MATHEMATICAL MODELING FOR INDUSTRY AND ENGINEERING

Thomas Svobodny
Wright State University

PRENTICE HALL, Upper Saddle River, New Jersey 07458

To my family

Library of Congress Catalogue-in-Publication Data

Svobodny, Thomas.
   Mathematical modeling for industry and engineering/Thomas Svobodny.
      p.   cm.
   Includes bibliographical references and index.
      ISBN 0-13-260894-4
   1. Mathematical models.   I. Title.
QA401.S03   1998
511'.8—dc21   97-35547
                                              CIP

Acquisitions Editor: George Lobell
Editorial Assistant: Gale Epps
Assistant Vice President of Production and Manufacturing: David W. Ricciardi
Editorial/Production Supervision: Robert C. Walters
Managing Editor: Linda Mihatov Behrens
Executive Managing Editor: Kathleen Schiaparelli
Manufacturing Buyer: Alan Fischer
Manufacturing Manager: Trudy Pisciotti
Marketing Manager: Melody L. Marcus
Marketing Assistant: Jennifer Pan
Creative Director: Paula Maylahn
Art Director: Jayne Conte
Cover Designer: Bruce Kenselaar
Cover Photo: Architect's color rendering of building; James R. Steinkamp/Murphy/Jahn,
      Inc. Architects

 ©1998 by Prentice-Hall, Inc.
Simon & Schuster / A Viacom Company
Upper Saddle River, NJ 07458

Printed in the United States of America
10   9   8   7   6   5   4   3   2   1

ISBN 0-13-260894-4

PRENTICE-HALL INTERNATIONAL (UK) LIMITED, LONDON
PRENTICE-HALL OF AUSTRALIA PTY, LIMITED, SYDNEY
PRENTICE-HALL CANADA INC. TORONOT
PRENTICE-HALL HISPANOAMERICANA, S.A., MEXICO
PRENTICE-HALL OF INDIA PRIVATE LIMITED, NEW DELHI
PRENTICE-HALL OF JAPAN, INC., TOKYO
SIMON & SCHUSTER ASIA PTE. LTD., SINGAPORE
EDITORA PRENTICE-HALL DO BRASIL, LTDA., RIO DE JANEIRO

Contents

Preface

This book grew out of notes used in a modeling course attended by upper-division undergraduate mathematics, science, engineering, and economics students that I and others have taught over several years in the mathematics department at Wright State University. From the beginning I had to rely heavily on home-made lecture notes, as I could not find a text suitable for beginners, and yet dealt with models challenging to a third or fourth year student. Some books had good modeling examples, but the students didn't like to read them as they didn't develop the mathematical structure; however, I found the "easy-reading", generalized, modeling texts too light on material. After a couple of years of hearing from colleagues that they would like the same sort of text that I envisioned, I started to write this book.

The technical prerequisites for a reading of the text are minimal: a solid calculus course, some exposure to differential equations (such as is sometimes found in calculus courses), and a little matrix algebra.

Although the use of the whole book would be ideal for a one-year introduction to applied math, the form of the book reflects to some extent a couple of one-quarter courses that are taught at Wright State University: MTH 306/606 Mathematical Modeling, attended by junior-level and senior-level mathematics majors and graduate students in math education, biomedical engineering, biophysics, economics, mechanical engineering, electrical engineering, computer science, and statistics; and MTH 333/533 Partial Differential Equations, attended by junior-level and senior-level math and physics majors, as well as graduate students in electrical, mechanical, and biomechanical engineering. A student can take the courses in either order. The modeling course covers, typically, Chapters 2 through 6 with supplementary material from Chapters 7, 8, and 9. The PDE course is based on Chapters 9, 10, and 11, with supplementary material from chapters 3, and 5,6.

At the same time, I have tried to make it as suitable for self-study as I possibly could. Besides making the subject accessible to the large number of students who find themselves without a relevant course, this should be of help to a teacher who wants to devote a large amount of classtime to the discussion of projects. The wide range of examples and mathematical techniques will broaden the student's vista and help get across the idea that there is no fixed set of tools for modeling.

In fact, the chapters are largely self-contained, although there are definite connections and built-in redundancies so that the student can see the same idea used in a different context. In this way, a clear stream can be followed (see the figure at the end of this preface). As the graphic indicates, Chapter 3 is the foundation of the book. It is natural to proceed from there to Chapter 4. One can go directly to Chapter 6 if probabilistic methods are to be the main content of the course. Chapters 1 and 2 have found use in surveying and exploring with students some of the simple ideas of applied mathematics, but their quick pace may dismay the insecure student. In particular, the first half of Chapter 1 is meant as entertainment; if the student doesn't find it so, it could be skipped. The rest of chapter 1 contains some background material. The chapters themselves are written in a sort of newspaper pyramid style so that one can either study a chapter thoroughly or simply read the first part of each chapter. Sections that are not necessary for later chapters and/or require more mathematical sophistication and/or ask for much classtime are marked with an asterisk (*).

The pedagogical intent of this book is to help develop in students a feeling for the use of mathematics as a tool in the understanding of the world. A common complaint of students when beginning the modeling course is that they "lack the physical intuition" to be able to model. The book is put together with the feeling that a "modeling intuition" can be nurtured in the mathematics student. It is mainly a matter of developing confidence, not just in problem solving, but in ability to approach complicated phenomena by asking a few simple questions. To this end and to encourage students to do much of the thinking on their own, exercises are built into the narrative. These exercises function as a governor: if they are trivial, the student can pick up speed; if they are not quite understood, a rereading of the text will be called for. Some require little work and may function simply to keep the student's pencil sharpened, but are designed to help the student take first responsibility for learning. Other exercises require some amount of thinking and/or search for data. There are also problem sections at the end of the chapters; these consist mainly of particular models, some of which may be suitable for a class project. Several independent trails can be followed through the problem sections. For example, chemical reactions and compartment models are introduced in the problems of Chapter 4 and reappear in problem sections in several later chapters.

It is clear to its teachers that modeling is not mathematics per se, but certainly the point of it is to use mathematics to show the underlying links between apparently disparate phenomena. Indeed, in many mathematics books, one often comes across a footnote remarking that the subject presently under discussion can also be clothed a different way, and in such and such a context. In this book these footnotes have been collected and expanded. I'd like to make the seemingly paradoxical statement that an engineer or scientist will want to take a modeling course, not to learn abstractions, but rather the opposite: to become better acquainted with concrete phenomena. Often in engineering texts one sees a formula derived and then some magical mathematics applied to it, and the result is a "theoretical rule of thumb" that appears in a box on the page. The student engineer accepts the boxed result as a substitution for the phenomenon and while becoming a practicing engineer will continue to do so. If he or shethe engineer ends up doing something more than paper work, he or she will notice a disparity and the boxed formula is thrown into the trash bin of "theory" which is disparagingly regarded as being unrelated to the "real world". On the other hand, the development of mathematical skills is necessary for a development of modeling skills. It may turn out that the mathematical technique needed for a particular model is yet to be found. Many problems started out as modeling problems and turned into areas of (pure) mathematics. Some of the great mathematicians spent an extraordinary amount of their time on modeling (Archimedes, Newton, Euler, Bernoulli, and others).

It is not possible to acknowledge everyone who aided in the development of this book. Special thanks must go to Jim Vance, Gloria Sickles, Masahiro Yamashita, Zdenek Kalva, Gabriel Svobodny, who carried out some of the experiments, and especially Anne-Marie Svobodny, who is responsible for much of the final art-work. Early partial versions of the text were class-room tested by David Miller and Larry Turyn. I benefited from the help of several institutions, including Wright State University, Center for Theoretical Study in Prague, and Dayton Museum of Natural History. I would like to thank the reviewers, Lester

Caudill, Ann Morlet, Walter Pranger, and Allan Struthers, and the editorial and production staff at Prentice Hall, especially George Lobell and Bob Walters. Of course I owe a debt to the authors of the many books that I have enjoyed reading and have found especially valuable in writing this one; the reader will find them in the recommended reading sections at the end of every chapter.

Thomas Svobodny
Dayton, Ohio

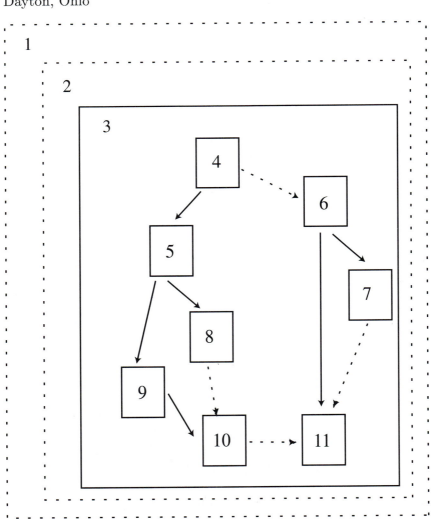

Chapter 1

The Modeling Adventure

1   What Is Modeling?

A first point that should be made is that we reflexively draw upon
models to aid in our understanding of and our dealings with the world
around us. Scientists, for example, must construct models of one sort or
another to make sense of their findings, to communicate these findings
to others, and to make comparisons with the work of their colleagues.
Models are an indispensable part of that thrilling aspect of scientific
endeavor: prediction making.

The models that we call upon may be

1. Pictorial

2. Analogical

3. Mathematical

Our concern in this book is, of course, with the overtly mathematical, but let us first see how the ontogeny of a mathematical model
sometimes replicates, at least in spirit, a phylogeny of scientific discov-
ery.

We will turn to the example of the evolution of models for the
resting potential in nerve cells.

Information is sent through nerves and muscles as signals that are
electrical in nature. These signals (nerve pulses) are a time record of
changes in a homeostatically permanent electric potential—a voltage—
that exists between the inside and the outside of the cells. It was realized that in order to understand and model more complicated aspects
of nerve-cell firing, one had to have a model of this resting potential.
The existence of the resting potential follows from a straightforward

1

observation. If one electrode of a voltmeter is placed inside the cell and the other electrode is grounded outside the cell, the needle of the voltmeter will be deflected. Let us denote the measurement of the voltmeter by V.

If the voltage reads -80 mV (a typical value), then we know the size of the electrical field, and the negative sign tells us that it is directed into the cell. An electrical field manifests a separation of charge: here, negative charge inside the cell and positive outside. How is this possible in the biological cell?

Biological fluids are *electrolytes*. That is, they are solutions of salts, whose molecules dissociate into charged atoms, or *ions*.[1] For example,

$$NaCl \longrightarrow Na^+, Cl^-$$
$$KCl \longrightarrow K^+, Cl^-$$

Here is a first model for the existence of V: when the cell is first formed, it is filled with an overabundance of negative ions that are sealed inside, while the leftover positive ions are in the surrounding extracellular fluid. This model was quickly laid low by experimental observation. First, to any measurable accuracy, both the fluid inside and the fluid outside the cell seemed to be electroneutral. In every spatial region of the fluid large enough to be measured, there are equal numbers of positive and negative ions. Second, the cellular membrane, the wall separating inside from outside, is known to be permeable to some ions, in particular the positive potassium ions, K^+. When a substance is dissolved in another, if all outside forces are equal, the dissolved substance, or *solute*, tends to a uniform or homogeneous concentration. High concentrations of the solute disperse of their own accord. The time rate of change of this dispersal is proportional to the spatial gradient of the concentration. The constant of proportionality is negative, to mean that the motion is down a gradient. (This is called *diffusion* and forms the subject matter for Chapter 11.) If the membrane were permeable to all ions, then any differences in concentration across the membrane would soon even out, and then

$$V \to 0,$$

in equilibrium.

---

[1]Negative ions are called *anions* because they are attracted to the positive electrode, which is called the anode. Positive ions are called *cations* because they are attracted to the negative electrode, which is called the cathode.

On the other hand, if the membrane could allow only the positive potassium ions to pass through and not any corresponding negative ions, then a voltage could be set up that just balanced the diffusive force of the positive ions. To see this, suppose that there were an excess of K^+ ions inside the cell. In other words, we assume a spatial gradient exists between inside and outside. The K^+ ions would diffuse with a net outward motion. However, because overall electroneutrality would thereby be disturbed and the inside would become negative relative to the outside, an electrical field would be set up that would oppose the emigration of positive ions. At first this voltage would be small, increasing as time went on to exactly balance the diffusive force. A static equilibrium would be reached (Figure 1.1):

$$F_{\text{electric}} + F_{\text{diffusion}} = 0.$$

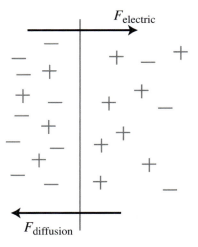

Figure 1.1 The balance between the electric and diffusive forces across the membrane

The situation is completely analogous to a common battery.[2] In a battery, the energy supplied by chemical reactions produces a separation of charge and so an electrical potential. In our case, diffusion is the driving force of the voltage. When we measure -80 mV, it is as though

---

[2] That chemical batteries are typically referred to as "cell" batteries is beside the point.

we had a 80-mV battery between the electrodes of our voltmeter, with the negative terminal on the inside and the positive terminal outside. Figure 1.2 describes this analogy.

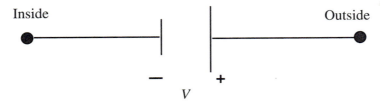

Figure 1.2 The battery analogue of the cell membrane.

Using the ideas of diffusion (Chapter 11), we can calculate this voltage from the assumption of static equilibrium. This is

$$V_{\text{static}} = E_K = 61 \log \frac{[\text{K}^+]_o}{[\text{K}^+]_i}, \tag{1.1}$$

where E_K is the equilibrium potential of K^+ ions (measured in mV), $[\ ]_o$ is the concentration on the outside, and $[\ ]_i$ is the concentration inside. (Batteries have a built-in resistance, but since we haven't provided for currents, we can ignore this for the time being.)

Our model has become mathematical. The output parameter, voltage, is determined by parameters that can, in principle, be measured. This is a good model. Except that it is wrong. When we look deeper, it doesn't match reality. First, the equilibrium is *not static*: V is affected by metabolic changes. If the cell is deprived of substances known to be used for metabolism, the resting potential fades to zero. Second, the cell membrane was found to be permeable to many other ions besides K^+, in particular to Na^+ and Cl^-.

The picture of an equilibrium of static charges must be thrown out. We could have a *dynamic* equilibrium, whereby moving charges, that is, currents, of potassium and other ions could be opposed by some energy producing mechanism, a pump. Each species of ion would have its own diffusive force (potential), and electroneutrality is maintained by the pump. Figure 1.3 shows the picture that modifies Figure 1.1.

The picture is a model of sorts, a pseudomechanical cartoon analogue. The highlights of the picture are: it portrays the pumping action (1 Na^+ for 1 K^+), it shows the relative conductivities of the ions by the widths of their channels, and the net force on each ion is given

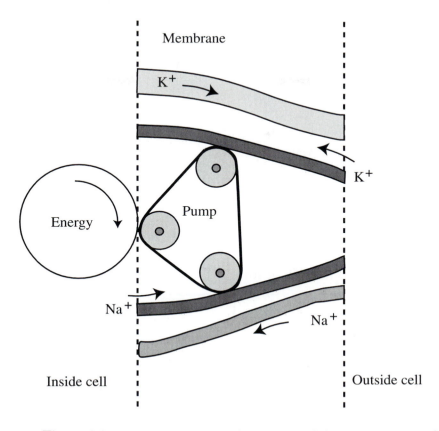

Figure 1.3 Mechanical cartoon of the origin of the resting potential.

by the slopes of the channels. But the picture doesn't really explain things as well as an analogical model can. To extend our electrical analogue, each ion species is to be represented by a battery powered by diffusive flux. Currents must flow through these batteries, and in order for equilibrium to be achieved, there must be a capacitive current across the membrane. The measured membrane voltage, V, is to be thought of as a capacitative potential, that is, the voltage across a capacitor (Figure 1.4). Conservation of charge requires an equilibrium of currents,

$$I_{\text{ions}} + I_{\text{cap}} = 0.$$

This electrical circuit analogue is easily converted into mathematical equations; these (algebraic) equations are also easily related to mechanical balance principles. But the dynamics of action potentials need to be described by differential equations and these equations can easily be written down from the electrical analogue or the mathematical equations; the same dynamical equations would only with great difficulty be translated directly from a mechanical explanation.

Most importantly, even though the electrical analogue would not seem to be as "real" as a mechanical description of the movements of ion concentrations, it is set at the level of our measurements, that is, at the level of verifiability. Although ion concentrations are the reality, the "truth," the movements of ions are too small to be measured. The separation of charge required for $1\,\mu F$ of membrane capacitance is effected by the movement of 0.005% of the cellular ions. One last point before moving on: A mathematical model should not just copy reality, but should aim for an explanation of a particular aspect; the ultimate worth (usefulness) of a mathematical model depends on our comprehension of it.

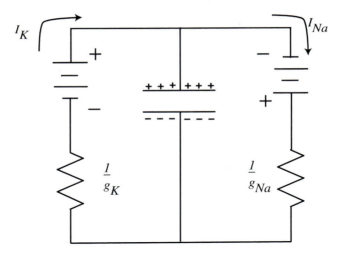

Figure 1.4 Electrical circuit analogue of membrane resting potential

2   Multiple Models: Shopping Around

A squat tube with a radius of about 25 mm, the aorta carries oxygenated blood from the heart bound for the extremities. It has two branches, and each of the branches has two branches. This branching continues until the blood arrives at the capillary, a structure about 0.008 mm in size, where oxygen diffuses into tissue.

Is there a simple relationship between the radius of the mother pipe and the radii of the daughter pipes? (See Figure 1.5.) Is there a way of explaining the observed structure? Can we be led to making quantitative predictions? We want a mathematical model.

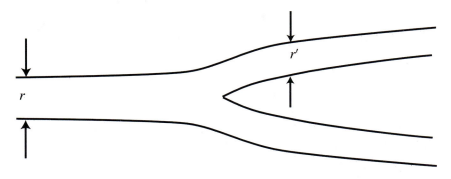

Figure 1.5 A tube splits into two.

We can first account for the idea that blood flows at a constant velocity. The one principle that *must* hold is conservation of matter: what flows in any section of the tubing must flow out of that same section. In mathematical terms, the flow rate (in volume per unit time) must be constant.[3] If A is the area at any cross section and V is the velocity, then the flow rate is

$$Q = AV.$$

If V is assumed constant, then we must have that

$$A = \text{ constant.}$$

This means that the branching should follow the law

$$r^2 = 2(r')^2.$$

This ratio

$$r/r' = \sqrt{2} \approx 1.4,$$

is higher than what is observed. We must throw out the idea that the velocity is constant over the length of the tube. Instead, let us suppose that the hydrodynamic force transmitted to the artery wall is constant.

---

[3]We are assuming that the fluid is incompressible so that the mass density is constant. Conservation of mass is more fundamental than volume conservation.

This is reasonable from the biomechanical point of view, since the cells of the artery wall can sense this shearing force. The shear force is assumed proportional to the rate of shear strain,

$$F_S = \mu \left| \frac{dv}{ds} \right|_{\text{wall}},$$

where v, the flow velocity along the direction of the tube varies only in the radial coordinate, s, and μ is the viscosity of the blood (Figure 1.6).

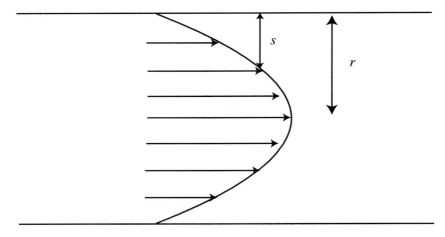

Figure 1.6 The Poiseuille flow has a parabolic profile.

If we assume the perfectly laminar flow known as Poiseuille flow (Figure 1.6), the flow through the mother pipe is

$$v(s) = \frac{A}{\mu}(r^2 - s^2),$$

and the flow through each daughter pipe is

$$v'(s) = \frac{B}{\mu}((r')^2 - s^2).$$

The constants A and B are proportional to the pressure gradient in the respective branch. The wall shears are easily computed:

$$\left| \frac{dv}{ds} \right|_0 = 2Ar, \qquad \left| \frac{dv'}{ds} \right|_0 = 2Br',$$

so that

$$B/A = r/r'. \tag{1.2}$$

We can eliminate B/A through the use of the conservation law. The volume flow rate through the mother pipe is

$$Q = \frac{2\pi}{\mu} \int_0^r A(r^2 - s^2)s\,ds = \frac{\pi}{2\mu}Ar^4.$$

Likewise,

$$Q' = \frac{\pi}{2\mu}B(r')^4.$$

But $Q = 2Q'$ means that

$$B/A = \frac{1}{2}r^4/(r')^4. \tag{1.3}$$

Together, (1.2) and (1.3) give

$$r^3 = 2(r')^3, \tag{1.4}$$

which is known as Murray's law.

Before we discuss what conditions can be drawn from this, let's turn to a different model for the same structure. With efficiency as our guiding light, let us assume that we are dealing with an optimal structure. The overall efficiency would represent both the cost of structural materials and the effort needed to pump the blood. Big tubing would have higher construction costs, but the blood could be pumped with less resistance.

Let us write down an objective functional, f, that represents costs per unit length:

$$f = \frac{\text{work of}}{\text{pumping}} + \text{cost of materials}$$

The first term is the power output of the heart, $\bar{w}$, and its form can be found by a simple dimensional argument: The heart works to convert energy to kinetic energy of the blood; thus

$$\bar{w} \sim v^2,$$

where v is the blood velocity, but

$$Q = \text{area} \cdot v \sim r^2 v,$$

and so

$$\bar{w} \sim Q^2/r^4.$$

(This also follows from the expression for the power generated by Poiseuille flow; see problems at the end of this chapter)

Contributions to the second term in f can come from various sources. The cost of the structure itself is proportional to the amount of the material, that is, to the volume, so the cost per unit length is proportional to r^2. There is also the cost associated with replacing cells that wear away; this is proportional to the inside area, and so this cost per unit length is proportional to r. Which one is dominant? This may depend on the function (see problems at the end of this chapter); let's compromise and say that the cost per unit length is proportional to

$$r^a,$$

where a is to be treated as a parameter between 1 and 2. So the total cost per unit length is

$$f = k\frac{Q^2}{r^4} + Kr^4, \quad 1 \le a \le 2.$$

The condition for a minimum is

$$\frac{Q^2}{r^{a+4}} = \frac{aK}{4k},$$

that is,

$$Q \sim r^{a+4},$$

so that, by the conservation of matter condition, $Q = 2Q'$, we have

$$r^{a+4} = 4(r')^{a+4}.$$

When $a = 2$, this is exactly Murray's law. How should we judge the different models? We can measure the relative sizes of the branches. We can count the number of branches and compare this to predictions of the scaling of the branching structure. We will leave the investigation of these things to the problems and recommended reading section. In passing, the important thing to note is how the underlying structure of different effecting mechanisms may dovetail in applications (as may have actually happened in the process of evolution). It is interesting that we didn't actually use any features of the blood itself except that it is a viscous fluid. Thus everything should be portable to other natural plumbing systems. In particular, when put into service in the case of the lung, the present models can be extended to "explain" the observed scaling in the lung (see problems and recommended reading).

3 An Example of the Modeling Process

Here I'll set out an example of the type of analysis that will be within the capability of anyone who will have studied this book. The purpose is simply to show how the modeling process might be carried out by such a dedicated, indefatigable reader. It will serve to illustrate how such a person will go about making decisions regarding problem formulation, choosing modes of mathematical analysis, and distilling the important ideas.

When we set out to model something, we should follow the steps:

1. Find a *specific* question of interest. This should be the first thing we do, even though the particular question may not be the last word or even of any lasting interest. The more specific we can make the question, the better our chances are of following a fruitful path.

2. Find the relevant variables and parameters. We might want to start with the output variables: *if* we knew the value of such and such parameter, or *if* we knew the shape of such and such a function, would our question be answered?

3. Write down any applicable relations between the variables. Are there any laws? Is it reasonable to require that one of our variables be a conserved quantity? It is likely that this will introduce new variables. Are we going to have trouble closing our model? We might start thinking about the relative numerical size of some of the variables. It doesn't hurt to be as complete as possible here, even if some quick estimates show that we can eliminate most of the relations and variables. Don't do anything yet with all these equations.

4. Make a guess as to the form of the solution. What are the important scales? Go back to the last step. Can you see something that can be thrown out? Can you simplify any of the relations? Get something that you can solve. Go back to the first step. Can you clarify the question? Can you make it more specific? Do you have to throw it out and start all over?

5. Solve. Approximate, scale, and extrapolate your solution to make its relation to step 1 clear.

Let us see how this works on a real problem.

Free Radical Formation by Ultrasound

We start with an experimental observation: When frog skin is exposed to therapeutic levels of ultrasound radiation (at an energy level of 300 $mWcm^{-2}$ and frequency of 1 MHz), an increase in ionic conductance is measured.[4] The introduction of antioxidants such as cystamine, cysteamine, and vitamin C reduce this increase [14]. Since these substances are free radical scavengers, this would seem to indicate that the increase in conductance is caused by the formation of free radicals (ions). How can sound waves of so little energy ionize the molecules of solution?

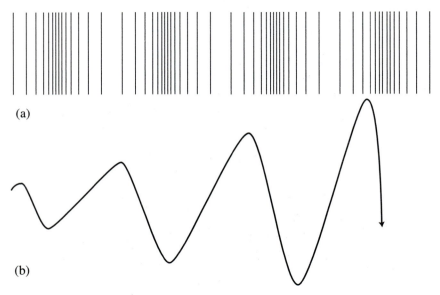

(a)

(b)

Figure 1.7 (a) The alternating compression and rarefaction of pressure in a longitudinal sound wave (see chapter 10). (b) The radius of a bubble oscillates in magnitude due to the periodic pressure variations in a steady train of sound waves.

Liquids can withstand tensions, but not too much, particularly if there are impurities such as ions present. A sound wave in its simplest form consists of a compression followed by rarefaction, and so on (Figure 1.7). During the rarefaction phase small seed bubbles are

---

[4]Ultrasound radiation refers to the production of sound waves having frequencies above that of human hearing (usually above 100 kHz). It is widely used in medicine for diagnosis and therapy. Epidemiological studies have indicated that it is not dangerous.

released from surfaces and grow in size; during the compression phase they shrink. If the match with frequency is right, the bubbles can grow bigger at each consecutive wave, eventually to many times their original size, and then collapse. This whole process of cavitation and its destructive possibilities has been studied in many different contents. It is believed that the collapsing bubble can ionize molecules [58]. This is interesting, and given its importance, you have been asked to evaluate this with a mathematical model. Let us follow the program.

1. Let us start with the obvious question: Can collapsing bubbles ionize molecules? You might feel that this is too vague. Can you think of something more specific at this stage?

2. The bubbles are pockets of gas and vapor in a liquid acted on by sound waves, so there are many components. Let us deal with each one in turn.

   (a) *The sound wave* This clearly involves input parameters. If the ultrasound is applied in extremely short bursts, then we might have to take into account the form of the wave. In the experiment referred to, the sound was applied as a continuous sinusoid, so we only need to know the two parameters, amplitude, F, and frequency, f.

   (b) *The bubble* What is the size of a bubble? What is its shape? Inside the bubble is a gas. The gas is described by its density, ρ_b, pressure, p_b, and temperature, T_b. The surface of a bubble involves surface tension, σ, possible electrical charge, q_b, and, possibly, dynamical variables that describe its motion.

   (c) *The liquid* All this takes place in a biological liquid. Does it have the properties of water? It has a viscosity, μ, and a velocity field, $\mathbf{u}$, as well as the scalar fields ρ, p, T. Is it compressible? Is heat being conducted?

   (d) *The molecules* Where do the free radicals come from? Are the water molecules ionized? Are there other reactions?

3. We might start with a calculation to establish a length scale. We can substitute the experimental value for the frequency,

$$f = 10^6 \text{ Hz,}$$

and, using the sound speed for water of 1500 ms^{-1}, calculate the
sound wavelength,

$$\lambda = 1500 f^{-1} = 1.5 \times 10^{-3} \text{ m}.$$

Longer than a millimeter. We can then decide how to treat things
based on their size relative to this length. The bubbles in a bi-
oliquid are nowhere as big as this. At this point we have to decide
if we want anything more to do with the sound field. Remember, the function of the sound field is to make the bubble grow
and then compress it. The compression of the bubble by the
sound wave (as is easily calculated) cannot be responsible for the
cavitation damage; so if we accept that the bubble has grown
to a certain radius, we can forget the sound field for the time
being, except for keeping a record of the wavelength and using
the frequency to define a characteristic time scale. These same
length scale considerations lead us to assume that bubbles are far
enough apart and not interacting with each other so that we can
concentrate on just one bubble.

What is the shape of the bubble? Until we have reason to think
otherwise, we will suppose it to be a sphere of radius a.

Is it an ideal gas inside? This will depend on its constitution,
amount of vapor present, and whether it can be considered rarified. The ideal gas law is an equation of state:

$$p = \rho R T.$$

Under conditions of thermodynamic equilibrium, there is a relation between the pressure field inside the bubble and the pressure
field outside the bubble:

$$p_b - p = 2\sigma/a, \tag{1.5}$$

where a is the radius of the bubble and σ is the surface tension. Is
the mass of the bubble conserved in its motion? This is a likely
assumption since evaporation and condensation at the surface
take place much slowly than the time scale imposed by the sound
wave. But if so, then the two equations above are incompatible.
Since ρ is inversely proportional to the volume of the bubble, as
the bubble grows, it can happen that $p_b < p$. The relation (1.5)
is an equilibrium relation, but what we are decribing is not at
equilibrium.

Let's turn to the liquid. Is it compressible? Ever so slightly—otherwise there would be no sound waves; but we can separate the motion of the fluid in the sound wave and the motion local to the bubble. The flow near the bubble (which, remember, is going to be very small compared to the wavelength) is expected to be small enough that we can ignore compressibility (chapter 3). For virtually the same reason, as well as considerations of symmetry, we will probably be able to ignore the viscosity of the fluid. There is one source of asymmetry: the gravitational force field. The bubble will experience a force

$$\frac{4}{3}\pi a^3 (\rho - \rho_b) g. \tag{1.6}$$

As the density inside, ρ_b, is less than the density of the fluid, this will be an upward (buoyancy) force. It is not obvious what to do about this, but since it breaks the perfect symmetry, let's put it on the back burner and, since we have no reason to suspect that it might be significant, we'll analyze it after we have some idea of related terms. The effect of surface tension is difficult to deal with: from (1.5) we can see that there is not going to be a well-defined surface tension for the extremely low bubble pressure that the conservation of mass requires. Let's ignore this, simply because our modeling powers are not up to dealing with it; then we'll come back in the end. The variables can be related by considering the equations that express conservation of mass, momentum, and energy. Since the liquid is incompressible, the energy flux is by conduction. By simply comparing time scales, we can eliminate any consideration of heat conduction.

We have not yet mentioned the chemical reactions. What do we really need? To ionize molecules. Where are these molecules? If, as we believe, the gas in the bubble is rarified with ordinary amounts of vapor, then we should bet on the liquid. (It really is a matter of relative numbers.) Free radicals are formed when a molecule receives a certain amount of energy to break the covalent bonds between atoms. That is, a large amount of energy must be available in a very small space.

4. The last remark gives us a path to follow: account for energy production. In general, good fiscal policy in modeling requires a clear accounting of energy expenditures. Thus, we can refashion our question in step 1 as follows:

$1'$. Can collapsing bubbles concentrate a lot of energy in a small volume?

Since we are only investigating the concentration of energy, we can ignore what really happens physically to a collapsing bubble; whether it collapses to a singularity, or breaks up into smaller bubbles, or whatever. In fact, many physical things can happen before a singularity develops, such as shear instabilities or shock waves radiating pressure. It is easy now to imagine what might happen: a large amount of energy comes to be concentrated in a small volume during a short time period. A little thinking about dimensions (Chapter 3) leads us further: the quantity of importance here is clearly energy density; but energy density has the same dimensions as pressure. Pressure is a mechanical variable, and so, since it is clearly the energy and pressure in the liquid that we are dealing with here, we are led to a consideration of the mechanics of the liquid. First, we can rephrase the question in step 1 as

$1''$. How high can the pressure in the liquid phase become as the bubble collapses.

Then we can figure out the mechanics of the liquid. The conservation of mass (Chapter 10) for an incompressible fluid is simply

$$\text{div}\,\mathbf{u} = 0. \tag{1.7}$$

The conservation of momentum (force balance) for an inviscid fluid, neglecting gravity, is

$$\frac{\partial \mathbf{u}}{\partial t} + \mathbf{u} \cdot \nabla \mathbf{u} = -\nabla p/\rho. \tag{1.8}$$

We must solve these equations for the unkowns $\mathbf{u}$ and p on the spatial domain exterior to the sphere of radius a (Figure 1.8). What are the boundary conditions? We can assume that most of the relevant fluid motion is going to be concentrated near the bubble as liquid rushes in to fill the volume vacated by the receding bubble. Far away from the bubble, the velocity can be assumed to be zero, and the pressure there is the mean pressure in the liquid. These values are assumed to hold at infinity. What about the interior boundary—the surface of the bubble? By continuity, the velocity of the fluid at the boundary is the speed at which the bubble is shrinking,

$$\mathbf{u}\big|_{r=a} = \dot{a}.$$

The collapse of the bubble is due to the rarified condition inside, so let's assume that

$$p\big|_{r=a} = 0. \tag{1.9}$$

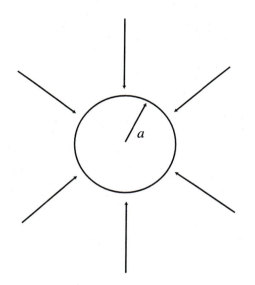

Figure 1.8 Radially symmetric potential flow.

Now we just have to hypothesize suitable initial values for the variables and we can proceed to solve (1.7, 1.8) subject to the boundary conditions. To solve a boundary-value problem for a system of partial differential equations is, in general, a daunting affair. Let us see if we can guess a reasonable form for the solution. From the spherical symmetry of the set up (Figure 1.8), we can look for a solution that is spherically symmetric. This by itself doesn't help a lot (you can go ahead and try it), but there is a simple trick that makes this easy to solve. If we assume that the initial velocity is conservative, or irrotational,

$$\operatorname{curl} \mathbf{u} = 0, \tag{1.10}$$

then it will always be irrotational. But (1.10) means that **u** is a gradient of some potential:

$$\mathbf{u} = \nabla\phi. \tag{1.11}$$

Furthermore, the condition (1.10) implies that the velocity field satisfies

$$\mathbf{u} \cdot \nabla\mathbf{u} = \nabla(u^2/2), \tag{1.12}$$

just as though it were a scalar field.[5] Thus, our system of partial differential equations reduces to the three scalar equations (1.11), (1.7), and (1.13):

$$\nabla\left(\frac{\partial\phi}{\partial t} + \frac{1}{2}u^2 + \frac{p}{\rho}\right) = 0. \tag{1.13}$$

Actually, we can combine (1.11) and (1.7) to get an equation for ϕ alone:

$$\nabla^2\phi = 0.$$

This is the Laplace equation and it is easy to find a spherically symmetric solution that satisfies the boundary condition (Chapter 11):

$$\phi(r) = \frac{Q}{4\pi r},$$

where Q is a source term,

$$Q = \int\int_{r=a} \nabla\phi \cdot d\mathbf{S} = \int\int_{r=a} \mathbf{u} \cdot d\mathbf{S} = \text{ volume flux } = \dot{V}.$$

But

$$V = \frac{4}{3}\pi a^3,$$

so that

$$Q = \dot{V} = -4\pi a^2\dot{a},$$

which allows us to find

$$\phi = -\frac{a^2\dot{a}}{r},$$

and

$$\mathbf{u} = \frac{a^2\dot{a}}{r^2}\mathbf{r}.$$

---

[5]The magnitude of the vector **u** is u.

From this, we compute

$$u^2 = (\frac{a}{r})^4 \dot{a}^2,$$

so (1.13) becomes

$$-2\frac{a\dot{a}^2}{r} - \frac{a^2\ddot{a}}{r} + \frac{1}{2}(\frac{a}{r})^4\dot{a}^2 + \frac{p}{\rho} = \frac{p_\infty}{\rho}. \tag{1.14}$$

At the bubble surface, $r = a$, this equation becomes

$$-\frac{3}{2}\dot{a}^2 - a\ddot{a} = \frac{p_\infty}{\rho}.$$

This latter equation admits $a^2\dot{a}$ as an integrating factor and, after integrating, we arrive at

$$a^3\dot{a}^2 = \frac{2}{3}\frac{p_\infty}{\rho}(a_0^3 - a^3). \tag{1.15}$$

5. It sometimes helps to clarify analysis to express an equation in terms of dimensionless variables (Chapter 3), particularly if you want to make sense of graphical output. A characteristic length scale is the initial radius of the bubble, a_0, while a characteristic time scale is the period of the sound wave, f^{-1}. Accordingly, we define the dimensionless variables, x and s, as

$$x = \frac{a}{a_0}, \quad s = ft.$$

In terms of these variables, (1.15) becomes

$$(\frac{dx}{ds})^2 = A^2(x^{-3} - 1), \tag{1.16}$$

where A^2 is the dimensionless parameter

$$A^2 = \frac{2p_\infty}{3\rho a_0^2 f^2}.$$

Notice that (1.14) gives p as a function of a and r and thus, via (1.15), as a function of (r, t). It turns out that the pressure maximum is found at a single radius that tends to the bubble surface as t increases. Thus, the only work we have to do is to solve the initial-value problem for the ordinary differential equation,

$$\frac{dx}{ds} = -A^{1/2}\sqrt{x^{-3} - 1},$$
$$x(0) = 1,$$

and use this to compute p_{max}/p_{∞}. Now, of course, the solution to this initial value problem is simply $x \equiv 1$. But it is unstable, so for numerical computations we take an intial condition something like

$$x(0) = 1 - 10^{-6}.$$

Figure 1.9 shows the result of the numerical calculations using the following values:

$$p_{\infty} = 1 \text{ Atm} = 10^5 \text{ Jm}^{-3}$$
$$\rho = 1 \text{ kg m}^{-3}$$
$$a_0 = 10^{-7} \text{ m}$$
$$f = 1 \text{ MHz} = 10^6 \text{ s}^{-1}$$

It is clearly seen that the pressure blows up during a small fraction of the oscillation period. It can be shown analytically that

$$p_{max} \propto x^{-3},$$

which helps to explain the results of Figure 1.9.

4   Approximations

When, in our work as applied mathematicians, we have diligently set about to model a phenomenon, when we have completed painstaking observations or studied the experimental results, when we have identified the relevant factors and variables, when we have accounted for all the mutual influences and interactions between these variables, it is almost inevitably the case that we have included too much in our model to make it amenable to a straightforward analysis. It is often necessary to find a simplification or approximation to a complicated function or expression. This simplification is typically only useful for a smaller range of parameters or variables than the full expression. It is fair to say that a large part of the work of an applied mathematician is

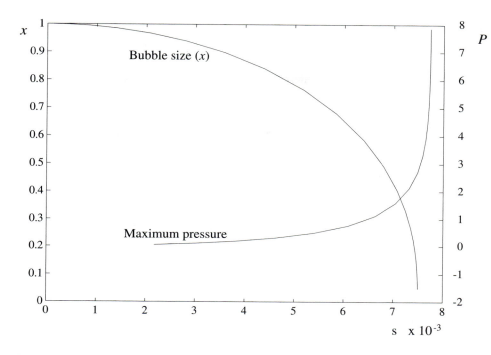

Figure 1.9 After a short time there is a catastrophic collapse of the bubble, and an extreme rise in the maximum pressure. The label on the pressure scale is reduced pressure, $P = \log\left(\frac{p_{max}}{p_\infty}\right)$.

involved with approximation. It will take the rest of the book to hint at the possibilities, so I will just make some observations here.

Much of the time we don't need an exact value of something, but only an order of magnitude estimate; furthermore, the relative accuracy is what is important. For example, we might say, with an understood accuracy, for some quantity whose true value happens to be 1020, that "it is *about* 1000," the error being 20. It would clearly be less accurate to say of some quantity whose value is 30 that "it is about 10," even though the error in each case is 20. In the first case, the relative error is

$$\frac{1020 - 1000}{1000} = 0.02,$$

or 2%, while in the second case the relative error is

$$\frac{30 - 10}{10} = 2,$$

or 200%.

It is important to keep in mind the relative sizes of different parameters in a model and not just the *limits* of actual sizes. That is, we must know the relevant scale. Consider the function

$$f(x) = \frac{10^{10}}{1 + x^2}.$$

It is easy to see, for example by expanding f in a geometric series, that, for $x < 1$, f has constant order of magnitude. A special scale has to be chosen for the dependent variable if we want to see any variation (Figure 1.10).

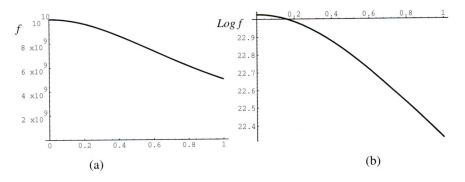

(a)   (b)

Figure 1.10

For larger values of x, the logarithmic derivative of f,

$$\frac{f'}{f} = -\frac{2x}{1 + x^2},$$

indicates that f decreases two orders of magnitude for every order of magnitude increase of the x-scale. The function

$$g(x) = \frac{10^{10}}{1 + 10^{10} x^2},$$

has, like f, the limit 10^{10} as x approaches 0. It is in fact a rescaling of f, and its graph for $x < 1$ shows different behavior (Figure 1.11).

It is important to always be aware of the appropriate scale for considerations of your variables.

Polynomials are the simplest functions, and polynomial approximations of functions are thus very important. There many different ways to use polynomials as approximations to more complicated functions;

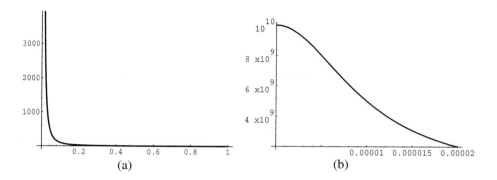

Figure 1.11

for the time being we will only consider the simplest. If we are interested in the behavior of a (smooth) function near a particular point, then we can generate a polynomial approximation by Taylor expansion:

$$f(x) = f(a) + \sum_{k=1}^{N} \frac{f^{(k)}(a)}{k!}(x-a)^k + R_N,$$

where the error, R_N, satisfies

$$|R_N| \leq \frac{B}{(N+1)!}|b-a|^{N+1},$$

and $B \leq \max_{a \leq x \leq b} |f^{(N+1)}(x)|$. This estimate shows that the error, over each fixed interval, decreases roughly by an order of magnitude for each term added to the sum, as long as the required derivative is bounded on the interval. However, this gives us no easy method to find the order, and thus the complexity, of an approximation when we require a certain accuracy. For example, if for the interval $[a, b]$, the error should be less than a specified amount, say, $R_N \leq \epsilon$, then the number of terms N needed may be very large. A straightforward Taylor approximation is not well-suited for such tasks. There are other types of finite expansions (that don't form a convergent series) that do a more efficient job. However, we are sometimes interested only in the action of a function at a point; we might need to know the instantaneous direction of the function, whether it is increasing or decreasing, or the shape of the graph. For instance, the critical points where the function changes its course are special. These critical points are the points where the derivative is zero:

$$f'(x_0) = 0.$$

These include local maxima and minima. The Taylor expansion about
such points takes the form

$$f(x) - f(x_0) = \frac{f''(a)}{2!}(x - a)^2 + \ldots, \tag{1.17}$$

and the quadratic term is the most significant. If $f''(x_0) = 0$ at a critical
point, the point is said to be **degenerate**, and the most significant
terms will be of higher order. For critical points that are nondegenerate,
the sign of $f''(x_0)$ gives the quality of the extremum. If it is positive,
we know that the point is a local minimum for f, while if it is negative
it is a local maximum. This is an easy consequence of (1.17) in taking
into account the form of the quadratic.

5 Curve Fitting and Parameter Estimation

Graphs are an important part of modeling because visual patterns are
easily processed by our brains. Graphs are great for presenting the
quantitative and qualitative reults of a model, but they can also be used
in the beginning to help make sense of much data for which the theory
is missing. A graph can clarify the validity of a model by immediately
(and viscerally) showing the match to data:

$$\begin{array}{ccc} \text{model} & \longleftrightarrow & \text{data} \\ \text{(theory)} & \text{graph} & \text{(reality)} \end{array}$$

In the simplest case, a model results in a law, or algebraic relation,
between two variables,
$$Q = F(P).$$
If a graph of the experimental results of Q versus P,

$$\{(P_i, Q_i)\},$$

is available, then it is possible that a glance at the two graphs can tell
us if our model has substance. We can also compare the predictions
with reality, directly,
$$F(P_i) = Q_i,$$
but it is unlikely that our model would be so exact. Then we could
compare the errors,
$$\text{error} = |Q_i - F(P_i)|. \tag{1.18}$$

Typically, our model will leave some parameters unspecified, to be determined by a match to the data. If the errors (1.18) are nontrivial, then we could choose the parameter (or parameters) that minimize the total error:

$$\sum |Q_i - F(P_i)|.$$

The advantage to minimizing the sum of the squares of the errors,

$$\sum |Q_i - F(P_i)|^2,$$

is that the objective function in this minimization problem is differentiable. This is the *method of least squares*. It is a straightforward procedure when working with linear "laws":

$$F(p) = \alpha p + \beta,$$

then the function to be minimized is

$$\epsilon(\alpha, \beta) = \frac{1}{2} \sum (Q_i - \alpha P_i - \beta)^2.$$

The least-squares solution is found as the solution to the linear equations:

$$0 = \frac{\partial \epsilon}{\partial \alpha} = -P_i(Q_i - \alpha P_i - \beta),$$

$$0 = \frac{\partial \epsilon}{\partial \beta} = -(Q_i - \alpha P_i - \beta).$$

6 The $O(\cdot)$ and $o(\cdot)$ notation

If $\lim_{x \to 0} f(x) = A$ exists then we can distinguish three cases: $A = 0$, $0 < A < \infty$, $A = \infty$. In case $A = 0$, we can measure the rate at which $f(x) \longrightarrow 0$ by comparing $f(x)$ to a known function, $g(x)$. The notation, $F(x) = O(G(x))$ is shorthand for $|F(x)| \leq C|G(x)|$, for sufficiently small x; equivalently,

$$\lim_{x \to 0} |\frac{F(x)}{G(x)}| < \infty.$$

The notation, $f(x) = o(g(x))$ means that

$$\lim_{x \to 0} \left| \frac{f(x)}{g(x)} \right| = 0.$$

Most commonly, the power functions $g(x) = x^p$ are used. Here are some examples:

$$
\begin{array}{cc}
\text{as} & x \longrightarrow 0, \\
\sin x = O(x), & \sin x = o(1), \\
\cos x = O(1), & \cos x = o(x^{1/2}), \\
\sinh x = O(x), & \sinh x = o(1).
\end{array}
$$

7   Pictorial Introduction to the Fourier Transform

It is useful, particularly when dealing with vibrations and waves, to describe a function in terms of its Fourier transform. Briefly stated, a function's Fourier transform describes the distribution of frequencies of the function. This distribution is sometimes called its *spectrum*.[6] For example, the Fourier transform of a function with exactly one distinguished frequency,

$$ f(t) = \sin kt, $$

is a point mass at a single point (Figure 1.12a). A function that is periodic in time will always have a spectrum that is supported at discrete points, even though there may be infinitely many frequencies (Figure 1.12b). A basic idea of Fourier analysis is that the original function can be recovered as a superposition of simple sinusoids of frequencies that are present and having as coefficients the values of the distribution:

$$ f(x) = \sum_{k=1}^{\infty} c_k \sin kt. $$

A nonperiodic function will have a continuously varying frequency distribution. Figure 1.12c shows the frequency spectrum of a sinusoidal signal that is turned on and off in a smooth way. Figure 1.12d shows the spectrum when the signal is turned on and off abruptly. Figure 1.12e shows the spectrum of a signal that is turned on and off too quickly for the sinusoidal to develop. Some functions are only easily defined in terms of their Fourier transform. A signal process with a constant distribution of frequencies is referred to as *white noise* (Figure 1.12f).

Figures 12a and 12c illustrate an important property of the Fourier transform: the more spread out is a signal, the more compact is its

---

[6]The term frequency is used here in the sense of elementary vibrations or oscillations (see Chapter 5).

frequency spectrum, and vice versa. This is called the *uncertainty principle* and can be succinctly expressed as

$$\Delta t \, \Delta k \geq C,$$

where Δt is the deviation in the "position" of the time signal, Δk is the deviation in the "position" of the spectrum, and C is a constant independent of the signal.

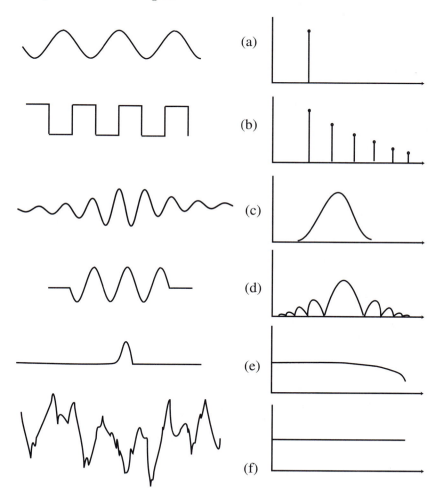

Figure 1.12 Time signals (on left) and their corresponding frequency spectra (on right)

8   Problems and Recommended Reading

0. You can come back to these problems better prepared after reading a few of the ensuing chapters. For an easy introduction to nerves, see [32]. Beautiful pictures of fluid flows tagged with appropriate parameters are found in [17]. The book [77] gives an entertaining introduction to biophysical modeling. A pleasant introduction to modeling at the precalculus level is [68].

1. *Branchings*   A big difference between the arterial system and the lung is the constraints place on the arteries (having to wrap around bones, go down limbs, etc.), while the lungs are in a wide-aspect cavity. Which system will depend more strongly on hydrostatic pressure?

   (a) Calculate the number of branches in the arterial system. In the lung.

   (b) Find the total surface area of the capillaries.

   (c) *Fractal scaling* What is the relative variation of the pressure in the branching system? Is it reasonable to assume that pressure is equalized throughout the system? Under this system, show that the branching is a self-similar structure. What is its fractal dimension? (*Hint: What is the pressure in the Poiseuille flow?*)

   (d) Find the power generated in the Poiseuille flow. Verify the expression for $\bar{w}$.

2. Ohm's law expresses a linear relation between current and voltage across a resistance R:
$$IR = V.$$

Conductance is the reciprocal of resistance,

$$g = 1/R.$$

When a capacitor has been charged up, the current through the capacitor is zero. The circuit is in a steady state. Show that when the membrane circuit is in a steady state:

   (a) We have
$$V_{SS} = \frac{\sum g_i E_i}{\sum g_i},$$

where g_i is the conductance of the ith ion, and E_i is the Nernst potential of the ith ion. (Why is this analogous to a balance beam?)

(b) If the conductance of an ion is increased, then the effect of the ion is increased. (It "weighs" more.)

(c) If the gradient of an ion is increased, then the effect of that ion on the steady state voltage is increased.

3. Verify that (1.15) follows from (1.14).

4. Is the assumption that the buoyancy force on the collapsing bubble is small consistent with the solution that we have computed? Compare the relative force on the bubble of our solution.

5. Evaluate the effect of surface tension on the collapsing bubble.

6. Is (1.9) a consistent assumption?

Table 1.1

Planet	Semimajor axis (10^6 km)	Sidereal period (days)
Mercury	57.9	87.97
Venus	108.2	224.7
Earth	149.6	365.26
Mars	227.9	686.98
Jupiter	778	4332.4
Saturn	1427	10759

7. Table 1.1 gives orbital data for six planets. Fit this data to a power law of the form $y = Cx^p$. The largest asteroid, Ceres, has a period of about 4.5 earth years; about how far far from the sun would you place it? (*Hint:* Kepler used data similar to this to formulate his third law. What was his basic model of planetary motion? Was it pictorial, analogical, or mathematical?)

8. Why can one see through rain much farther than through fog?

9. How is the force needed to open a twist-off jar related to the area of contact of the top with the jar? If the force used to put on the tops is normally distributed, how many will be too difficult to get off?

10. Why do some birds fly in a characteristic V-formation? Aerodynamic models have been invoked, but is there a much simpler explanation?

11. What are the amounts of coffee and water that it takes to brew the perfect cup of coffee?

12. *Product evaluations*
    (i) A manufacturer claims that its cereal is not wetted by milk because of the ridgelike surface. The ridges are about 0.5 cm in size. If the cereal is in fact not wettable, is it due to the ridges?
    (ii) Can a *two-stage* cigarette filter make for a "cleaner" smoke?

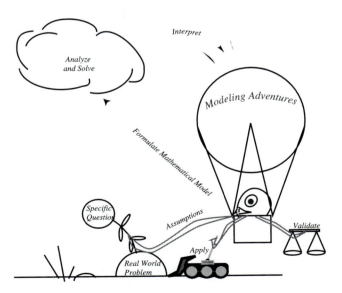

Figure 1.13 Illustration of the modeling process showing that the modeler needs many arms.

Chapter 2

Stability and Bifurcation

You may have noticed the way things respond differently to the wind or other fluid flow. An anemometer consists of three or four cups arranged symmetrically about an axis. As the wind blows, the arrangement of cups turns about the axis with a speed that is proportional to the wind velocity. When the wind blows soft, the vane turns slow; when the wind blows hard, the vane really goes. A different type of behavior is seen while riding a bicycle to school with a notebook in the front basket. The notebook will lie flat in the basket until the speed reaches, say, 15 mph; then it will flip up against the back of the basket and stay there. It will stay there even when the bicycle slows to below 15 mph, in fact to much slower than that.

1 Potentials

A system will be specified by a certain number of variables. These variables are sometimes called the degrees of freedom of the system. A system is said to be in **equilibrium**, or in an equilibrium state, if there is nothing that acts to displace it from that state. In a physical system, this corresponds to a state upon which no net force is acting. A system may be defined by a **potential**, a function whose critical points are the equilibria of the system. The reason is that the difference in potential between two states gives the **work** necessary to transfer the system from one state to the other. In other words, the work, W, needed, *in opposition to the force F*, to move the system from state x_0 to state x_1, is

$$W = -\int_{x_0}^{x_1} \mathbf{F} \cdot \mathbf{dr} = \phi(x_1) - \phi(x_0), \qquad (2.1)$$

where ϕ is the potential. From calculus we know that such a potential exists if and only if the above integral can be defined independently

31

of path. Physical systems with a potential are called **conservative**, because the integral around any closed path is zero. This is not the case for physical systems that exhibit friction or other dissipative processes; they are accordingly referred to as nonconservative. The statement that the equilibria of the system are given by the critical points of the potential, the points where

$$\nabla \phi(x_0) = 0,$$

is sometimes known as the principle of virtual work: zero work is done by infinitesimal (virtual) displacements of the state at an equilibrium. This is easily understood if one keeps in mind the useful picture of a ball rolling around on a surface (Figure 2.1). On any sloped surface, gravity will change the ball's position. If the ball is placed in the bottom of the depression, then it will stay there. Furthermore, it will return to the bottom if displaced from it a small (finite) distance. This is a **stable** equilibrium.

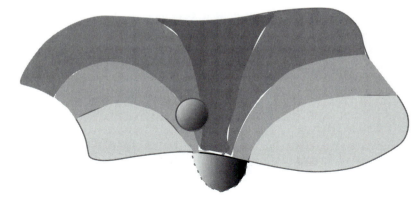

Figure 2.1 Ball rolling on a surface where there is a depression. The bottom of the depression is the only stable point.

If the ball is placed very carefully on the exact summit of a hillock (Figure 2.2), then it will stay there. But this is an **unstable** equilibrium: subjected to any finite push, the ball will roll away.

Combining the expression for work (2.1) with the Taylor formula at a critical point, one obtains a criterion for the stability of an equilibrium, which is a nondegenerate critical point of the potential. In Chapter 1, we found that, for a function of a single variable, this crite-

Figure 2.2 Ball rolling on a landscape where there is a hill. The summit is an equilibrium but it is not stable.

rion is

$$\phi''(x_0) > 0 \implies \text{stable}$$

$$\phi''(x_0) < 0 \implies \text{unstable}$$

A simple example of a system with a potential is given by a (point) mass attached to a steel spring. Let us assume that the mass is lying on a frictionless surface (Figure 2.3) so that the only force acting is the force exerted by the spring. What is the nature of this force? Perhaps we can turn to experiment and observation. (We'll detach the mass, not to be confounded by its inertial effect.) After some essays with this system we find that there is a certain position of the block, say at a distance l_0 from the wall, where it is in equilibrium. We find that if we displace the mass from this equilibrium, then the force will act in the direction opposite to the displacement in order to restore the equilibrium; for this reason, we call it a *restoring force*. We find that (as long as we don't pull too hard) the magnitude of the force is proportional to the amount of the displacement from equilibrium, and we can measure this. The displacement is $x - l_0$, where x is the (variable) distance measured from the wall, and so our observations can be summarized by writing

$$F_{\text{spring}} = -k(x - l_0). \tag{2.2}$$

The constant of proportionality is known as the spring constant or *stiffness*.

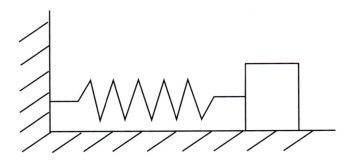

Figure 2.3

Can we figure out the value of k without having to experiment with
this particular type of spring, knowing what the spring looks like and of
what material it is made? Perhaps, in principle; but to give an idea of
what would be involved in such a calculation, note that k will depend
mainly on two things. First, the geometry of the spring configuration.
For instance, a tightly wound spring will be stiffer than a more loosely
wound spring. Second, the properties of the spring material. Our
spring is made of steel, which is an example of an elastic material.
This means that the material will return to its equilibrium state after
being deformed. This would not be the case if the material was, say,
caramel candy. Suppose we take a solid piece of size l of a material and
we deform it by pulling it straight in one direction by an amount Δl.
The material is **elastic** if satisfies

$$\sigma = E\frac{\Delta l}{l}, \tag{2.3}$$

where σ is the restoring force per unit area normal to the direction of
deformation. This force per area is called **stress**; (2.3) says that, in
an elastic material, stress is proportional to the relative displacement.
This relative displacement, $\frac{\Delta l}{l}$ is called the **strain**. The constant of
proportionality, E, is known as the modulus of elasticity, or Young's
modulus. The modulus is typically arrived at by experiment. Since E
depends only on the material, can it not be found theoretically from
knowledge of the atomic lattice structure of the material? No, in general the mechanical properties of materials are determined more by
defects in the crystal structure than the lattice structure itself. However, the effects of these defects can sometimes be estimated.

By combining (2.3) with (2.2), we can write

$$k = fEd,$$

where d is some characteristic size of the spring, and f is a geometric shape factor (f is large for tightly wound springs).

The potential for this system is found from the work integral

$$U_{\text{spring}}(x) = \int_{l_0}^{x} k(s - l_0)ds = \frac{1}{2}k(x - l_0)^2.$$

We have taken the potential level of the equilibrium to be zero. Since $U''_{\text{spring}}(l_0) > 0$, the equilibrium is stable. We are not surprised, since the force is a restoring force.

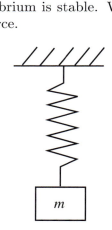

Figure 2.4

Suppose now that we let gravity act on the mass that is attached to the spring by hanging the spring from the ceiling (Figure 2.4). If the variable x represents distance (down) from the ceiling, then the gravitational force is

$$F_{\text{gravity}} = mg,$$

where m, the mass, and g, the gravitational acceleration, are constants. That is, the gravitational potential is

$$U_{\text{gravity}} = -mgx,$$

and

$$U = U_{\text{spring}} + U_{\text{gravity}}$$

is the total potential. (Notice that since the potential is defined to be the work done in opposing the force, its sign will be opposite to that of the force.)

EXERCISE 1   (i) Show that the addition of this constant force only changes the position of the equilibrium, but not the stability of the equilibrium. (ii) Show that any quadratic can be written in the form $y = Ax^2$ by a suitable choice of coordinate origin.

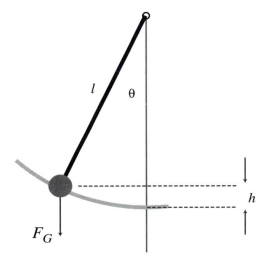

Figure 2.5

Another simple but important example is afforded by the gravitational potential of a simple pendulum (Figure 2.5). This system has one degree of freedom: the angle of displacement, θ. Note that two values of θ that differ by a multiple of 2π represent the same physical displacement.

The work that one does against the constant force, mg, in elevating the mass a vertical distance, h, is

$$U = mgh = mgl(1 - \cos\theta);$$

this is the potential, which is bounded and periodic. There seem to be an infinite number of equilibria, corresponding to the solutions of

$$0 = U'(\theta) = mgl\sin\theta,$$

but there are only two physical equilibria. It is easy to see that the "down" equilibrium is stable and the "up" position is unstable.

EXERCISE 2 (i) Use the second derivative to verify these statements about stability. (ii) Suppose that one approximates the potential by its Taylor expansion about the downward equilibrium. Keep the first two or three terms. Where does this predict the "up" equilibrium to be?

A true pendulum will not return to an equilibrium directly after being displaced, since the mass has inertia. It will, in fact, swing about the equilibrium: the inertia will convert the change in potential to kinetic energy. In the present chapter, we are ignoring dynamics for the most part; but, having studied the forms of the equilibria we will not be surprised to find, when the full dynamics are included, that the mass of the pendulum can spend time in its oscillation near a stable point, but not near an unstable one.

We now turn to a more striking example of the use of potential arguments to make predictions, this time in a situation where the actual dynamics are much more complicated than in the previous cases. When we watch air bubbles rising in a water tank or a pitcher of beer, or see rain drops falling, or dewdrops on a cool morning, we are not presented with a very wide range of sizes for the drops. Why are particular sizes of drops more likely to be seen than others? This is a very complicated problem and there are very many different variables (degrees of freedom) that play a role, but let us see how far we can get with a simple model, constructed with potentials, that only takes into account one independent variable that represents the size of a raindrop. Since we are ignoring shape, let us suppose that the drop is spherical, and the variable of interest will be the radius, r.

What work is involved in building a raindrop? One must build the surface between the air and the water. The work is proportional to the surface area

$$E_{\text{surface}}(r) = \gamma S = \gamma 4\pi r^2.$$

The constant of proportionality, γ, is the *coefficient of surface tension*. By considering this as our potential, it seems that, $r = 0$ being the only equilibrium and stable, we shouldn't see drops at all! This is certainly not what is observed, so let's see if we have left something out. If we confine our attention on a fixed portion of space at a fixed temperature, then there is a given saturated vapor pressure. This means that there will be, in our region, a certain amount of liquid water in thermodynamic equilibrium with the vapor. Thus, there is a certain amount of liquid water mass that must be present regardless of the size of the individual drops. We can take into account the conservation of mass by considering not the absolute energy of a drop, but the *energy per unit mass.* If ρ is the density of water, then the mass of a spherical raindrop is

$$M = \rho \frac{4}{3} \pi r^3,$$

and the surface energy per unit mass is

$$U_{\text{surface}} = E/M = \frac{3\gamma}{\rho r}.$$

According to this potential, there are *no* finite equilibria, but as the potential is a decreasing function of radius, the only stable equilibrium should be $r = \infty$. This, also, is clearly wrong; it would seem to imply that the water should form only in extremely large lumps. We'll have to consider more factors. One possibility is that to make drops we will have to do work against ambient forces, such as a gravitational force. The work against constant gravitation to build a drop of of mass M with radius r is Mgr, so the potential is

$$U_{\text{gravity}} = Mgr/M = gr.$$

The effect of this potential is to penalize the formation of large drops. The graphs of the two potentials, U_{surface} and U_{gravity} are drawn with dotted lines in Figure 2.6. The total potential

$$U_{\text{surface}} + U_{\text{gravity}} = \frac{3\gamma}{\rho r} + gr$$

is the solid line in Figure 2.6. Notice that, as in the case of the mass-spring system, the addition of a term in the potential to account for a constant (gravitational) force simply changes the position of the equilibrium. The effect is more dramatic in this case because a finite equilibrium is established.

EXERCISE 3 Use the stable equilibrium of this potential to estimate the size of a raindrop. (For water at room temperature (293 K), $\rho = 10^3$ kg/m^3, $\gamma = 7.25 \times 10^{-4}$ J/m ^{2}.) How does the equilibrium depend on variations in the surface tension? According to this model, should we expect larger drops at high altitudes or low? What is the effect of temperature on the equilibrium size of drops?

In this example, as in the previous ones, we have ignored the possible mechanism for approach to equilibrium; this involves thermodynamics; the interested reader is referred to the suggested readings.

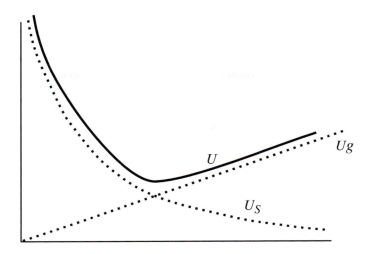

Figure 2.6 Potential energy curves for waterdrop sizes. The combined energy curve has a minimum.

2 Bifurcation

It may happen that as a parameter changes the equilibrium position doesn't change, but changes its stability. For example, if you take a thin card, made of some fairly elastic material, between thumb and forefinger and squeeze gently, you will find that the card doesn't stay flat but assumes a curved shape.[1] The card could be squeezed while retaining the flat shape if constrained to do so, but the fact that, without the constraint, one never sees this, indicates that the flat solution is unstable (for large enough compressions). There are two possible stable configurations for the stressed card: it can bend up or it can bend down. With some cards, one sees up or down with about equal chance, but more commonly a card will almost always tend to a favored one of the two. Yet both solutions *are* stable; if, while keeping constant compression, you guide the card with your other hand to the unfavored position, it will stay there. Another way to say this is that with a strong enough transverse force the card will *jump* from one stable equilibrium to the other. An interesting effect can be observed with a biased card. If the card is placed in the unfavored equilibrium and the compression is slowly released, the card will not tend continuously to the flat shape, but will first *jump* to the opposite (stable) curved state.

---

[1]Do this with several different samples.

An elastic card, or beam, or wire has, in principle, infinitely many degrees of freedom, since to specify it exactly we have to give its complete shape. However, taking advantage of the symmetry of our setup, we can represent the state of the card by the position of its midpoint, which we call W in Figure 2.7. For all values of the compression pa-

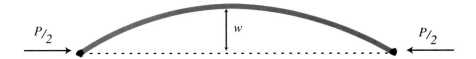

Figure 2.7 Compressed Euler strut. W is the height of the midpoint, P is the compression.

rameter, P, the zero state exists, but it becomes unstable when P reaches a certain value, P_c. For values of P much larger than P_c, there are two stable states, but one of them is favored, and the other can only be reached by jumping directly to it from the favored state. Our experiments can be summed up, by plotting the observed equilibrium values of W versus P, as in Figure 2.8. Note that there is no other equilibrium, stable or otherwise, than the zero state, for $P < P_c$. Our

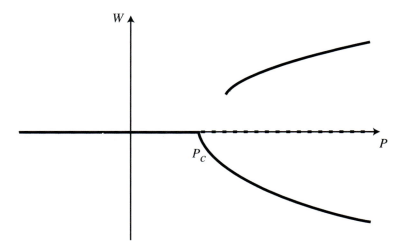

Figure 2.8 Equilibrium values of W plotted against compression P. The dashed line indicates an unstable equilibrium.

qualitative experiments do not allow us to be very precise about the situation near the point $(P_c, 0)$.[2]

Another setup that is mechanically almost identical to the example of the compressed card is that pictured in Figure 2.9.[3] Indeed, this

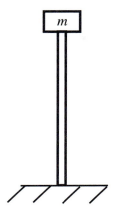

Figure 2.9 Vertical Euler strut, compressed by a mass.

is a compressed elastic beam or column, the horizontal compression P being replaced by mg, the vertical compression. As the mass is increased, it eventually attains a value, m_c, so that when $m > m_c$ the column is tipped (Figure 2.10). As before, we have a symmetric bifurcation, where, as a parameter increases, the trivial solution loses stability and two new stable equilibria come into being. This, again, is an elastica that can bend into various shapes and requires, in general, infinitely many degrees of freedom (modes) for representation.

Let us see if we can glean some understanding of the bifurcation phenomenon by modeling the elastica as a rigid rod attached to a spring (Figure 2.11). We will assume that the spring is unstressed when the rod is in the vertical position. This will ensure that the vertical position is an equilibrium for all values of the parameters. Later, we will look at the effect on the model if this condition cannot be controlled precisely.

The spring potential is $\frac{1}{2}k\theta^2$, and the gravitational potential is

---

[2]The reader is encouraged to set up an apparatus to make more precise measurements

[3]Either of these is known as an Euler strut, named after Leonhard Euler (1707–1783), who first studied these examples and generally began the study of stability.

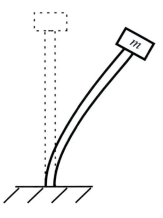

Figure 2.10 For $m > m_c$, the vertical position is unstable.

$-mgl(1 - \cos\theta)$; thus the total potential is

$$U = \frac{1}{2}k\theta^2 - mgl(1 - \cos\theta). \qquad (2.4)$$

The equilibria are given by the solutions of

$$0 = U'(\theta) = k\theta - lmg\sin\theta.$$

The zero state is an equilibrium for all values of the parameters, in particular for any value of m. If we eliminate this solution, then we can solve for m,

$$m = \frac{k\theta}{lg\sin\theta} = \frac{k}{lg}(1 + \frac{1}{6}\theta^2 + \cdots) \qquad (2.5)$$

to find the other branch of solution, which can be represented as a curve in the (θ, m) plane. Figure 2.12 is the bifurcation diagram for this system. It shows how the features of the equilibria depend on the control parameter, which in this case is the mass m. The heavy solid lines are stable equilibria, and the heavy dashed line is an unstable equilibrium. The zero state is shown as changing its character with respect to stability at the lowest point of the nontrivial equilibrium branch given by (2.5). This can be seen via the Taylor expansion of U about $\theta = 0$:

$$U \simeq \frac{1}{2}(k - lmg)\theta^2 + \frac{1}{24}lmg\theta^4 + \cdots \qquad (2.6)$$

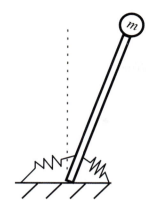

Figure 2.11 Point mass m, on a massless rigid rod of length l, constrained by a spring.

The sign of the quadratic term reverses when $m = m_c = \frac{k}{lg}$; the stability of the nontrivial solution branch is guaranteed by the fact that the coefficient on the quartic term is always positive.

EXERCISE 4 Graph

$$U_{ss}(x) = ax^2 + cx^4, \quad c > 0,$$

for values of a positive, zero, and negative. This potential and the examples just considered exemplify the form of the *stable symmetric transition* or, more picturesquely, the *supercritical pitchfork bifurcation*.

The ideal Euler strut is a physical example of this pitchfork bifurcation. It can be associated with a potential of the form given by U_{ss} in Exercise 4, even in the context of nonlinear elasticity theory. I'd like to give a taste of this theory here. Consider the vertical Euler strut, shown in Figure 2.13 labeled with the relevant variables. The basic relation of the bending beam states that bending moment is proportional to the curvature:

$$M = (EI)\kappa. \tag{2.7}$$

The constant of proportionality is the *bending stiffness*, and it is the elasticity modulus, E, times the cross-sectional moment of inertia, I. In Figure 2.13 the bending moment, about a point a distance s from the join, is

$$M = mg(w_{max} - w),$$

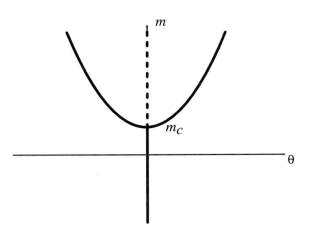

Figure 2.12 Bifurcation diagram for the system of figure 13. This is called the supercritical pitchfork bifurcation.

and the curvature is

$$\kappa = \frac{d\phi}{ds}.$$

The shape of the bent column is specified by giving either w or ϕ as a function of s. Notice that the springed-hinge model corresponds to the case with a curvature singularity at $s = 0$. Since

$$\frac{dw}{ds} = \sin \phi, \tag{2.8}$$

we can differentiate both sides of the equation (2.7) and make the required substitutions to arrive at the equation that describe the deformation of the column:

$$(EI)\frac{d^2\phi}{ds^2} + mg \sin \phi = 0.$$

We are not going to solve this equation;[4] instead we would like to construct the potential energy in order to investigate the bifurcation. To this end, it is more convenient to work with w as the variable. We can write κ in terms of w using (2.8):

$$\kappa = \frac{\ddot{w}}{\sqrt{1 - \dot{w}^2}},$$

---

[4]We will in Chapter 5.

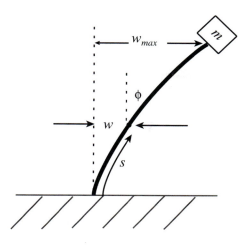

Figure 2.13 Variables needed for large displacements of the one-dimensional elastica: s is arclength, measured from support, ϕ is angle of deflection of tangent from vertical, w is horizontal displacement.

where dots refer to differentiation with respect to arc length. Two types of potential energy are present. There is the strain energy stored in the bent beam, and there is the gravitational potential energy of the mass. An increment of strain energy is

$$dU = \frac{1}{2}M\,d\phi = (EI)\frac{\ddot{w}^2}{1 - \dot{w}^2}\,ds,$$

and so the strain potential is

$$U_{\text{strain}} = \int_0^l (EI)\frac{\ddot{w}^2}{1 - \dot{w}^2}\,ds \simeq (EI)\int_0^l \ddot{w}^2(1 + \dot{w}^2 + \dot{w}^4 + \cdots)\,ds$$

Any incremental deformation along the length contributes to a vertical displacemnt of the mass, so we can write an incremental potential as $dU = mg\,dy = mg\cos\phi\,ds$, and the gravitational potential is

$$U_{gravitational} = \int_0^l mg\sqrt{1 - \dot{w}^2}\,ds \simeq mg\int_0^l (1 - \frac{1}{2}\dot{w}^2 + \frac{1}{8}\dot{w}^4 + \ldots)\,ds$$

This seems to be a formidable expression, but we simplify it considerably by taking an educated guess as to the form of the bent column. From experimental observation, we notice that curvature of the bent column doesn't change sign, but, indeed, decreases monotonically from the base to the mass. (This can also be deduced from the basic equation (2.7).) A simple (but not *too* simple) function that has this form is

$$w(s) = W \cos \frac{\pi s}{2l}.$$

If we keep quadratic and quartic terms for the potential, we can now perform the integration to arrive at an expression in terms of the single degree of freedom W:

$$U(W) \simeq \frac{\pi^2}{8l}(EI(\frac{\pi}{2l})^2 - \frac{1}{2}mg)W^2 + \frac{\pi^4}{128L^3}(EI\frac{\pi^2}{4L^2} - \frac{1}{8}mg)W^4$$

EXERCISE 5 What is the critical value, m_c, at which the stability of the equilibrium changes? By referring to Exercise 2, show that this is a case of a supercritical bifurcation. Compare to (2.6).

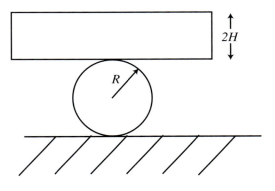

Figure 2.14

A different but related type of bifurcation is displayed by another class of phenomena. The simplest example is the block and cylinder of Figure 2.14. Shown is the equilibrium position, where the center of mass of the block is directly over the center of the cylinder. In a small displacement from equilibrium, the block rolls along the cylinder (without slipping). The potential governing the system is proportional to the vertical distance that the center of mass of the block travels as it rocks and rolls.

EXERCISE 6 Show that $(V/mg) \simeq \frac{1}{2}(R - H)\theta^2 - \frac{1}{24}(3R - H)\theta^4$.

The symmetrically placed equilibrium is stable as long as $H < R$, but when $H \geq R$ the equilibrium is unstable. In this example, there are no other stable equilibria.

EXERCISE 7 Graph the potential

$$U_{us}(x) = ax^2 + cx^4, \quad c < 0,$$

for values of a positive, zero, and negative. This potential expempifies the form of the *unstable symmetric transition*, or the *subcritical pitchfork bifurcation*.

The bifurcation diagram in this case looks like Figure 2.15.

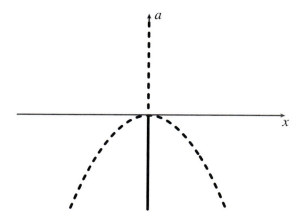

Figure 2.15 Bifurcation diagram for subcritical pitchfork.

Another example of this type of bifurcation is afforded by the shape of a raindrop that is electrically charged. In our previous study of drops we considered one degree of freedom; this was the size of the raindrop. Only spherical drops were considered. In the present example we wish to consider whether the spherical shape itself is stable. If we perturb the shape of the drop slightly, will the effect of the perturbation be minimal (stable equilibrium), or will it turn out that the drop seeks a completely different equilibrium shape? This is the question that Rayleigh asked. To make the analysis fit with the present discussion, we will consider only particular deformations of the sphere. More precisely, we consider deformations that are ellipsoids with two equal minor axes (prolate spheroid). This will allow the use of only one parameter (degree of freedom) to represent the class of deformations. This parameter is the eccentricity of the ellipsoid.

If a drop has a net electrical charge Q, then the electrostatic energy is the work needed to put all these charges together in the shape of the drop; this energy is given by $Q^2/2C$, where C is the capacitance of the drop.[5]

The size of this electrostatic energy is the important thing here. The sphere is a minimal surface, which means that among all surfaces of a fixed volume the sphere has the smallest surface. Since surface energy is proportional to surface area, we can say this a different way by stating that the sphere is a stable equilibrium of the surface potential with respect to shape. The gravitational energy is ignored because it is independent of shape. Of course, the anisotropy of the gravity field will make some perturbations more likely than others but won't affect the actual stability.

Recall that the eccentricity of an ellipsoid of revolution is

$$e^2 = 1 - \frac{b^2}{a^2},$$

where a is the semimajor axis and b is the semiminor axis. We want to treat e as a small parameter (since we are perturbing a sphere) so both the surface area and the capacitance are to be written in terms of the eccentricity:

$$S = 2\pi ab[\sqrt{1 - e^2} + \frac{\arcsin e}{e},$$

$$C = 8\pi\epsilon_0 a(e/\ln \frac{1+e}{1-e}).$$

Neither of the relevant forces depends on the mass; this tells us that the density is not important here, but we still have to conserve volume. This can be done by requiring that the ellipsoidal perturbations have the same volume as the equilibrium sphere,

$$ab^2 = r_0^3.$$

With this latter equation and the expression for the eccentricity, we can write a and ab, and thus C and S, in terms of e and then expand in power series:

$$E_s = \gamma S = \gamma 4\pi r_0^2 (1 + \frac{4}{90}e^4 + \cdots),$$

---

[5] Capacitance is the ability to hold charges. If, when a charge Q is placed on an isolated conductor, its potential is increased by V, then $C = Q/V$. The capacitance depends only on the geometric shape and size of the conductor. See the first appendix.

$$E_e = Q^2/(2C) = \frac{Q^2}{8\pi\epsilon r_0}\left(1 - \frac{1}{45}e^4 + \cdots\right).$$

EXERCISE 8 Verify these steps. Use the total potential to show that the sphere is stable to such perturbation for $Q^2 < Q_c^2 = 64\pi^2\epsilon_0 r_0^3$, but that it is unstable for greater values of charge. By considering higher-order terms, decide whether this bifurcation is subcritical or supercritical.

3 Catastrophe

Sometimes, under smooth change of a certain parameter, the position of a stable equilibrium will change in an abrupt and discontinuous way. The equilibrium seems to disappear, or the state jumps from one equilibrium to one that is far away, as we have seen in the experiments with the Euler strut and recorded in Figure 2.8. This is an example of what is called a *catastrophe*. Before we examine the catastrophe in the Euler strut, let's turn to another interesting and well-known example whose catastophe has the simplest topological structure.

The soap film that forms between two coaxial rings is a surface of revolution known as a catenoid (Figure 2.16). How does this shape

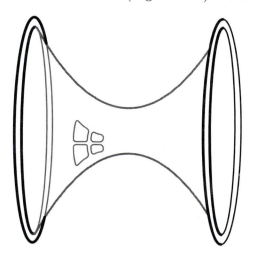

Figure 2.16 Soap film between two coaxial rings.

change as the distance between the rings is varied?[6] The shape of a soap

---

[6]The reader is encouraged to experiment using a solution of detergent in water. Glycerine helps to slow evaporation.

film is determined by minimizing surface energy subject to constraints. In this case, the constraint is that the boundary of the surface must be the two rings. Since surface energy is proportional to surface area, soap films are area-minimizing surfaces. If a is the radius of the rings and b is the distance between them (Figure 17), the area of this surface of revolution is

$$S = \int_0^b 2\pi r \sqrt{1 + (\frac{dr}{dx})^2} \, dx. \qquad (2.9)$$

If we minimize this integral (see problem section), we are led to the

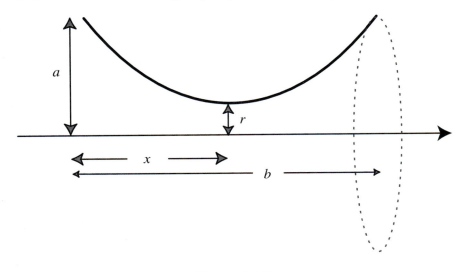

Figure 2.17

differential equation

$$\frac{r^2}{c_1^2} = 1 + (\frac{dr}{dx})^2,$$

where c_1 is a constant of integration. The solution to the differential equation is

$$r = c_1 \cosh(\frac{z - c_2}{c_1}),$$

where c_2 is another constant of integration.

EXERCISE 9 Fix c_2. (Use $r(0) = a$.) Use the other boundary condition to get an equation for c_1. Show that c_1 has two real solutions (only one is minimal) as long as b is less than some critical value. For b larger than this critical value, there is *no* solution.

This is an example of what is known as a **fold catastrophe** (often it is called a *limit point instability*). The word *catastrophe* is simply a technical term that refers to certain types of singularities whose manifestation is a discontinuous change brought about by a continuous variation of a parameter.

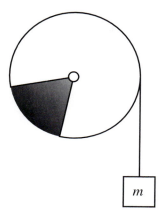

Figure 2.18

Another example of a system that exhibits the fold catastrophe is pictured in Figure 2.18. An asymmetrically weighted wheel of radius R has a mass hanging from it in a pulley arrangement. Suppose that, the mass of the wheel is M and that of the hanging is m. Suppose that without m attached, there is a unique equilibrium position of the wheel (call that $\theta = 0$). When a small m is attached to the rope, the equilibrium position is shifted.

EXERCISE 10 (i) Suppose that the mass of the wheel is concentrated at one point. Show that the relevant potential for the system is $U(\theta) = MgR(1 - \cos\theta) - mgR\theta$. (ii) Find the equation for equilibria and show that none exist for $m > M$. (iii) How is the analysis changed if we consider other asymmetric mass distributions for the wheel?

The critical angle is $\theta_c = \frac{\pi}{2}$ (as you found in the exercise). If we expand the potential about that critical point, then, using $\phi = \theta - \frac{\pi}{2}$,

$$U(\phi) = MgR + (M - m)gR\phi - \frac{1}{6}MgR\phi^3 + \cdots$$

The constant term is irrelevant.

EXERCISE 11 Show that there is an equilibrium when $m < M$, which disappears for larger values of m.

This potential displays the canonical form of the fold catastrophe

$$V(x) = \epsilon x + bx^3.$$

The effect of varying ϵ is shown in Figure 2.19.

EXERCISE 12   (i) The graphs in Figure 2.19 are drawn for $c > 0$. Draw the analogous graphs for $c < 0$. (ii) Show that any cubic can be written as $y = ax + bx^3$ by proper choice of origin.

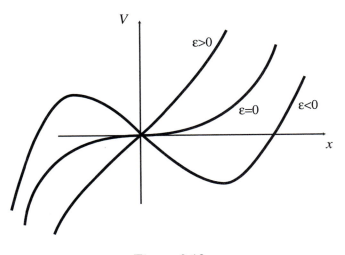

Figure 2.19

The fold catastrophe gets its name from the surface that is the graph of the potential V considered as a function of the two variables (x, ϵ). The bifurcation diagram for the fold is drawn in figure 20. Notice that, as ϵ increases to the critical value, two equilibria, one stable and the other unstable, come together and annihilate each other.

The diagram for the fold catastrophe shown in Figure 2.20 is different than the bifurcation diagrams seen previously in that, strictly speaking, there is no bifurcation, that is, there is no forking of the paths of equilibria. It is this feature that distinguishes the elementary catastrophes, of which the fold is the simplest example.

The qualitative features of either the graph of the potential or the plot of the equilibria shown in Figure 2.20 are not affected when small perturbations, respresenting noise or imperfections, are introduced in the underlying system. The fold catastrophe is said to be *structurally stable*. This is not the case for the bifurcations studied in the previous

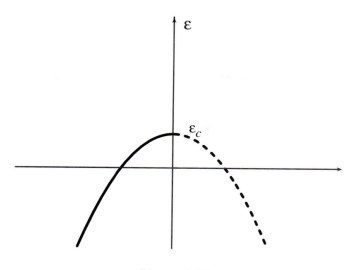

Figure 2.20

section. To see this, suppose that we introduce a perturbative linear term in the potential of the supercritical bifurcation:

$$V(x) = \epsilon x - ax^2 + cx^4, \quad a, c > 0$$

EXERCISE 13 Show that the potential of the spring-strut (2.6) has this form if the unstressed state of the spring is slightly off.

The equilibria are the zeros of the cubic

$$V'(x) = \epsilon - 2ax + 4cx^3,$$

which is graphed in Figure 2.21 for different values of ϵ. Notice that there is a range of values, $-\epsilon_c \leq \epsilon \leq \epsilon_c$, for which there are three equilibria. Otherwise, there is only one.

EXERCISE 14 (i) How many equilibria are there at $\epsilon = \epsilon_c$? (ii) Show that $\epsilon_c = \frac{4}{3}a\sqrt{a/6c}$.

The important thing to realize is that

$$\epsilon_c = \tilde{C}a^{3/2}.$$

If we plot ϵ_c and $-\epsilon_c$ versus a in the (a, ϵ) plane, the graph shows the form of a cusp (Figure 2.22).

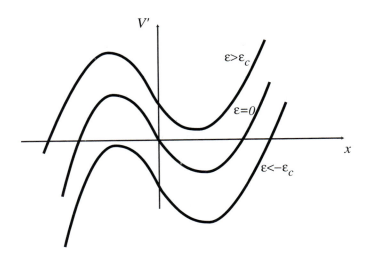

Figure 2.21

Inside the cusp, there are three equilibria, and there is one equilibrium for values of a, ϵ outside. This diagram gives its name to the second elementary catastrophe, the **cusp catastrophe**. This catastrophe, as we have seen, arises from the pitchfork bifurcations when we take into account certain perturbations. To see this, we must consider both parameters, a, ϵ, as varying. In this way, the bifurcation diagram, that is, the plot of the equilibria x_e versus the two parameters, is given by the folded surface of Figure 2.23. If we have a built-in, fixed, imperfection ϵ, then the restriction of this surface to ϵ fixed gives us the usual bifurcation diagram where x_e is plotted versus a. If $\epsilon = 0$, then we recover the usual supercritical pitchfork diagram. On the other hand, for any other value of ϵ, the bifurcation diagram looks like Figure 2.24.

With this diagram we can more easily understand the behavior of the nonideal Euler strut (as recorded in Figure 2.10). As a is increased gradually from negative values, we always obtain the equilibrium of the lower branch. The upper branch is stable, but it does not connect to the lower branch. If the state is placed in the upper stable equilibrium, then, as we decrease a to a^*, the strut snaps through to rest in the equilibrium of the lower branch. (The actual value of a^* will depend on ϵ.) If we fix a and use ϵ as the control parameter, we have two limit point instabilities or what amounts to a double fold catastrophe. By allowing ϵ to range over any interval containing $[-\epsilon_c, \epsilon_c]$, we observe a *switching* from one stable equilibrium to the other. This is illustrated

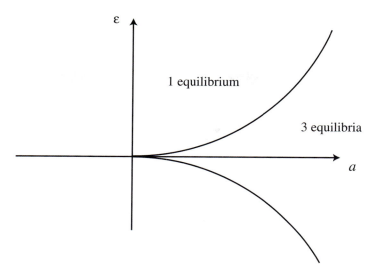

Figure 2.22

by the path $ABCD$ in figure 25. At the points B and D we have limit point instabilities. In the type of behavior illustrated, where a parameter, here ϵ, varies in a continuous way from a value and back to that value, the return to a state, here A, is not the reverse of the forward. (This is commonly referred to as hysteresis.) In this way, the cusp catastrophe can be seen to incorporate via two control parameters, a bifurcation to multiequilibria as well as a mechanism for switching between stable equilibria. Indeed, the cusp has often been used in models of switching phenomena. In the problem section of the chapter we will mention examples of the cusp catastrophe and will refer to a as the bifurcation parameter, λ_b; this determines the number of stable equilibria. We will refer to ϵ as the switching parameter, λ_s. The reader should keep in mind that there may be nothing in a certain switching phenomenon that corresponds to the bifurcation parameter λ_b. The switching may proceed by a series of limit point instabilities, or there may not be a continuous equilibrium surface. One might also mention that bifurcation theory, while being based on constructions that are not structurally stable, is still useful, especially in engineering where one can fine-tune control parameters at will.

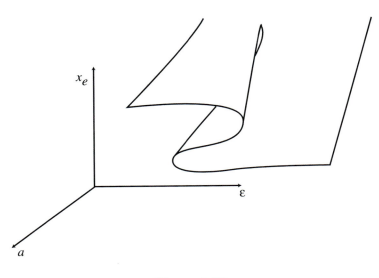

Figure 2.23

4   Problems and Recommended Reading

1. What is the Taylor expansion for a function of several variables? What is the criterion for stability of an equilibrium?

2. Notice that the equilibrium in Exercise 3 can be found at the place where the forces balance. Relate this to the intersection of the marginal cost and marginal income curves in the economics of the firm.

3. For the method of minimizing integrals such as (2.9) see the discussion of the Lagrange method in chapter 5.

4. A capacitor plate is fixed horizontally and is charged to voltage V relative to the grounded plate that is suspended from a spring above it. The plates are separated by a distance, d. As V increases, d gradually decreases, but then, at some critical value V_c, d jumps to zero. What kind of bifurcation or catastrophe is this?

5. *Further discussion of the charged droplet* can be found in [53] and [71].

6. *Approach to equilibrium* In the raindrop example, the actual mechanism responsible for driving the size of the drop to the

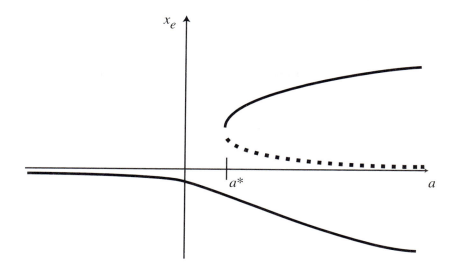

Figure 2.24

equilibrium size is governed by how the saturated vapor pressure above a surface is affected by the curvature of that surface. The saturated vapor pressure (SVP) is the pressure of a vapor that is in thermodynamic equilibrium with its liquid. It depends on the temperature. If the pressure above a liquid surface is less than the SVP, there will be a net evaporation; if the pressure is greater than the SVP, there will be a net condensation. We will denote the SVP above a flat surface by $p^0(T)$. If r is the radius of curvature of the liquid surface, the SVP is given by

$$p(r) = p^0(T)e^{(2\gamma/r)(M/\rho RT)}.$$

(M is the molar mass of the liquid, and R is the universal gas constant.) The reason for this is that, if the surface has positive curvature (like a raindrop), then molecules just below the surface are not held as strongly as they would be just below a flat surface. Thus, there will be a higher rate of evaporation and the vapor pressure can increase in equilibrium (SVP). What is the critical size radius for a nucleus in supersaturated vapor?

7. *Phase transitions* Matter can be arranged in various distinct organizational states. These different states are called *phases*. When a substance passes from one of these ways of arranging itself to

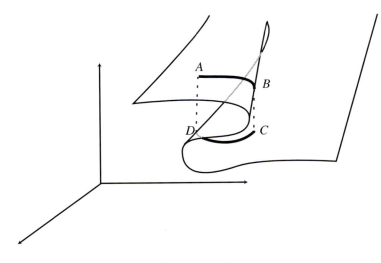

Figure 2.25

another it is said to undergo a *phase transition* or *phase change.*
A phase transition is characterized by the co-existence in equilibrium of two phases at the same set of values of the thermodynamic parameters. This bistability of equilibrium is one reason
why some people have looked to the the cusp catastrophe as a
possible model for phase transitions. Another reason is that the
Van der Waals equation, a popular equation of state for the gas–
liquid transition, gives qualitatively the equilibrium surface of the
cusp catastrophe.

(a) In the cusp catastrophe, as we have seen, there are several
ways to pass from one equilibrium branch to the other. If
we follow the path of Figure 2.25 where ϵ is changing but
a is fixed, that is, we obey the *delay convention*, then we
jump from one branch to the other. There is a discontinuity
of the equilibrium position and also a discontinuity in the
potential. This corresponds to what is known as a *zeroth-
order phase transition.* If the process is to have bistable
equilibria, but the value of the potential must change continuously, then one should follow the *Maxwell convention.*
Draw a diagram to illustrate the Maxwell convention. If
we obey the Maxwell convention, then the path from one
equilibrium to the other is at a constant value of the poten-

tial. This is *first-order phase transition*. Justify these terms. The Landau theory of second-order transitions is based on considerations of the supercritical pitchfork bifurcation.

(b) The Van der Waals equation of state is

$$(P + \frac{a}{V^2})(V - b) = nRT, \qquad (2.10)$$

where P is the pressure, V is the volume, and T is the temperature. This equation is sometimes used as a crude model for the first-order phase transition that occurs between the liquid and gas state of a substance. When $a = b = 0$, (2.10) reduces to the ideal gas equation of state. The parameter b expresses the volume taken up by the gas molecules, while a is a first order term of the interaction energy. Show that the surface defined by the equation (2.10) in (V, P, T)-space has the form of a cusp catastrophe. (*Hint*: Use the density $\rho = 1/V$ as the state variable. Introduce scaled dimensionless variables

$$\frac{\rho - \rho_c}{\rho_c} = \rho - 1, \qquad \frac{P - P_c}{P_c} = p - 1, \qquad \frac{T - T_c}{T_c} = t - 1,$$

where (ρ_c, P_c, T_c) is the so-called critical point of the phase transition — there is only one phase for $T > T_c$. One difficulty you will have is that the parameters $p - 1$ and $t - 1$, do not correspond directly to the parameters of the cusp catastrophe. In fact these will be found as linear combinations: $\lambda_b = \frac{1}{3}(p - 1) + \frac{8}{3}(t - 1), \quad \lambda_s = -\frac{2}{3}(p - 1) + \frac{8}{3}(t - 1)$.

In fact, the cusp catastrophe is not a very good predictive model for the gas–liquid transition. The form of the cusp gives

$$\Delta\rho = 2\sqrt{\lambda_b} \sim (T_c - T)^{1/2},$$

but experimental measurements of the densities of liquids and gases on either side of the transition in a neighborhood of the critical point give

$$\frac{\rho_l}{\rho_c} - \frac{\rho_g}{\rho_c} = \frac{7}{2}(1 - \frac{T}{T_c})^{1/3}.$$

Another problem with models such as the Van der Waals equation is the incompatibility with a potential. Consult the references [22], [53], and [80], for further discussion of models of phase transitions.

8. *Plasticity* When subject to a stress that is subsequently relieved, a solid may return to its previous equilibrium. As mentioned in the text, this sort of deformation is known as an elastic deformation. If, however, the solid does not return to its prestressed state, we say it has undergone a *plastic deformation*. What would stress-strain relations look like for materials that undergo both elastic and plastic deformations? The explanation of plastic deformation hinged on supposing certain defects in the crystal structure known as dislocations, which were later observed ([24], [69]).

9. *Financial Panics* The usual mechanisms of investment can be quickly drained of capital in a panic due to a loss in the collective confidence of investors in the security of their investments. A panic is usually experienced in several ways. There may be a run on the banks; people are afraid that the bank, in the future, won't be able to back up their deposits with hard currency and, rather than take the risk, they make withdrawals. People sell their securities, prices fall, and further sales can be made by the issuing corporation only at very low prices. The precipitation into a panic is quick and, once in, seemingly inevitable. Small causes can have great effects. Irresponsible (though not necessarily unintentional) comments were supposedly responsible for much of the ruin in the panic of 1907. Can one think of wealth in terms of an order parameter that accounts for the extent to which wealth is concentrated? Thus, heavy investment in a certain market would be a high order parameter, and so on. What does this parameter depend on? Disposable wealth, public confidence?

10. *Business folds* Can a small change in operation bring about the catastophic collapse of an enterprise? What factors are involved in the health of shopping malls competing with each other? Can you relate failure to the fold catastrophe?

    (a) Suppose that the firm produces some commodity, x, with a cubic cost function. The price, p, is set by market conditions. Show that if the firm follows a profit maximization scheme it can experience catastrophic losses. For similar types of models, see [74].

    (b) The catastrophic decline in shopping centers has been modeled in [83].

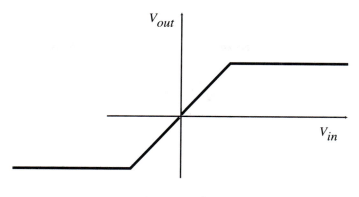

Figure 2.26

11. *Jumps in perception* Lately, "magic-eye" 3D pictures are very popular. There is a background picture, and if one stares in the right way, a new picture jumps out. What are the variables and parameters that describe this phenomenon? Can this be modeled as a catastrophe?

12. *Bistability and hysteresis* The equation for the input voltage in a circuit is

$$\dot{V}_{\text{in}} = \frac{V_{\text{out}} - V_{\text{in}} - V_{\text{control}}}{RC},$$

where $V_{\text{out}} = A(V_{\text{in}})$. Suppose that the gain of the amplifier is given as in Figure 2.26. Show that, depending on the values of V_{control}, there are either two stable equilibria or just one. This change in the equlibrium set takes place at critical values of the control voltage. Show that, by moving the control voltage, we can observe hysteretic behavior, such as with the cusp catastrophe under the delay convention. However, this circuit cannot be modeled as a catastrophe.

Chapter 3

Dimensions

What would you need to know to estimate the amount of gamma radiation passing through you every second? How would the computation change if you need to know the rate of neutrinos from the sun going through you? Is there an optimal running speed through the rain in order to minimize the drenching? Is there any connection between all these questions?

1 Dimensions

The discipline of mathematical modeling necessarily straddles both the "real world" and the abstractions of mathematics. This is fundamentally manifested in the different ways that we use numbers. In mathematics it is accepted that one deals with numbers as entities unto themselves, abstracted as pure sign and independent of ulterior significance. More often than not, in contrast, the objects of our models are not related in a canonical way to any absolute number. Except on those apt occasions when numbers are natural, and these are when we find ourselves counting integral quantities or expressing a proportion of quantities, the numbers that are assigned to quantities make use of standard comparisons. These comparisons define our standards or systems of units. Thus, the numbers representing physical quantities come to be attached to units, and units have dimensions.

This concrete use of numbers is typically associated with some kind of measurement. For example, if we want to measure the distance between two points A and B, we assign a number to this distance by comparing it to another distance that is, presumably, well known or easily reckoned. This latter is the standard and defines the unit of length. Suppose that our standard is a certain wooden bar that we have on hand. If we lay out the distance from A to B, laying down the bar n times end to end, we say that this distance is $n\,bars$ (long). There are three components to this statement: a pure number (n), a unit (the *bar*), and the implication that this unit measures length, which is its dimension. The use of units must accord with the following property of measurement. If we cut our bar in half and agree to use one of the new pieces as the unit, we will find by measurement that the distance from A to B is now $2n$ *half-bars* (long). We have chosen a new scale; our measurements have been suitably rescaled.

The comparison of two quantities that have the same physical dimensions must be independent of the scale we use. If the measured distance from C to D is m *bars*, then we know, without measuring, that the distance is $2m$ *half-bars*. Furthermore, the ratio of these two (physical) distances,

$$CD : AB,$$

will be the pure number m/n independent of which scale is used. We expect the ratio of two physical quantities of the same dimensions to be a physically meaningful pure number, although either one of those quantities will not be expressible without the use of units. We can refer to this observation as the fundamental scaling rule.

Areas could be measured using a certain square tile as the standard. A certain floor space is covered by n tiles, and we would say that the area is n *tiles*. The unit of *tile* has the dimension of area. Since the area of a tile is always the product of the lengths of its sides, the dimension of area is not as fundamental as the dimension of length. Indeed, in any circumstance, the dimension of area is $length^2$. Among the quantities that one uses in modeling, such as length, time, energy, entropy, intensity, product turnover, and so on, some must be distinguished as **fundamental quantities**, and others, expressible in terms of these chosen ones, as area is in terms of length, will be called **derived quantities**. The chosen fundamental quantities will be referred to as **dimensions**. The dimensions of a quantity Q will be denoted by $[Q]$.

It is common in mechanics to take as dimensions the quantities *mass, length, time* (M, L, T) and consider *force, energy*, and the like, to be derived quantities. The distinction between fundamental and derived is not straightforward and is to a large degree a matter of convention. Furthermore, although a minimum number of dimensions will always be required, we can admit as many fundamental quantities as we wish. It might prove useful to have two different length scales, one to measure *east–west* and another to measure *north–south* (Figure 3.1a), or else one to measure the length of waves and the other to measure their amplitude (Figure 1b). To be useful, fundamental quantities must be related to a standard of measurement.

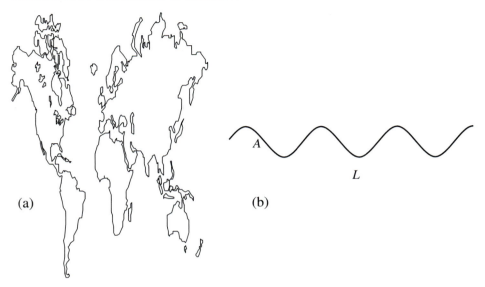

(a)   (b)

Figure 3.1

A basic (almost obvious) property of the fundamental quantities is that which we have called the fundamental scaling rule: the ratio of two measurements of the same quality must be dimensionless. The ratio of distances $AB : CD$ was independent of the measurement standard. If we require that this scaling law should hold for all quantities, fundamental or derived—and we mean by this that if

$$(p_1, \ldots, p_n), \quad (q_1, \ldots, q_n)$$

are the same fundamental quantities in two different scales and

$$f(p_1, \ldots, p_n)$$

is a derived quantity, and we require that

$$\frac{f(p_1, \ldots, p_n)}{f(q_1, \ldots, q_n)} = \frac{f(r_1 p_1, \ldots, r_n p_n)}{f(r_1 q_1, \ldots, r_n q_n)}$$

—then the derived quantity must be proportional to a product of powers of the fundamental quantities,

$$f(p_1, \ldots, p_n) = C p_1^{a_1} \ldots p_n^{a_n},$$

the constant C being dimensionless. If the powers $a_1, \ldots, a_n$ are all zero, then the derived quantity is dimensionless.

A nice property of a model or physical law is that it should hold independent of scale. An equation should be **dimensionally homogeneous**. Thus, any terms that are to be added together in the equation must have the same dimensions. This says a little bit more than that the equation should hold independent of the particular units. Notice that the equation

$$-kx + x = m\ddot{x} + \dot{x}t$$

could be true in any system of units, but it is not dimensionally homogeneous.

Newton's law states that force is equal to mass times acceleration:

$$F = MA.$$

In terms of dimensions, then, $[F] = [MA]$. If M, L, T are fundamental quantities, the dimensions of force are MLT^{-2}. On the other hand, if F, L, T are to be the fundamental quantities, the dimensions of mass can be found as

$$[M] = \frac{[F]}{[A]} = FT^2L^{-1}.$$

A standard system of units for physics and engineering is provided by the *Système International* (SI). The fundamental quantities (dimensions) in this system are: mass, length, time, temperature, electric current, and luminous intensity. This system and its units are listed in Appendix B. A feature of this system is that the *units* of derived quantities are formed in exactly the same way as these quantities are formed from the fundamental quantities. For example, since

$$[\text{force}] = MLT^{-2},$$

the unit of force is kg m s^{-2}. This unit is called the newton (N).

EXERCISE 1 (i)Suppose that the fundamental dimensions are force, length, and velocity. What are the dimensions of mass and time?
(ii) What are the fundamental quantities (dimensions) of the Imperial or "English" system of units? (*Hint*: The mass of an object is not considered fundamental, but rather its *weight* is.)
(iii) What are the standards of measurement that define the Imperial system?
(iv) What is a slug?

Before we involve ourselves in any formal analysis of dimensions, let's look at the dimensions of some derived quantities that we have seen. We will start with the elastic modulus. If a solid with a characteristic length l is deformed by an amount Δl, then the fractional deformation is called the *strain*:

$$\epsilon = \frac{\Delta l}{l}.$$

Notice that ϵ is a pure number; it has no dimensions. The reason for this is that it is a geometric description of the "shape" of a deformation, independent of the absolute magnitude of the deformation. The two (longitudinal) deformations of Figure 3.2 have the same strain.

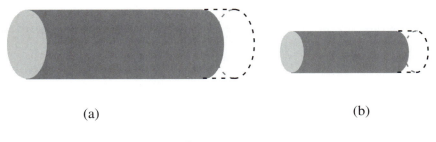

(a)   (b)

Figure 3.2

There are corresponding dimensionless numbers to represent types of strain other than the longitudinal strain pictured in Figure 3.2. Both blocks in Figure 3.3 are *sheared*, and they both have the same *shear strain*, which is represented by the angle of shear.

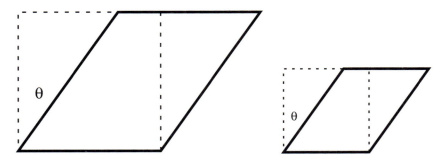

Figure 3.3

In elastic materials, the *stress* is proportional to the strain:

$$\sigma = E\epsilon. \tag{3.1}$$

Since the strain is dimensionless, the elastic modulus E has dimensions of stress. Stress itself is a *force per unit area*, and so

$$[E] = [F/A] = ML^{-1}T^{-2}.$$

As E has the dimensions of stress, we should be able to relate it to some characteristic stress or force per unit area. Indeed, as (3.1) shows, the modulus is the stress needed to effect a strain of 1, or a 100% deformation. (Of course, most materials fail to retain their elasticity when subject to such large stresses. They break.)

EXERCISE 2 Figure 3.4 shows a portion of a bent beam or rod, at some point along its length. The fundamental equation of the bending beam, supposed to hold at that point, is

$$\frac{M}{I} = \frac{S}{d} = \frac{E}{R}. \tag{3.2}$$

In this equation, R is the radius of curvature, I is the moment of inertia of

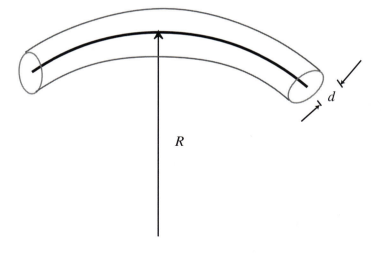

Figure 3.4

the beam section, M is the bending moment, S is the stress, d is the radial distance from the neutral axis, and E is Young's modulus of elasticity. Check this equation for dimensional homogeneity.

The simple idea of requiring that our model equations be dimensionally homogeneous is the basis of the method of *dimensional analysis*. A formal presentation of this method will be given later. For now, we can give a flavor of the method via some simple examples.

How does the speed of sound in a solid depend on material parameters? Sound is propagated through a solid as waves of longitudinal
elastic deformations. The nature of the phenomenon is seen in pulling
and releasing one end of a slinky spring (Figure 3.5): the coils in each
portion of the spring periodically contract and expand as compression
waves move the length of the spring.

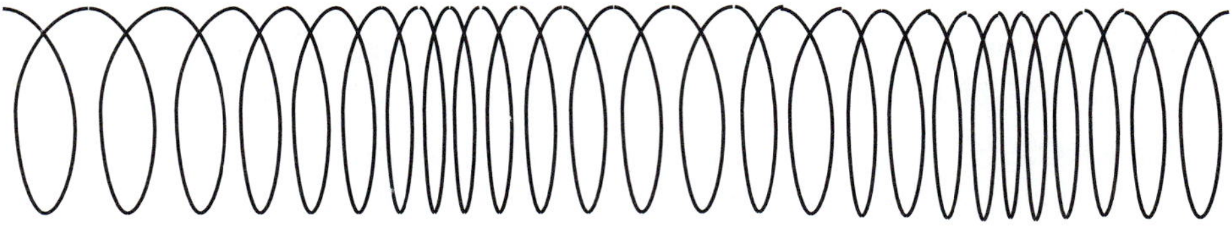

Figure 3.5 Compression waves in a slinky.

Clearly, the elastic modulus, E, will play a role, since it is what
governs the dynamics of each contraction in the material. What else
must be included? Let's compare the dimensions of E to those of the
sound speed, a:

$$[E] = ML^{-1}T^{-2}, \quad [a] = LT^{-1}.$$

E involves the M dimension, but a does not, so whatever relates them
must include mass as a dimension. We can divide a by $\sqrt{E}$ to form a
timeless combination:

$$[\frac{a}{\sqrt{E}}] = L^{3/2}M^{-1/2} = (M/L^3)^{-1/2}.$$

But mass density has dimensions ML^{-3}; so if ρ is the mass density of
the solid,

$$[\frac{a}{\sqrt{E}}\sqrt{\rho}] = 1,$$

then

$$a = \text{const.}\sqrt{E/\rho}. \tag{3.3}$$

How constant is the constant? We know that it is dimensionless, which
means that it doesn't change if we change our units. Yet it may be a
function of dimensionless combinations of other variables of which the
importance escapes us in our ignorance. But, it turns out, in this case
that is all there is to it, and (3.3) gives the speed of sound with the
constant equal to 1 !

From the example, it is clear that dimensional analysis is an accounting procedure, a method, similar to that a financial auditor might employ, intended to establish the logical checks and balances, without inquiring into the precise nature of each entry. Its use can sometimes keep one from making a mistake, or clarify a certain relation among variables, or even, as the example shows, lead to a physical understanding of the phenomenon at hand.

EXERCISE 3 You can increase the "sound speed" of a slinky by holding it stretched out. Why?

Dimensional analysis can lend a hand in the discussion of any equation. For example, the speed of water waves depends on their wavelength. The wavelength is the distance from crest to crest. The speed can refer to a couple of different quantities, but for now you can think of it as the speed of any particular crest. For very long so-called gravity waves, such as one would see far out in the ocean, the relationship is

$$c = \text{const.} \sqrt{g\lambda}, \tag{3.4}$$

where c is the wave speed, λ is the wavelength, and g is the acceleration due to gravity (see Chapter 10). Such a relation is known as a *dispersion relation*.

EXERCISE 4 (i) Show that the constant in (3.4) is dimensionless.
(ii) Find the form of the dispersion relation for small ripples on a quiet pool. The relevant variables (besides the wave speed and wavelength) are the density of the water and the surface tension.

The fact that the constant appearing in the equations (3.3) and (3.4) is dimensionless allows us some confidence that we have included all the appropriate variables. Yet dimensional thinking is sometimes useful even when we know that we have left out something. Suppose that we ask for the form of the dependence of the pressure in a planet's core on the mass, M, and radius, R, of the planet. Dimensional arguments lead us, via the universal law of gravitation,

$$F = G\frac{Mm}{r^2}, \tag{3.5}$$

to

$$P = \text{const.} \frac{M^2}{R^2}. \tag{3.6}$$

The constant appearing here depends on G and is not invariant under changes of scale.

EXERCISE 5 (i) What are the dimensions (using M, L, T) of G?
(ii) If the dimensions (fundamental quantities) are chosen to be M, L, G, what are the dimensions of time?
(iii) Show that energy per volume has the same dimensions as pressure.
(iv) Show that, in a parallel-plate capacitor, the pressure is equal to the energy density.

Problems involving heat flow require a fundamental dimension other than the three of classical mechanics. The new fundamental quantity is temperature, and the unit is the degree. The temperature of a body is a measure of the quantity of the average kinetic energy of the particles of which it is composed. On the macroscopic scale, it determines the flow of heat by the second law of thermodynamics: heat flows from high temperature to low temperature.

EXERCISE 6 (i) The heat capacity, C, of a body is the amount of heat (energy) that is needed to raise its temperature by 1 degree. What are the dimensions of C?
(ii) The equation of state that defines an *ideal gas* relates the pressure, volume, and temperature. It states that for 1 mole of gas

$$pV = RT,$$

where R is the universal gas constant. Find the dimensions of R. (One mole (of anything) is 6.022169×10^{23}. Although the mole is just a counting number and is really dimensionless, it is convenient, and universally accepted, that we include it in the dimensions of any thermal quantity that is expressed in terms of absolute quantity. That is, *enumeration* is a dimension. Think of counting as another way of measuring. In this way, any quantity that depends on the amount of something can be expressed as a quantity per mole.)

*Dimensions in Electricity and Magnetism

To combine the laws of mechanics with electricity and magnetism requires the use of more than three fundamental quantities. As a matter of fact, since current, voltage, charge, magnetic flux, and so on, can all be expressed in terms of one another, we need precisely one more dimension. There are several possibilities for the choice of a fundamental quantity. Electric charge could be a dimension. Coulomb's law,

$$F = k\frac{Q_1 Q_2}{r^2}, \tag{3.7}$$

relates charge to the mechanical dimensions.

Current is time rate of change of charge and thus has dimensions of charge/time. The magnetic field strength or flux can be related to current by way of the law of Biot–Savart. This law states that the magnetic field strength due to charges flowing in a straight wire is proportional to current and inversely proportional to the distance from the wire.

Charge is used as the fundamental quantity in the cgs (centimeter-gram-sec) system so popular with physicists. In the scheme of that system, (3.7) is used with $k = 1$ (dimensionless) and the unit of charge is the esu (electrostatic unit).

To illustrate the use of charge as a dimension, consider the following situation. Light, negatively charged ions (or electrons) are moving around in a matrix of ponderous, positively charged ions, so the charge overall is zero. The static situation is unstable at nonzero temperatures, and what can happen is that a bunch of negatives are attracted to a positive, which makes the net charge in the immediate vicinity of the positive negative, and the negatives thereby experience mutual repulsion, and scatter (Figure 3.6a). Later they are back and the cycle repeats itself (Figure 3.6b). What is the frequency of these oscillations? Frequency is the number of cycles completed in unit time. The strength of the repulsion depends on the charge concentration, and it must overcome the inertia of the small charges. The charge concentration, the amount of charge per unit volume, is equal to the charge of a single particle times the concentration of particles. Thus, the oscillation frequency should depend only on the charge e, the mass m, and the concentration n of the negative charges. So we will write

$$f = \Omega(e, m, n)$$

and carry forth an investigation based solely on dimensional homogeneity. The frequency, f, has dimension T^{-1}. The mass, m, and concentration, n, do not involve time, but we can express charge in terms of M, L, T by use of Coulomb's law (3.7) (with k dimensionless):

$$[Q^2] = L^2[F] = L^3 M T^{-2}.$$

But T^{-2} is the dimensional form of f^2, so

$$[\frac{f^2}{e^2}] = M^{-1}L^{-3} = [m]^{-1}[n],$$

and our result is

$$f^2 \propto \frac{e^2 n}{m}. \tag{3.8}$$

Such oscillations are known as *plasma oscillations*. A plasma is a highly ionized gas, which means that most of the particles of which it consists are charged, but that overall the charges cancel. Plasmas are found in stars and neon tubes and inside some nuclear reactors. It is popularly called the fourth state of matter. Plasma oscillations are common in plasmas, but they can also occur in other physical systems that fit the model presented here. This is a popular model for conduction electrons in a metal.

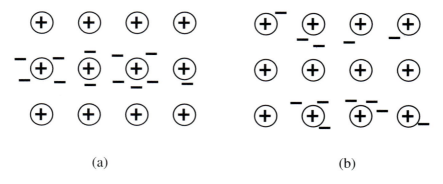

(a)   (b)

Figure 3.6 Plasma oscillations. Opposite phases in the oscillation cycle are represented by a and b

A different choice for the fundamental quantity in electromagnetism is the magnetic pole strength. The force between two magnetic poles of strengths m_1, m_2, respectively, is

$$F = \frac{m_1 m_2}{\mu r^2},\qquad(3.9)$$

where μ is the magnetic permeability. One can express μ in terms of F and m_1, m_2, or one can take it to be dimensionless, or μ itself can be taken to be the fundamental quantity.

EXERCISE 7 Taking μ as a fundamental quantity, find the dimensions (M, L, T, μ) of the following (note especially which ones have the same dimensions).
(i) Magnetic moment (magnetic pole strength · length)
(ii) Intensity of magnetization (magnetic moment per unit volume)
(iii) Magnetic field strength, H (force per unit magnetic pole)
(iv) Magnetic induction $(B = \mu H)$
(v) Magnetic flux $(\int B \cdot dA)$
(B is sometime called the magnetic flux density.)

The electrical quantities (charge, current, and so on) can be related to these magnetic quantities by use of the Biot–Savart law. In terms of dimensions, this is expressed as

$$[H] = [I]L^{-1}. \tag{3.10}$$

Since the force (per unit length) between two parallel wires that are a distance d apart, each conducting a current I, is

$$F = k\frac{I^2}{d}, \tag{3.11}$$

another possibility presents itself: we can use the current as the fundamental quantity. This is what is done in the SI (mks) system. (All the units of SI are in Appendix B.) The unit of current is the ampere (A) and is defined by (3.11) with $d = 1$ m and $F = 2 \times 10^{-7}$ Nm^{-1}. The unit of magnetic flux density is the tesla (T) and is defined by the fundamental force law

$$F = q(v \times B).$$

2 Scaling and Life

Our use of the term scaling refers to the effects produced in models by a change of the scale used for any of the fundamental quantities (dimensions). The basis for a study of scaling effects is the ancient notion of similarity or proportion. If two triangles have the same shape, then the ratio of any corresponding lengths is the same (Figure 3.7), so

$$\frac{l_1}{l_2} = r = \frac{h_1}{h_2}.$$

We know that the ratio of their areas will be the square of the ratio of their sides:

$$\frac{A_1}{A_2} = r^2.$$

We say that "area scales as length squared."

Jonathan Swift was an educated geometer and used the idea of proportionality in *The Voyages of Gulliver*. As Gulliver was in length equal to 12 Lilliputians, and the volume of any corresponding organ thus being 12^3 times that of a Lilliputian, he was offered a feast of meat and wine that was 12^3 times as much as a typical Lilliputian repast. We shall have occasion in the following to question Gulliver's

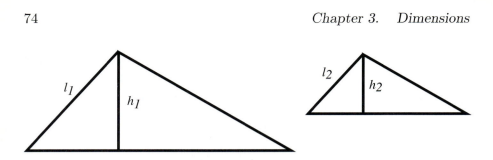

Figure 3.7 Geometrically similar figures.

need for such sustenance. For, even given the fact that Gulliver was structurally geometrically similar in every way to his hosts, one has no reason to expect that any physiological function, such as energy requirement, should scale in the obvious way. On the other hand, what else should we do but base our arguments on geometric similarity if and when we are faced with a lack of experimental data?

Most things that come under the rule of physical laws do not scale easily. The properties of matter do not. The picture of water flowing at a speed of 1 ms^{-1} in a small tube is not just a geometrically similar picture of the flow at the same speed through a much wider channel (Figure 3.8).

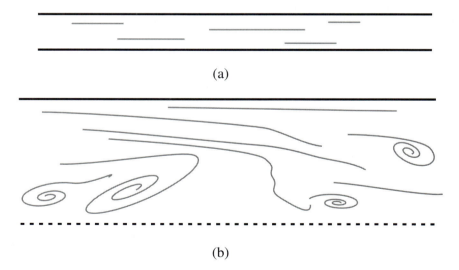

Figure 3.8 (a) Water flow in a thin tube. (b) Water flow in half of a wide channel (after a photograph).

We can consider length, time, or energy scales that are character-istic of some physical objects. Raindrops do not come in all sizes. The wave speeds of water waves do not scale evenly with wavelength; for long gravity waves, speed is proportional to the square root of wave-length, while for ripples, speed is inversely proportional to the square root of wavelength.

Our interest in the present section is to study how similarity and scaling can help us to understand how biological structure and function can depend on size. The interesting thing is that even some structural features of organisms do not scale in the way that one would think that they should. For example, the size of the cell is almost constant. The whale does not have whale-sized red blood cells and the mouse have wee ones, but they are the same size in each case (about 10^{-3} mm in diameter). The whale just has a whole lot more. One explanation for this can be found by combining the mathematics of geometric similarity with a regard to the function of the cells. The metabolic contents of the cell need oxygen to do their work. The content of the cell scales as the volume of the cell (thus, as L^3, where L is a characteristic length scale of the organism). One the other hand, the oxygen is received through the cell wall and unless the cell could somehow speed up diffusion, the availability of oxygen scales as L^2. If we define *efficiency* as the oxygen available per amount needed, we see that efficiency scales as L^{-1}. Large structures are less efficient. There is a size limit above which a cell becomes too big to supply itself with oxygen. A scaling argument can lead us to conclude that there is a cell-size limit and that we might expect all cells to be within a couple orders of magnitude in size. It only goes a small way to explaining why cell sizes are so nearly the same.

A few points, related to the assumptions that we make, must be borne in mind when sallying forth with this type of argument. We made implicit use of geometric similarity so that, strictly speaking, a conclusion should only be drawn for cells of one shape. In this way, a cell could escape the ban on large size by choosing a shape with a higher ratio of surface area to volume. Indeed, there exist very long nerve cells. In these arguments we are to some extent guided by as-sumptions that nature works according to an economy of material and an optimality of function. From this, one can then go quite a ways with volume/area considerations. For, although the load on a biolog-ical structure usually scales as the cube of a characteristic length, its strength will be proportional to its cross-sectional area. In broadest strokes, the reason for this is that although all elementary forces act pointwise, only gravity and electromagnetism are long range, and only gravity "adds up" in being proportional to matter. The effects of the electromagnetic forces are often experienced when two pieces of matter strike against each other. This is the origin of the usual "mechanical forces." (The reader should not get the impression that when a structure fails it is because the accumulated gravity forces have overcome the electromagnetic forces. Rather, the failure of a material is due to defects in its internal organization. Readable accounts are provided by [69] and [24].)

We have made reference to the fact that blood cells and nerve cells have considerably different shapes. The diversity of geometries in biological organisms makes the choice of a characteristic length to represent size a difficult matter. All animals are mostly water and so have roughly equivalent mass densities. This means that volume will scale as mass, and as it is an easy matter to weigh an organism, mass is more often than not taken to represent the size. Mass measures the amount of elemental matter present. Quantities that are to be considered per unit mass are said to be *specific*. For example, specific entropy is entropy/unit mass.

As a good illustration of scaling in biology and especially the dangers of a too quick reliance on geometric similarity, let us analyze the height of an animal's jump. How does the high jump of an animal depend on its size? Let h be the height of the jump and L a length scale. Clearly, we have to equate the energy needed to make the jump to the energy available. If we are careful to scale these two quantities according to the size of the animal, then we can solve for h in terms of the size. Let us set down our scaling assumptions in a series of propositions.

1. A jump of height h requires a change in potential energy of mgh.

2. The work done by the animal is the energy production of the muscle, being proportional to its metabolic content, and thus proportional to its volume.

3. The volume of a muscle is proportional to its cross sectional area times L.

4. The cross-sectional area of muscle is proportional to the cross-sectional area of the bone.

5. The cross-sectional area of the bone gives the strength of the bone, which must be proportional to the load, which is proportional to the mass of the animal.

6. In combining 1,3,5, we must conclude that

$$h \sim \frac{mL}{mg} \sim L.$$

Our conclusion should be that an animal's jump scales proportional to its size. Unfortunately (well, perhaps, fortunately) this does not fit the facts. The reader has perhaps noticed that dogs can jump as high as horses. In fact, as Table 3.1 shows, animals on all size scales jump roughly to the same height. Actually, the flea has a higher surface area to volume ratio and thus suffers more from air resistance; in a vacuum the contest between flea and human would be even closer. While the jumps listed in the table are of the same order of magnitude, the size scale spans *10 orders of magnitude*. The evidence is unambiguous: the conclusion we came to is wrong.

Table 3.1 Jump Heights of Various Animals

Animal	Jump Height (m)
Horse	3
Dog	3.5
Man	2
Flea	.3
Beetle	1
Frog	2
Antelope	1.5
Dolphin	4.5
Squirrel	1.5
Rat	2

Let us go over our track to see where we have have erred. There can be doubt as to the veracity of propositions 1 and 3. Also, as we have mentioned, studies in the strengths of beams and columns lend support to proposition 5: the strength of a bone will depend on its cross-sectional area. From a naive viewpoint, this is perhaps the most suspect of the lot, since one would expect a "characteristic area" to scale as the square of a "characteristic length," not as the cube. If we compare the skeletons of genetically similar animals that differ in size, we see that the large animal has bones that are thicker than geometric similarity would allow (Figure 3.9). This seems to be in line with an optimality assumption. Thus, we can with confidence state that bone thickness scales as L^3, while muscle thickness scales as L^2. If this were not so, large animals would be strong enough to break their own bones. This may eventually turn out to be the case with animals bred for performance.

Statement 4 is then the weak link. (Statement 2 is essentially correct, although it may seem to contradict our earlier discussion of cell size. We will return directly to the problem of scaling the metabolism.) Briefly, our argument is then modified to run as follows: The strength of a muscle scales as L^2 and so the work available for jumping $\sim L^3$, and as energy required $\sim L^3$, the height of the jump must be constant.

EXERCISE 8 (i) Explain why it is to be expected that small animals such as ants are proportionally stronger than large animals.
(ii) Notice that our modified argument at some point relies on geometric similarity. Do long lean animals jump higher than short squat ones?
(iii) If geometric similarity is not allowed, then we cannot assume that $m \sim L^3$. Show that this point is not essential.

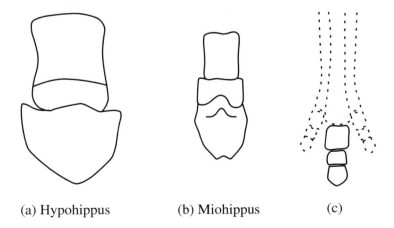

    (a) Hypohippus   (b) Miohippus   (c)

Figure 3.9 The corresponding bone in two extinct closely related species of horse are contrasted. Hypohippus was a significantly larger animal. The location of the bones on the lower leg is indicated in (c). Early recorded observations of this sort are in the writings of Galileo.

You'll recall that Gulliver was delivered his victuals in an amount dictated by a scaling based on geometric similarity. Whether G. actually requires that much depends on the rate at which he can convert it all to energy; this is his *metabolic rate*. One might think, subsequent to our arguments in limiting cell size, that, since oxydation occurs across surfaces and any heat that is produced in metabolism must be carried away through surfaces, the metabolic rate should scale as L^2, which is the same thing as saying $\sim m^{2/3}$. This is what was generally believed until 1932 when a worker at a California agricultural station made some measurements of oxygen uptake and plotted metabolic rate versus mass for a variety of animals. Others soon made measurements on many other (warm-blooded) animals. The plot of the congregrate results is known as the "mouse-to-elephant curve" (Figure 3.10). It is clearly seen in this log–log plot that the proper exponent is not 2/3, but 3/4. Thus, if P is metabolic rate, then

$$P \sim m^{3/4},$$

or

$$P \sim L^{5/2}.$$

There are perhaps many explanations for this. Although energy transfer transpires at the surface of the capillary or alveoli, an organism can do better than geometric similarity would seem to allow; this is most easily done by increasing the density of these energy transfer structures. Indeed, as the example in Chapter 1 shows, the ideal form of these structures comes close to space-filling. This would imply that the energy transfer rate could, in the perfect limit, scale as L^3. It is not unreasonable to suppose that compromise due to finiteness of the branching and conflict with other structures have conspired to reduce the scaling power. Another explanation [46] is based on a formula from the theory of elastic columns that relates the critical length for buckling, l_c, to the diameter, d:

$$l_c^3 = k\frac{E}{\rho}d^2 \tag{3.12}$$

Then, as weight should scale as ld^2,

$$d^2 \sim m^{3/4}.$$

If we assume that power output of the muscle $\sim d^2$, then

$$P \sim m^{3/4}.$$

We are accustomed to thinking of small animals as having elevated rates of metabolism. The metabolic rate P as used here accounts for the organism's total consumption of oxygen. An elephant inhales almost 10^5 times as much air per breath as does a mouse. To better compare different organisms, we should compare specific metabolic rates (per unit mass). Thus,

$$P_{\text{spec}} = \frac{P}{m} \sim m^{-1/4}.$$

It is interesting to note that empirical measurements give a heart rate (beats/unit time), f_h, that scales as

$$f_h \sim m^{-1/4}.$$

Can this frequency or its implied period be correlated to life expectancy? The interested reader is referred to the literature (cf. [62]).

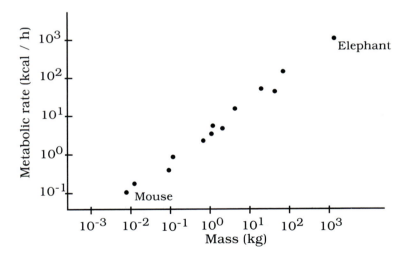

Figure 3.10 The mouse-to-elephant curve. Metabolic rate is plotted versus body mass for various warm-blooded animals. Notice that the data points lie on a line with slope 3/4 in log–log coordinates.

3   Dimensional Analysis and the Pi Procedure

Earlier in this chapter, we had deduced, using simple assumptions about dimensions and rather minimal physics, that (3.3) gives the speed of sound in a solid. Another way of stating this conclusion is that we have identified a certain combination of the variables,

$$\frac{a^2 \rho}{E},$$ (3.13)

that is *dimensionless*. If we were to rescale one of the dimensions, say length, and were careful to express all quantities in terms of the new units, quantity (3.13) would not be affected. The same can be said for the combination

$$\frac{V^2}{g\lambda},$$ (3.14)

formed from the parameters of long gravity water waves. However, these two examples are very different in one respect. The three derived quantities that form the combination (3.14) involve only the two dimensions L, T, while the three derived quantities that form (3.13) involve the three dimensions M, L, T. We should not expect to be able to form a dimensionless number from three quantities that involve three dimensions. For instance, the three variables P, M, R of (3.6) cannot form a dimensionless number. On the other hand, it is natural to expect to form the number given by (3.14).

A number such as (3.13) or (3.14) that is the product of powers of derived quantities is known as a **dimensionless group**. A formal process has evolved, called here the Pi-procedure, for finding dimensionless groups using dimensional analysis, and has been codified by a couple of statements known as the Buckingham Pi theorem. (Pi refers to the greek letter Π, traditionally used to designate dimensionless groups.) As a more attractive alternative to simply stating the theorem and then applying it, we will let it develop through some illustrative examples.

In Chapter 1 we looked at a steady velocity profile for a fluid flow in a pipe. This Poiseuille flow with its quadratic profile is known as a **laminar flow** because we can think of it as thin layers of fluid flowing, or sliding, one on top of the other (Figure 3.11a). This smooth type of flow, which you might see in a slow-moving river in late summer, is to be contrasted with the chaotic motion of the springtime stream, with its burbling breakers and swirling whirlpools that come and go and sometimes merge together (Fig 11b). This latter sort of flow is known as **turbulence**. In turbulent flow, the velocity field is a complicated function of both space and time. Turbulence is very important in technology because many phenomena of interest, such as mixing and heat transfer, are greatly enhanced in the presence of turbulence. The

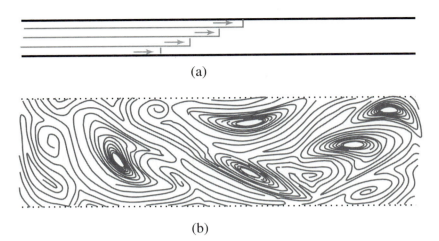

(a)

(b)

Figure 3.11 (a) Laminar flow and (b) turbulent flow.

modeling of turbulence presents one of the most formidable challenges facing science.

It is observed that the flow in any situation will become turbulent if the average flow velocity is great enough. You may have noticed this. When the water flows slowly from a faucet, the jet is smooth and transparent; but if we open the tap, we see the jet become rough, frosty, confused. The laminar flow becomes *unstable* for average flow velocities that exceed a critical speed, V_c. How does V_c depend on the parameters that describe the flow in a pipe? These are ρ, the density of the fluid, μ, the viscosity, and d, the diameter of the pipe. The dimensions of these quantities are

$$[V] = LT^{-1}, \quad [\mu] = ML^{-1}T^{-1}, \quad [\rho] = ML^{-3}, \quad [d] = L.$$

Time appears as a dimension only in V and μ, and so

$$[\frac{V}{\mu}] = M^{-1}L^2.$$

Also,

$$[\rho d] = ML^{-2},$$

and so

$$\frac{V\rho d}{\mu} = \text{const.}$$

In other words,

$$V_c = R\frac{\mu}{\rho d}, \tag{3.15}$$

where R is a dimensionless number.

EXERCISE 9 If the transition from laminar to turbulent flow for water in a pipe of diameter 1 cm occurs at $V_c = 2$ m s^{-1}, what is the velocity at which blood flowing in the same pipe becomes turbulent?

The objective of the Pi procedure is to find all the dimensionless groups we can form from a given set of variables. We will illustrate, using the example just above. Having set

$$\Pi = \mu^\alpha \rho^\beta d^\gamma V^\delta, \tag{3.16}$$

we look for all $\alpha, \beta, \gamma, \delta$ so that Π is dimensionless. Substituting into (3.16) the appropriate dimensions of the variables, we get

$$\begin{aligned}
\Pi &= (ML^{-1}T^{-1})^\alpha (ML^{-3})^\beta L^\gamma (LT^{-1})^\delta \\
&= M^\alpha L^{-\alpha} T^{-\alpha} M^\beta L^{-3\beta} L^\gamma L^\delta T^{-\delta} \\
&= M^{\alpha+\beta} L^{-\alpha-3\beta+\gamma+\delta} T^{-\alpha-\delta}.
\end{aligned}$$

Since Π is to be dimensionless, we are led to the system of linear equations for $\alpha, \beta, \gamma, \delta$:

$$\begin{aligned}
\alpha + \beta &= 0 & (3.17) \\
-\alpha - 3\beta + \gamma + \delta &= 0 & (3.18) \\
-\alpha - \delta &= 0. & (3.19)
\end{aligned}$$

It is easy to see that this system is rank 3 and thus there is a one parameter set of solutions:

$$\begin{aligned}
\alpha &= -\delta, \\
\beta &= -\delta, \\
\gamma &= \delta.
\end{aligned}$$

With $\delta = 1$, we get the parameter $R = \frac{\rho V d}{\mu}$ that we found above. All possible dimensionless groups that we can form from these variables are found by letting δ range over all real numbers; these are all powers of R. The choice of $\delta = 1$ specifies a basis for the solution set of the system of equations (3.17–3.19), and so we can think of R as generating all dimensionless groups in this example. R is called the Reynolds number, after O. Reynolds who studied the stability of flow in pipes. (His dimensional analysis is essentially that leading to (3.15).) In experiments with water flowing through glass tubes of various diameters, he demonstrated that the transition to turbulent flow, when the laminar solution becomes unstable, occurs at a critical value of R. Thus, the description of this aspect of the flow, which seemed to require that we use the variables ρ, μ, d independently, essentially depends only on R. In this way, any homogeneous relation of the form

$$F(\rho, V, d, \mu) = 0$$

can be reduced to a relation involving only R:

$$G(R) = 0.$$

Even in this simple case, there are factors acting to foil any predictions based only on R. These include the roughness of the pipe surface and the shape of the pipe cross section. One could model these independently (see problem 13 in the last section of this chapter). Notice that the physical properties of the flow only enter into the Reynolds number through the ratio $\frac{\mu}{\rho}$. This ratio is called the **kinematic viscosity** and is usually denoted by ν. We will have more to say about this in the next section, but for now simply note that a large density could cancel a large viscosity. That is, a heavy viscous fluid could have a flow pattern that resembles a light fluid that is less viscous. This is because those tumbling whirlpools have inertia, and there is only viscosity to tear them down. In this way, the flow of heavy metals in large planetary cores is often modeled as inviscid flow.

We now turn to a related example that will illustrate more fully the Pi procedure. A sphere moves through a viscous incompressible fluid. In what manner does the resistance that it experiences depend on its velocity and any other relevant factors, such as its size? (The resistance felt by a body in relative motion with a fluid is called a *drag force*.)

One possible way to answer this question is to do experiments. We can attach the sphere to a spring balance to measure the force that is registered for many different values of the velocity. That would be fine, but the experimental results would only apply to this particular sphere. To find drag force as a function of velocity for the general sphere, we would have to repeat the whole series of assays for many different sizes. This is a great increase in labor, and when we are done we would still only know the answer for spheres moving in a particular medium. If we wanted to get the lowdown for a different fluid, we would have to repeat the experiments again. Is anything to be done about this?

Dimensional analysis can help to simplify matters.[1] We follow the Pi procedure. What are the relevant variables? We want to relate the velocity, V, to the drag force, F_D; the fluid is viscous and incompressible, so that μ and ρ are important; and the diameter of the sphere D can represent its size. (If the fluid were compressible, then we should include thermodynamic parameters. Some fluids, water for example, are nearly truly incompressible. Others, such as air, are compressible, but at low enough speeds the effects of the compression are negligible. At airspeeds of less than 200 ms^{-1}, a flow can safely be taken to be incompressible.)

With these variables in hand we try to form a dimensionless parameter:

$$\Pi = V^a \rho^b \mu^c D^d F_d^e.$$

We write in terms of the dimensions

$$[\Pi] = L^a T^{-a} M^b L^{-3b} M^c L^{-c} T^{-c} L^d M^e L^e T^{-2e}.$$

The requirement that Π be dimensionless leads us to the system of linear equations:

$$a - 3b - c + d + e = 0, \tag{3.20}$$
$$-a - c - 2e = 0, \tag{3.21}$$
$$b + c + e = 0. \tag{3.22}$$

The first equation gives the exponent of the dimension L, the second the exponent of T, and the third the exponent of M.

EXERCISE 10 Solve this system.

The system (3.20–3.22) is rank 3 and thus has a two parameter set of solutions. By fixing any two of a, b, c, d, e, we get a dimensionless group. For example, if we set $e = 0, d = 1$, then $a = 1, b = 1, c = -1$; and we obtain

$$\Pi_1 = V \rho \mu^{-1} D,$$

which is just the Reynolds number appropriate to this situation. By choosing any solution of (3.20–3.22) that is (linearly) independent of

---

[1] The rampant complexity that is described in the previous paragraph is sometimes called the *curse of dimensionality*. The cure is through a completely different use of the same root word.

$(1, 1, -1, 1, 0)$, we arrive at a dimensionless number that is independent of Π_1.

EXERCISE 11   State precisely what independence of two dimensionless numbers means.

As an example, if $e = 1, d = 0$, then $a = 0, b = \frac{1}{3}, c = -2$, and we get

$$\Pi_2 = \rho^{1/3}\mu^{-2}F_D.$$

This number is independent of Π. This follows from the fact that the corresponding solution vectors of (3.20–3.22) are independent.

Neither of the numbers Π_1 or Π_2 gives us directly what we first set out to discover, which is a relation between drag and velocity. Can we find a dimensionless number that includes both drag and velocity? Of course: simply choose an appropriate solution of (3.20–3.22), say $(a, b, c, d, e) = (-2, -1, 0, -2, 1)$, which leads to

$$\Pi_3 = V^{-2}\rho^{-1}D^{-2}F_D.$$

This is the dimensionless group that we would have found if we had neglected viscosity in our original formulation.

This number clearly can't be independent of the first two, since the set of solutions of (3.20–3.22) is two dimensional, so we can write

$$\Pi_3 = g(\Pi_1, \Pi_2), \tag{3.23}$$

where (3.23) reflects the fact that Π_3 can be expressed as a product of powers of Π_1 and Π_2. As we can choose any two linearly independent solutions as a basis for the solution set of (3.20–3.22), let them be those that lead us to Π_1 and Π_3. The latter is interesting because it says that F_D is proportional to V^2:

$$F_D \propto \rho D^2 V^2. \tag{3.24}$$

This can only be the case when we neglect viscosity; if viscosity is included, the actual expression cannot be found directly from dimensional analysis. What can we do?

The Buckingham Pi theorem states, in effect, that any rationally homogeneous relation among the variables V, ρ, μ, D, F_D is equivalent to a relation involving just the two dimensionless variables Π_1 and Π_3. That is, if

$$F(V, \rho, \mu, D, F_D) = 0$$

is dimensionally homogeneous, it is equivalent to

$$G(\Pi_1, \Pi_3) = 0.$$

If we go ahead and apply the implicit function theorem, we can write

$$\Pi_3 = f(\Pi_1),$$

or

$$F_D = \rho D^2 V^2 f(R), \tag{3.25}$$

where $R = \Pi_1$ is the Reynolds number. This is all that we can conclude by use of dimensional analysis, yet it is quite a boon. Although, we do not get directly a relation that gives the drag force—the function f is not known—our experimental labor will be greatly reduced. Essentially, our only experimental parameter will be the Reynolds number R, and we can vary that by any variation of the flow parameters that we choose. Experiments like this have been carried out, of course, and results typical of these efforts are displayed in Figure 3.12. The function f is plotted vs. R. (Some experiments may still be difficult to carry out; cf. Section 3.4.)

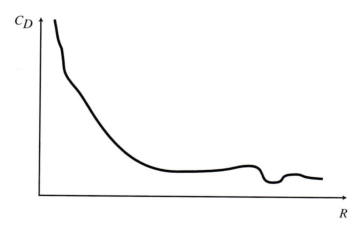

Figure 3.12 Measured drag coefficient versus Reynolds number.

In the low R region the graph indicates that

$$f \sim C/R,$$

so

$$F_D \propto \mu D V.$$

In this regime of "creeping flow," the drag is proportional to both viscosity and velocity. For larger values of R, there are some regions where

$$f \sim \text{const.},$$

and there

$$F_D \propto \rho D^2 V^2.$$

In this "high Reynolds number flow" the drag is proportional to the square of the velocity and independent of viscosity. This regime is sometimes called *inertial* because the drag or resistance depends on the mass density. Notice that the graph exhibits a little "blip" at R_c that separates two inertial regimes. This is the value where the thin layer of fluid about the sphere becomes turbulent, resulting in a lower drag coefficient. The number R_c is known to be dependent on the roughness of the sphere surface.

EXERCISE 12 Why is a golf ball dimpled?

*The Pi Theorem

With these examples in hand, we proceed with a formalization of the Pi procedure. The reader is encouraged to read the present section while referring to the previous examples.

Suppose that we are given n fundamental quantities (dimensions), $p_1 \ldots p_n$, from which we derive m quantities, $q_1, \ldots, q_m$, as follows

$$q_i = p_1^{a_{1i}} p_2^{a_{2i}} \ldots p_n^{a_{ni}}, \quad i = 1, \ldots, m. \tag{3.26}$$

A choice of scale can be represented by a vector

$$\boldsymbol{\lambda} \in R_+^n = R_+ \times \cdots \times R_+,$$

so that the scaling itself is

$$\boldsymbol{\lambda} \cdot \mathbf{p} = (\lambda_1 p_1, \ldots, \lambda_n p_n),$$

which induces, via (3.26), a scaling of the derived quantities:

$$\boldsymbol{\lambda} \cdot \mathbf{q} = (\lambda_1^{a_{11}} \lambda_2^{a_{21}} \ldots \lambda_n^{a_{n1}} q_1, \ldots, \lambda_1^{a_{1m}} \lambda_2^{a_{2m}} \ldots \lambda_n^{a_{nm}} q_m). \tag{3.27}$$

Another way of saying that a quantity is dimensionless is that it be left unchanged under a scaling of the form (3.27). We are going to derive an algebraic criterion for the number of dimensionless groups that can be formed from the quantities q_i.

Among the functional relations of the q_i,

$$F(q_1, \ldots, q_m) = 0,$$

we are interested in those that are left unchanged by scalings of the type (3.27). This would be the case for any dimensionally homogeneous relation. The conclusion of the Pi theorem is that any such invariant relations are equivalent to relations expressed only in terms of the dimensionless parameters. This results in a considerable simplification when there are few such "pure numbers."

When the Pi procedure is followed to find dimensionless numbers,

$$\Pi = q_1^{x_1} \cdots q_m^{x_m},$$

we are led to n linear equations for the m unknowns:

$$\sum_{k=1}^{m} \alpha_{jk} x_k = 0, \quad j = 1, \ldots, n. \tag{3.28}$$

Let $A = \{\alpha_{ij}\}$ be the $n \times m$ coefficient matrix, with $s = \operatorname{rank} A$. Solving the system of homogeneous linear equations

$$A\mathbf{x} = 0$$

is equivalent to computing the nullspace, $N(A)$, of the matrix A. Since rank $A = s$, the dimension of this nullspace is $m - s$. And so $N(A)$ has a basis of $(m - s)$ vectors $\mathbf{v}^1, \ldots, \mathbf{v}^{m-s}$, in other words, there are $m - s$ independent solutions to (3.28). Figure 3.13a shows that in x-space the $\mathbf{v}^k$ provide a basis for the $(m - s)$-dimensional horizontal coordinate space. The s-dimensional leaves are constant $\mathbf{v}^k$-coordinate surfaces. This gives rise to a corresponding picture in q-space with non-linear leaves that are the constant-Π_k surfaces (Figure 3.13b), where

$$\Pi_k = q_1^{v_1^k} q_2^{v_2^k} \cdots q_m^{v_m^k}, \quad k = 1, \ldots, m - s.$$

That is,

$$\Pi_k(q) = \Pi_k(q') \iff \mathbf{q} = \boldsymbol{\lambda} \cdot \mathbf{q}',$$

forall k.

EXERCISE 13 (i) Prove this last statement.
(ii) Show that $\Pi(q)$ is dimensionless if and only if $\Pi(q)$ is a product of powers of the $\Pi_j, j = 1, \ldots, m - s$.

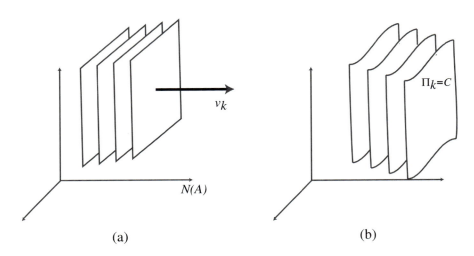

Figure 3.13

We conclude that the effect of the scaling transformation is to move a point $\mathbf{q}$ along the s-dimensional leaf to which it belongs. If any functional relation of the $q_1, \ldots, q_m$,

$$F(q_1, \ldots, q_m) = 0,$$

is invariant under scaling, then it is constant along these s-leaves. But the

$$\Pi_1, \ldots, \Pi_{m-s}$$

provide coordinates orthogonal to these s-surfaces. In other words, such relations can be written using only the $\Pi_1, \ldots, \Pi_{m-s}$:

$$G(\Pi_1, \ldots, \Pi_{m-s}) = 0.$$

Limitations and Extensions

We should mention the limitations of dimensional analysis. One is unlikely to learn a lot about any physical phenomenon if any formal mathematical procedure is approached in a mechanical manner. On the other hand, used as part of a balanced attack, dimensional analysis can be enlightening, and it can point the way toward other avenues of investigation. As an example of this, we refer the reader to Problem 18 and Komogorov's dimensional analysis of turbulent energy scales.

Other facets of dimensional analysis will be encountered later in the book; these include nondimensionalization and similarity solutions of differential equations (look under *scaling* in the index).

4 Scale Modeling, or How to Be Understood at the Fluid Dynamicist's Convention

What are the factors that govern the motion of an incompressible fluid in a given situation—through a culvert, or about a sharkfin? Let us be concrete and refer to the flow about the obstacle in Figure 3.14a. Besides the specification of U, the speed of incoming flow, and the material properties, density (ρ) and viscosity (μ), what is of obvious importance is the "geometry" of the situation, illustrated here by the shape of the projectile. It is this latter variable that one expects to be the deciding one in the specific flow structure and the hardest to model. For one thing, we cannot use a scalar parameter to represent the shape of an obstacle; and even in cases where it is possible to do so, as with a disc or a sphere, the relation of this ("all points at a distance d from a point") to the flow is not easy, since the flow reacts to the obstacle through information about the curvature, and so on.

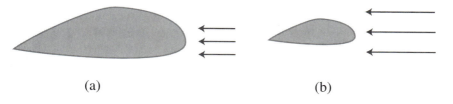

(a)   (b)

Figure 3.14 Scale modeling.

Suppose that we factor away precisely this geometry variable. In other words, we build a geometrically similar projectile, a *scale model*, and compare the flows about the two obstacles (Figure 3.14). Should we expect the flow itself to be similar? Recall that the flow with these variables is characterized by one dimensionless number, the Reynolds number,

$$R = \frac{\rho V l}{\mu}. \tag{3.29}$$

This number must be invariant under scaling of the dimensions. Any

scaling of the length
$$l' = rl$$
must be accompanied by an appropriate rescaling of the other parameters so that
$$R = R' = \frac{\rho' V' l'}{\mu'}.$$

This is *dynamic similarity*. When dynamic similitude is guaranteed with the Reynolds number, we call it Reynolds modeling. If you wanted to test a model in the same medium as the real thing, you'd have to scale the velocity inversely to the length. This makes scale modeling less practicable than would be imagined at first glance. If it is an airplane that we are modeling, then the airspeed needed in the wind tunnel can be prohibitive. For example, suppose that the airplane is a B-52 with a length of 48 m, and we want to study the flow at 100 m/s (220 mph). To use a 1-m model we would have to have a wind tunnel speed of 4800 m/s which is too high. In fact, it is well over 10 times the speed of sound; even if a wind tunnel were available with such a draft, the flow around the model would not be incompressible, and we would be out of the proper regime of Reynolds modeling. Even a large model of 5 m would require a tenfold increase in windspeed, which in this case is still supersonic. For these reasons and others, depending on the type of craft being modeled, whole craft are not typically modeled at once. Perhaps one is content with detailed knowledge of just the flow about a wing (which is still an experimental challenge for, say, the 25- m wings of a B-52).

It is possible to change the material medium in order to respect dynamic similarity. In the airplane experiment, if we switch from air, with a kinematic viscosity of $\frac{\mu}{\rho} = \nu = 15 \times 10^{-6}$ m^2 s^{-1}, to water, with a kinematic viscosity of $\frac{\mu'}{\rho'} = \nu' = 1 \times 10^{-6}$ m^2 s^{-1}, see that we can reduce the needed flow speed by a factor of 15. To model a 1-m wing of a B-52 at the speed mentioned would require a flow speed of 160 m/s, which still requires prodigious engineering.

EXERCISE 14 How could you simulate the dynamical environment of a spermatozoid or other microscopic swimmer using common household materials?

The physical quantities corresponding to the parameters in our model of the flow can be affected by factors not present in the model. The viscosity of many materials has a marked dependence upon temperature, decreasing as temperature increases. (On January 19, 1994, the temperature in Dayton was $-25°$ and the gasoline at the pumps was so viscous that the flow was noticeably slower.)

Scale modeling illustrates a fairly rigorous use of the Reynolds number. One would like to go further and use it as a description of flow. It is true that, in every sense, a flagellating bacterium is always in a low Reynolds number flow as opposed to the high Reynolds number flow of an attack submarine (the difference ranges over 15 orders of magnitude), yet a look at some of the pictures in the book [17] will convince one that it is not possible to categorize all flow situations as being either "low R" or "high R."

It is perhaps surprising that flow in water is always at a higher R than corresponding flow in air. We have already noted that this is because it is not the viscosity, μ, alone that matters in a dynamical situation, but the ratio, $\frac{\mu}{\rho} = \nu$, the kinematic viscosity. And water, while being more viscous than air, is much heavier. It is interesting, and instructive, to try to find a physical reason for this. In fact, the Reynolds number, as is the case for many physically relevant dimensionless numbers, is the ratio of two forces or energies. To see this, note that the kinetic energy density in the fluid, representing its inertia, is

$$\text{Kinetic Energy (per vol)} = \frac{1}{2}\rho V^2$$

and the work done by the viscous (friction) force is the rate of shear strain,

$$\mu \frac{V}{l}.$$

The ratio of these is

$$\frac{\rho V^2}{\mu \frac{V}{l}} = \frac{\rho V l}{\mu} = R.$$

(This might worry you, since the rate of strain should be V/d, where d is a characteristic length transverse to the flow (for example, the distance across the boundary layer), while the characteristic length, l, may be parallel to the flow. However, in scale modeling geometric similarity makes this irrelevant. For such reasons we also ignore such factors as the $\frac{1}{2}$.)

By use of such energy balancing, foregoing tedious dimensional analysis while obtaining some physical intuition, one can arrive at many of the "pure numbers" of fluid mechanics. (Of course, one could just as well balance the relevant forces.)

As another example, suppose we balance the kinetic energy density with the gravitational potential energy density (hydrostatic pressure),

$$\rho g h;$$

then

$$\frac{\rho V^2}{\rho g h} = \frac{V^2}{gh} = F$$

is a pure number, the Froude number.[2] Such a number is useful in building scale models of boats to study the effects of surface waves, which are gravitationally driven. Scaling according to constant F poses no problems: if the model is $1/100$ scale, then the towing speed should be $1/10$. Again, although h is a height, it is directly related to any characteristic length via geometric similarity. One difficulty with maritime modeling is that the boat has part of its hull submerged. For very small F, characteristic of large, slow boats, surface phenomena, such as wave drag, are negligible; viscous friction and thus the Reynolds number become important. These two numbers, F and R are often combined.

On the other hand, the surface tension becomes important for small waves, such as would be produced by a small model.

What happens at the free surface between water and air? The surface must be a constant-pressure surface. The pressure difference may change instantaneously due to inertial impingement of fluid on the surface due to some wave action. It will require a change in the shape of the surface to keep the surface of the water at atmospheric pressure. If the resulting wave is large enough, then gravity will act to restore equilibrium (this should remind the reader of a pendulum), but the restoring force for small waves is the surface tension.

Another situation where inertial and tension forces interact is when a bubble of air moves through a liquid, or even airborne bubbles. Recall that the surface energy is proportional to surface energy,

$$E = \gamma S,$$

where γ is the surface tension. Thus, if l is a typical length (the size of a bubble, perhaps), then γ/l is a surface energy density. If we balance kinetic energy with surface energy, then

$$\frac{\rho V^2}{\gamma/l} = \frac{\rho V^2 l}{\gamma} = We$$

---

[2]William Froude (1810–1879) was a naval architect.

is the Weber number. What of those phenomena without a characteristic velocity? Water waves (far from the boat) are an example: you'll recall (Exercise 4) that the speed of water waves depends on the wavelength, but in a different way depending on whether gravitation or surface tension is the dominating effect. If we compare the gravitational potential energy density to the surface energy density, then we get

$$\frac{\rho g l^2}{\gamma} = B,$$

the Bond number (no, it's not equal to .007). This would be relevant to a spherical drop, part of a surface wave, or a water strider—to any object with characteristic size l. Notice that

$$K = \sqrt{\gamma/\rho g}$$

is a length characteristic of the liquid; K^2 is called the *capillary constant*.

Another dimensionless combination arises when we balance kinetic energy with the elastic potential energy. Sound travels in longitudinal waves through a medium by repeated compression and decompression of the material. This is analogous to the situations in solids (3.3). There we saw that the sound speed was $a = \sqrt{E/\rho}$, where E was the modulus of elasticity. If we imagine that waves can only move in one direction (plane waves)[3] then E represents the elastic potential energy density, and so

$$KE/PE \sim \frac{\rho V^2}{E} \sim \frac{\rho V^2}{\rho a^2} = \frac{V^2}{a^2} = M^2,$$

which is known as the Mach number. That may be a roundabout way of defining the ratio of two velocities, but it gives some intuition as to the meaning of M. Incompressible fluids have $M = 0$; its significance is in high-speed flow in gases. In fact, the details of Mach modeling can be very complicated.

5   Problems and Recommended Reading

1. Suppose that the speed of light, the gravitational constant, and

---

[3]All we need is a quantity that expresses springiness of deformations in that direction; this is how we think of E. With compressible fluids we relate the pressure or stress to a change in the volume; this ratio is the bulk modulus or compressibility.

Planck's constant are all equal to 1:

$$c = h = G = 1.$$

What are the units of length, mass, and time?

2. How many different dimensionless groups can you make by mixing energy with M, L, T, I, θ?

3. Larger items in this supermarket cost less per unit weight than smaller items. Is this a marketing strategy, or is the manufacturer passing on savings in production, packing, and handling of larger items? Consider the various costs (production, materials, shipping, handling) and scale each in terms of weight. What is the wholesale cost per unit weight? What savings should be passed on to the consumer? This is discussed in detail in [4].

4. How does the running speed of an animal scale with size?

5. How does the maximum time between waterings for a desert animal scale to size?

6. How would life forms scale in the absence of significant gravity?

7. Is there an upper limit on the size of mountains?

8. Is there an upper limit on the size of trees? Trees are not observed to stand more than 100 m tall. Moreover, tall trees are thick; the height of large trees never exceeds about 50 diameters. Are the observed limits due to mechanical constraints? Might a too tall tree crush itself or buckle? Important parameters for a tree are density, $\rho \approx 500 \frac{kg}{m^3}$, elastic modulus, $E \approx 10^9 \frac{N}{m^2}$, and the breaking stress, $\sigma_{max} \approx 10^7 \frac{N}{m^2}$. How tall would a vertical tree have be before it buckled or was crushed? How much bend can a tree take? Show that there must be a bound on the ratio of height to diameter. Clearly, if trees were stiffer, they could stand taller. Are trees uprooted before they break? Vogel [77] seems to think so. What does that say about an optimal form for trees? Another reason for an extensive root system is that a very large amount of water must be absorbed by the roots to be pumped up the trunk to irrigate the leaves. Assuming a super subsystem and proper thickness is there an upper limit to tree height? It seems that a very tall tree must be very old. Assuming a constant height to

diameter ratio and constant rate of increase of mass, show that the height of the tree scales as the cube root of the time; thus, height rate slows considerably. These scaling notions are explored in detail in [48]; see also [47].

9. Is there an upper limit to the size of a flying animal? Relate the power required for flying, or just hovering, to the power available from metabolism. Relate your limit to the size of some of the pterosaurs? How much more power does a hovering helicopter need to lift twice as much weight?

10. Can you give an explanation for Bergman's rule, which states that *as we move from poles to equator, the size of warm-blooded animals of the same species decreases*? Compare, if you will, collies and chihuahuas.

11. How does the size of eardrums scale with respect to size in animals? How does the resonant frequency of a drum scale with diameter?

12. How does the size of eyeballs scale with size? The smaller animals don't have eyeballs. Why not?

13. How does volume flow in a pipe depend on roughness of the pipe?

14. A famous example of dimensional analysis is that which was carried out by G. I. Taylor at the time of the first atomic blast at Alamogordo. Deciding which variables are important, find the relationship between the size of the ball, the energy, and time. Can you find the energy released in the blast if you are given the figures in Table 3.2?

15. Show that the equilibrium shape of a (2D) sessile drop is given by the solution to the equation

$$h_{xx} = Bh,$$

where B is the Bond number.

16. How can the surface tension of a liquid be measured? One possibility is to measure the mass, m, of drops formed at the bottom of a tube of radius b. Use dimensional analysis to find a relationship between the relevant variables. How do you make sure that viscosity is not a relevant variable?

Table 3.2 Radii of the Visual Ball of the Atomic Explosion for Various Times (adapted from [70])

Time (msec)	Radius (m)
0.38	25.4
0.80	34.2
1.22	41.0
1.65	46.0
1.93	48.7
3.26	59.0

17. To see that one can quickly get to the limitations of the method of dimensional analysis, carry out the dimensional analysis of the dependence of the drag force on the flow parameters, but include a second length scale or perhaps a characteristic area.

18. *Turbulence energy dissipation scaling* Unless we watch a small twig or a bug, slow laminar flow presents no structure to the eye. In faster flow we see whirlpools here and there, sometimes big ones, sometimes small ones that quickly dissappear. What distinguishes true turbulence is the presence of whirlpools, or eddies, of all sizes and intensities. A generally accepted concept is that whatever is powering the torrent is pumping energy into the largest eddies, from which the turbulence mechanism distributes this energy down a "cascade" of eddy scales, from large inertial scales to the smallest sizes where viscosity acts to convert the energy to heat. Kolmogorov was able to derive some information concerning the distribution of energy on the cascade, using a little dimensional analysis, along with some simplifying assumptions about the flow. The relevant parameters are: ε, the total rate of energy dissipation per unit mass; λ, an eddy-size variable; and the density ρ and viscosity μ of the fluid. Suppose that the energy dissipation occurs predominantly on some range of wavelengths $\lambda_{min} \leq \lambda^* < \lambda_{max}$, and is given in terms of a density on that range,

$$\varepsilon = \int_{\lambda_{min}}^{\lambda^*} e(\lambda)d\lambda.$$

How should the density, e, depend on the flow parameters?

(a) Set
$$e \propto \mu^a \rho^b \varepsilon^c \lambda^d,$$

and show that dimensional homogeneity requires that $b = -a, c = 1 - a/3, d = -1 - 4/3a$.

(b) Argue, as follows, that $a = 1$. In viscous flow, where momentum is diffused, the relevant balance equation is

$$\rho \mathbf{v}_t = \mu \nabla^2 \mathbf{v},$$

where $\mathbf{v}$ is the flow velocity (see Chapter 7). But, since $\frac{1}{2}\rho V^2$ is the kinetic energy density, the dissipation rate is

$$\frac{d}{dt} \int \frac{1}{2}\rho V^2 dx = \int \rho v v_t \, dx.$$

This implies that e is proportional to μ. (*Hint:* Integrate by parts. The result follows if you treat the flow as one dimensional, though, of course, no real flow is.)

(c) We have, then,
$$e = C\mu/\rho \varepsilon^{2/3} \lambda^{-7/3}.$$

Argue that C is truly constant in the regime of interest and not a function of some dimensionless combination of variables.

(d) If $k = \frac{2\pi}{\lambda}$ is the *wavenumber* of the eddies (essentially, the eddy density), then

$$\varepsilon = \mu/\rho \int_{k_{\min}}^{k_{\max}} E(k)k^2 dk,$$

where $E(k) = C\varepsilon^{2/3}k^{-5/3}$.

19. *Bounds on turbulence energy dissipation* How fast is the energy being dissipated in a flow between two plates that are moving parallel to each other? (Figure 3.11a) Suppose that the plates are h apart and the top is moving relative to the bottom at V. The energy dissipation is greater in turbulent flow than in laminar flow. A laminar flow for (Figure 3.11a) is easily found. Use it to show that a lower bound for the energy dissipation is

$$\epsilon_{\text{laminar}} = \mu V^2/h.$$

The greatest amount of energy (kinetic) and thus the most power is contained in the inertial range where viscosity should be unimportant. What is a characteristic time scale for this problem? Find a power expression not involving viscosity, and propose the form of an upper bound. Exact bounds have been calculated for this simple case ([15]).

Chapter 4

Growth and Relaxation

How many people can the earth support? Should we worry that the population of the earth is growing too fast and the future is dim for large numbers of humans? To start to answer such questions, we need a good idea of how the earth's population will grow in the future. In 1800 there were 950,000,000 people on earth; in 1900 there were 1,650,000,000; and 5,890,000,000 in 1996. How many will there be in 2100? To predict, should we extrapolate with curves based on past data? Should our model change over time?

1 Exponential Growth

If we think of growth, in a general way, as any lengthening, broadening, or accumulation over the course of time, it is not to hard to find examples that warrant our attention as modelers. Each example quickly suggests the units of time in which we should measure the growth. The foetus grows by the week, a crystal grows by the hour, a tree grows by the year, a tumor grows by the month. Such growth seems continuous. There are other types of growth that we think of as taking place in lumps or jumps in time. A small colony of rabbits can increase by a noticeable number day by day; there may be 20 rabbits one day, 22 the next, and 26 a week later. However, if the colony is very large, in the thousands, then we are not pressed to keep accurate statistics and will instead keep rough estimates of the increase, which can be thought of as averages of the change taking place on finer time scales. As a matter of fact, this is universal: real-life phenomena are discrete processes, but seem continuous due to the scales of observation. Thus, a crystal grows by accretion of molecules from a solution onto a surface, a tumor grows by a sequence of discrete cellular divisions, and so on.

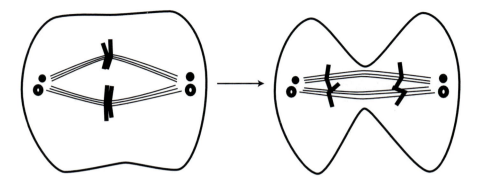

Figure 4.1 Mitosis.

The growth of a mass of cells by repeated division is simple. In the life of the cell, it goes through several phases. One of these phases is known as *mitosis*. In mitosis, the genetic material of the cell is divided, and the cell then cleaves to become two cells, each close to identical with the parent cell (Figure 4.1). Let us consider the simplest case of a population of cells with a single ancestor. Suppose that the time between divisions is constant, T_c, and let N_k denote the number of cells present after k divisions, which is a time $(k-1)T_c$ since the first mitotic division. The doubling at each time step means that

$$N_{k+1} = 2N_k. \tag{4.1}$$

This recursion states that the sequence N_k follows a geometric progression. We can easily find the solution as a function of k: this is

$$N_k = 2^k N_0 = 2^k, \tag{4.2}$$

since there was one cell at the beginning. As N_k is an exponential function of the discrete time, we call this *exponential growth*. This same recursion (4.1) and its solution (4.2) can model exactly the growth by a doubling process starting from any number, N_0, of ancestors as long as

1. the interdivision time for each cell is the same,

2. the divisions are *synchronized*: they occur at the same time.

EXERCISE 1 How would you modify the model or its interpretation if either of these conditions is violated?

The exponential growth is to be contrasted with the linear growth that occurs when a constant amount is added at each time step (Figure 4.2). This recursion is

$$A_{k+1} = A_k + d.$$

EXERCISE 2 (i) Why is this called linear growth?
(ii) How would you distinguish exponential from linear growth on a log–log plot?

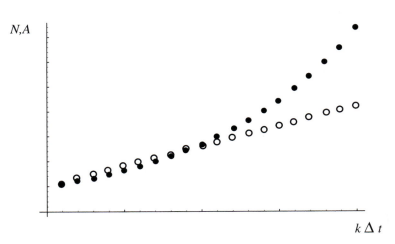

Figure 4.2 Exponential growth (solid circles) and linear growth (hollow circles).

Once a population gets very large, whether it takes the form of a crystal or a tumor or a swarm of mayflies, the discrete growth model becomes unwieldy. It is useful to think of change as unfolding in a continuous time stream and the population count as some continuous function of time. We can match the function of discrete time (4.2) to a function of continuous time by interpolation:

$$N(kT) = N_0 2^k,$$

or

$$N(t) = N_0 2^{t/T_c}. \tag{4.3}$$

How long does it take to double the population? Let's fix t and calculate the additional time we must wait for the population to be twice its present size. This is the *doubling time, $T_d(t)$*, given by

$$N(t + T_d) = 2N(t) = N_0 2^{t/T_c+1} = N_0 2^{\frac{t+T_c}{T_c}} = N(t + T_c).$$

Thus, T_d is independent of t and is equal to T_c. This is true even if the cells are unsynchronized and divide at different moments, as long as the interdivision time T_c is the same for all cells. It is natural to consider the rate at which the cells divide. In some small time Δt there will be a fraction, ρ, of the cells undergoing division. Thus, in Δt there are $2\rho N(t)$ new cells, but also $(1 - \rho)N(t)$ cells that did *not* divide. The number of cells that divide in a certain time interval will depend on the length of the time interval, that is, $\rho = \rho(\Delta t)$.

We will make the (not unreasonable) assumption that for small time intervals the fraction dividing is proportional to the duration of the interval, i.e.,

$$\rho = r\Delta t + \circ(\Delta t),$$

where r is a constant rate. (We can think of r as the birth rate.) Expressing the number in the population at the time $t + \Delta t$ as those who have not divided during the time Δt plus the cells born during Δt (remember that there are two new cells for every one that divides), we have

$$
\begin{aligned}
N(t + \Delta t) &= (1 - r\Delta t)N(t) + 2r\Delta t N(t) + \circ(\Delta t) \\
&= N(t) + r\Delta t N(t) + \circ(\Delta t).
\end{aligned}
$$

If we take the limit as $\Delta t \to 0$, we get the differential equation

$$\frac{dN}{dt} = rN. \tag{4.4}$$

The solution to this equation is

$$N(t) = N(0)e^{rt}. \tag{4.5}$$

If this is to match the previous growth expression (4.3), then

$$r = \frac{\ln 2}{T_c}.$$

EXERCISE 3 Show this explicitly.

Which of the quantities r, T_c, T_d is most fundamental? Typically, only T_d is directly measurable. If this number is observed to be independent of t, then the exponential growth with rate

$$r = \frac{\ln 2}{T_c}$$

is the correct model. The equation $T_c = T_d$ that we found earlier presupposed that such a number T_c exists as characterizing the entire population. In general, there is no reason to suppose that the inter-division time for all cells is the same. In that case, T_d is the *mean* or average time between divisions. As will be made clear in the sequel, especially in later chapters, the model of exponential growth is appropriate when there is very little interaction between individuals in a population. Clearly, then, whether the division is synchronized or not should be irrelevant.

Once we have moved from the discrete to the continuous, there is nothing sacred about *doubling*; one could just as well ask about a tripling time. Notice that, as $[1/r] = T$, $1/r$ is a characteristic time. It is in fact the time it takes for the population to effect an e-fold increase.

The Relaxation Response

If r in (4.4) is negative, then we do not have growth but rather exponential decline. Then r can be interpreted as a death rate. Figure 3 contrasts the cases $r > 0$ and $r < 0$. Exponential decay is observed often in nature; it is the simplest model of approach to equilibrium. In this case it is typically referred to as (exponential) *relaxation*.

In Chapter 2 we looked at the elastic spring with an unstressed length of l and a restoring force proportional to the displacement from l : $-k(x - l)$. The equilibrium state, $x = l$, is the only solution to the force balance equation,

$$-k(x - l) = 0,$$

and by consideration of this alone we would expect the spring, if displaced, to instantaneously choose this equilibrium. Suppose that we attach the spring to a dashpot (Figure 4.4) so that there is a viscous force proportional to the time rate of change of the length of the spring. Now the force balance is

$$-\eta \dot{x} - k(x - l) = 0. \tag{4.6}$$

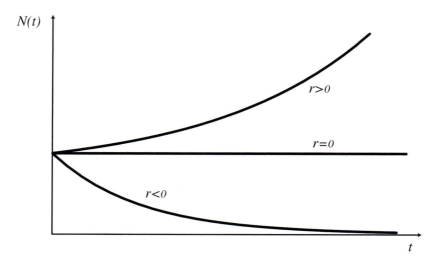

Figure 4.3 Exponential growth and decay.

EXERCISE 4 Decide when this type of rate proportional force law is appropriate. This is usually called a *viscous friction* law. How does it depend on the viscosity and other parameters of the apparatus of Figure 4.4?

From (4.6) we can see that, although $x = l$ is still the unique equilibrium, this position cannot be obtained instantaneously because the friction term would become singular. The differential equation (4.6) can be solved to get

$$x(t) = l(1 - e^{-\frac{k}{\eta}t}). \tag{4.7}$$

EXERCISE 5 Solve (4.6) to obtain (4.7). What is the initial condition? What is the solution for an arbitrary initial condition?

The equilibrium $x = l$ is reached asymptotically as $t \to \infty$ at an exponential rate (Figure 4.5). We say that the displacement "relaxes" to l. The time $\tau = \frac{\eta}{k}$ is known as the *relaxation time*.

No spring is perfectly elastic and there will always be some relaxation even without the added dashpot. If, in the basic stress–strain relation, we allow a rate-dependent term,

$$\sigma = E\epsilon + r\frac{d\epsilon}{dt}, \tag{4.8}$$

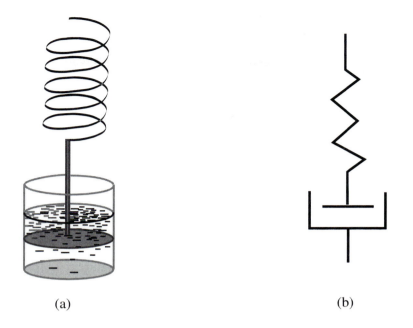

(a)   (b)

Figure 4.4 (a) Spring attached to plate that must be dragged through a liquid. (b) Corresponding schematic.

there will be a characteristic relaxation time $\tau = \frac{r}{E}$. The elastic limit is $r \to 0$. In some solids, such as steel, $r \ll E$, and so the relaxation time is negligible on most scales.

EXERCISE 6 What is the time-dependent response of the displacement to a constant applied stress?

The word "relaxation" is most evocative in the context of structures. A familiar example is rolled poster paper (Figure 4.6). If a rolled poster is unrolled, it will quickly resume its scroll form. If it is held flat for a long enough time, it will conform to the flat shape. This is an example of a long relaxation time (long compared to the elastic behavior). A relaxation response is always of the simple form

$$\sim e^{-t/\tau};$$

the *relaxation time* is τ.

Another example of a relaxation phenomenon is the charging of a capacitor in the presence of a resistive load (Figure 4.7a). The voltage

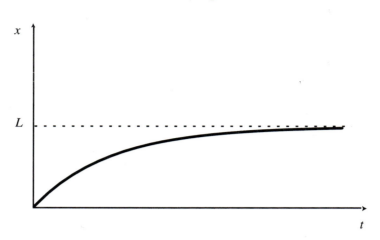

Figure 4.5 Relaxation to equilibrium.

across a capacitor is proportional to the charge,

$$V_C = Q/C,$$

and the voltage across a resistance is proportional to the current, the time rate of change of the charge,

$$V_R = RI = R\frac{dQ}{dt}.$$

If there is a constant applied voltage, V_0, it must be the sum of the voltages through each of the components:

$$V_0 = Q/C + R\frac{dQ}{dt}. \tag{4.9}$$

Notice the similarity between equations (4.9) and (4.8). Solving this equation gives the charging of the capacitor,

$$Q(t) = CV_0(1 - e^{-\frac{t}{RC}}),$$

and reveals that the relaxation time is RC. The response to the applied voltage is the charging of the capacitor.

EXERCISE 7 How long does it take to get within 10% of the full charge?

EXERCISE 8 The voltage through an inductive element, such as a coil, is proportional to the time rate of change of current,

$$V_L = L\frac{dI}{dt}.$$

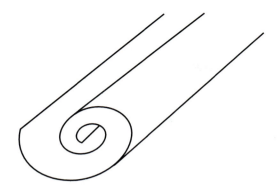

Figure 4.6 Rolled paper.

If a constant voltage is applied to an inductance and a resistance in series (Figure 4.7b), what is the response of the current? What is the relaxation time?

EXERCISE 9 If a capacitance and a resistance are in parallel (Figure 4.7c), what is the response of the voltage to an applied current?

2 Self-limiting Growth

We have remarked that exponential growth is a consequence of the lack of interaction among the members of a population. This interaction may take the form of physical crowding, competition for resources, the exposure to disease, and so on. Such interactions are observed to limit the sizes of populations. (To be sure, there are possible ameliorating interactions, such as sexual reproduction, and nurturing of the weak and young; these can certainly complicate a model.) The growth of a young deciduous forest, the size of which is measured by the verdure or leafmass, might initially be exponential because all the leaves have plenty of sunshine to carry out the necessary photosynthesis, and there is nothing to discourage new bud formation; but as shade develops, leaves must compete for sunshine and photosynthesis is reduced, resulting in the slowing of the foliage growth. This effect can be modeled by a growth rate that is a decreasing function of the size of the population, N:

$$\dot{N} = r(N)N. \tag{4.10}$$

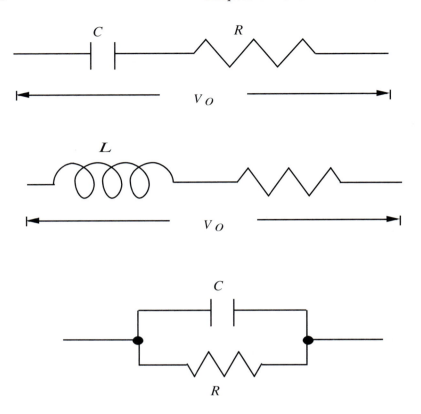

Figure 4.7 Simple circuit elements.

Besides the slowing down of growth rates, equilibria will be introduced; as many equilibria as there are positive zeros of the function r.

EXERCISE 10 What kind of growth corresponds to $r(N) = e^{-N}/N$?

A simple model is to take r to be linear. Let's examine the growth of some foliage, say a field of tea leaves. The growth equation is

$$\dot{\Lambda} = r(\Lambda)\Lambda,$$

where the dimensions of Λ, $[\Lambda] = ML^{-2}$, represent the mass of leaf-cover. The dimensions of r are T^{-1}, and we take r to be

$$r(\Lambda) = a - b\Lambda.$$

EXERCISE 11 What are the dimensions of a, b?

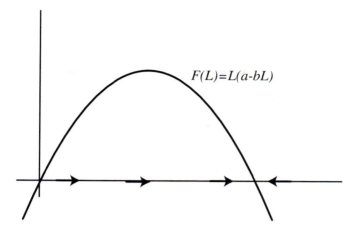

Figure 4.8

The equation with this growth law is known as the **logistic equation**. There are two equilibria for (4.10), $\Lambda = 0$ and $\Lambda = a/b$. The latter is stable and the former unstable. To see this, consider the graph of

$$F(\Lambda) = (a - b\Lambda)\Lambda,$$

which is plotted in Figure 4.8. Since $F(\Lambda)$ is the time derivative of any solution (trajectory), we just have to check the sign of F to find out if $\Lambda(t)$ is increasing or decreasing. F is positive for $\Lambda < a/b$, so Λ is increasing there: for any initial condition $\Lambda(0) < a/b$, we will approach a/b from below. Likewise, $F < 0$ for $\Lambda > a/b$, and so Λ is decreasing when above the equilibrium. Notice the arrows in Figure 4.8. These indicate the "flow" in the one degree of freedom available to Λ. This is an example of a one-dimensional *phase plot* (see also Figure 4.10). The basic idea is that the equation determines a *vector field* (in this case, the vector field on the one-dimensional space is

$$(a - b\Lambda)\Lambda),$$

and the solution must be such that this vector field gives its direction of "flow" at each point. The solution trajectories can also be plotted versus time (Fig. 9).

For Λ very small, we can ignore the quadratic term, which gives the equation

$$\dot{\Lambda} = a\Lambda,$$

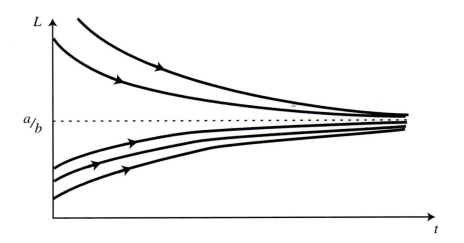

Figure 4.9

so that for small Λ we have exponential growth. To investigate the behavior near the stable equilibrium, we write a Taylor expansion about $\Lambda_\infty = a/b$:

$$
\begin{aligned}
\frac{d}{dt}(\Lambda - \Lambda_\infty) &= F'(\Lambda_\infty)(\Lambda - \Lambda_\infty) + \cdots \\
&= [a - 2b(a/b)](\Lambda - \Lambda_\infty) + \cdots \\
&= -a(\Lambda - \Lambda_\infty) + \cdots
\end{aligned}
$$

Since we are only interested in the behavior *near* equilibrium, we will only keep the linear terms. If $\lambda = \Lambda - \Lambda_\infty$, then the equation for the *linearized dynamics* about the equilibrium is

$$
\dot{\lambda} = -a\lambda.
$$

This is exponential relaxation to equilibrium. Thus, the logistic equation exhibits an initial exponential growth and then exponential relaxation. An analytic solution to the logistic equation is easily found:

$$
\Lambda(t) = \frac{\Lambda_0 a e^{at}}{a - bL_0(1 - e^{at})}.
$$

EXERCISE 12 Integrate the logistic equation to get this solution. Show explicitly that a/b is the unique stable equilibrium.

What will be the effect of harvesting? Suppose that the harvesting is done at a constant rate. Then

$$\dot{\Lambda} = (a - b\Lambda)\Lambda - c.$$

The rate c gives the harvest (mass) per unit area per unit time. The effect of harvesting can be gleaned from the graph of Figure 4.10 which is analogous to Figure 4.8. One effect is that small populations have no chance to grow and will die out quickly. Another effect is that the equilibrium leaf mass is less. The stable and unstable equilibria annihilate each other at a critical value of the harvesting rate:

$$c_{\text{crit}} = \frac{a^2}{4b}.$$

This is an example of a limit point instability or a fold catastrophe. For $c > c_{\text{crit}}$, all solutions decay. Since $L = 0$ is not an equilibrium, we do not have a relaxation, but rather a catastrophic decline in the leaf population.

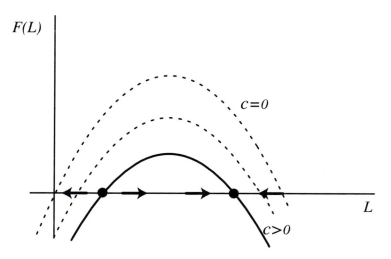

Figure 4.10 Effects of cropping.

What is the optimal cropping policy? Clearly, one cannot take $c > c_{\text{crit}}$. It might be thought that we should take $c < c_{\text{crit}}$ as close as we can to c_{crit}. The problem with this is that, when we are close to critical, the stable and unstable equilibria are very close; they are both about $L = \frac{a}{2b}$. Since there will be uncertainties in the values of a and b

as well as unmodeled fluctuations, this would place us perpetually on the brink of disaster. Suppose that we have some way of measuring the density of the foliage. Then we can crop in proportion to the density that we measure, obtaining a *feedback law* for harvesting,

$$c = -kL.$$

In this case the equation becomes

$$
\begin{aligned}
\dot{L} &= aL - bL^2 - kL \\
&= (a-k)L - bL^2,
\end{aligned}
$$

which is just a modified logistic equation. There is a fully stable equilibrium at $L = \frac{a-k}{b}$. Notice that if $k = a/2$ then the equilibrium is at $L = \frac{a}{2b}$, and the asymptotic cropping rate is

$$c_\infty = -kL_\infty = \frac{a^2}{4b},$$

which is precisely the maximum possible rate of constant cropping. The moral here is that, if we are willing to wait the short relaxation time it takes for the system to stabilize, we can expect in the long run to enjoy maximal profits. The foregoing is an example of *phase-plane analysis*, which is also useful in two dimensions (see the next section and chapter 5), but less useful in higher dimensions for obvious reasons.

Another limited-growth equation, originally used in actuarial studies and now popular with biologists, is found by taking the growth rate to be

$$r(N) = \alpha - \beta \ln N. \tag{4.11}$$

It is easy to verify that, as in logistic growth, there is a single stable equilibrium. The approach to the equilibrium is governed by the constant β, and not by the initial growth rate α.

EXERCISE 13 (i) Show that the equation governing small displacements from equilibrium is

$$\dot{\eta} = -\beta\eta. \tag{4.12}$$

(ii) What are the dimensions of β?

A problem with the rate law (4.11) is that the resulting equation is not dimensionally homogeneous, due to the logarithm. For that reason the rate is usually used in the form

$$r(N) = \alpha - \beta \ln \frac{N}{N_0},$$

where N_0 is some characteristic population level, typically the initial population. In that case, α and β have the same dimensions. The actual equilibrium value depends on N_0, but the approach to equilibrium (the linearized dynamics) is still given by (4.12). The differential equation with this rate law is the Gompertz equation:

$$\dot{N} = (\alpha - \beta \ln \frac{N}{N_0})N. \qquad (4.13)$$

EXERCISE 14 Show by explicit integration that the solution to (4.13) is

$$N(t) = N_0 e^{\frac{\alpha}{\beta}(1 - e^{-\beta t})}.$$

Autoregulation

The two models of self-limited growth presented so far are fine for situations, such as foliage growth, where one can directly relate the form of the equation to a competition for resources, such as sunshine energy. This is less tenable in more complicated situations. In the case of cell growth, the population is typically in an environment where the individuals are not in competition for nutrients. Of course, an individual could get caught in a corner or the population could grow into a ball that effectively keeps some of the individuals from feeding, but this needn't imply the cessation of growth of the population. Yet normal cells do not relentlessly divide and normal populations do not grow without bound. Unlimited growth of a cell population is known as *cancer*.

It is believed that cells follow a homeostatic control path by synthesizing a chemical that prevents mitosis from taking place. This is known as **mitotic autoregulation** ([82]). The control chemical is known as the **inhibitor**, because it inhibits the growth of the cell population.

Let n denote the cellular density (the number of cells per unit volume in some region). If α is the rate at which cells divide, and ω is the loss rate, then

$$\dot{n} = (\alpha - \omega)n. \qquad (4.14)$$

Let c be the concentration of inhibitor (the amount per unit volume). It is reasonable to assume that the inhibitor concentration should affect the cellular growth rate in an inverse way so that the greater the concentration of inhibitor, the lower the cellular growth rate. Thus, α should be replaced by

$$\frac{\alpha}{1 + \beta c}.$$

Meanwhile, a homeostasis should imply that the cells control production of the inhibitor, perhaps by synthesizing it themselves. Also, there should be a mechanism for catabolism of the inhibitor. This means that the chemical is to be broken down when the concentration is too high. Taking these comments into account, a growth of the following form seems to be justified:

$$\dot{c} = \gamma n - \delta c - L,$$

where L is the leakage of the chemical away from the physical domain of interest. This leakage is due to mechanical transport as well as diffusion and clearly depends on the spatial geometry of the cell population as well as that of its environment. If the population had some distinctive geometry, such as tumor growth, special care would be needed to model this term.

Here we will consider a population of cells distributed more or less uniformly in some fixed region of space. This is a model for the stem cells of the blood that grow in the bone marrow. Stem cells, also called *clonogens*, have the capacity, if not otherwise restricted, to carry out divisions in perpetuum. Differentiated cells, responsible for much physiological function, do not typically undergo mitosis. Uncontrolled growth of the stem cells of the blood is *leukemia*.

The leakage is due to cells finding exit at the boundary of the region, and this will be proportional to the total amount, and thus to the concentration. Thus the leakage effect can be incorporated in the parameter δ. We now have a system of equations for cell density and inhibitor concentration:

$$\dot{c} = \gamma n - \delta c, \tag{4.15}$$

$$\dot{n} = \left(\frac{\alpha}{1 + \beta c} - \omega\right) n. \tag{4.16}$$

It is easily seen that the system has an equilibrium ($\dot{c} = \dot{n} = 0$) at the intersection of two lines in the (c, n) phase plane. These lines,

$$n = \frac{\delta}{\gamma} c, \quad c = c_0 = \frac{\alpha - \omega}{\omega \beta},$$

are plotted in the phase plane in Figure 4.11. Is this a stable equilibrium? To investigate the local dynamics, we can write down the equation for small deviations from equilibrium. If these deviations, $\xi = c - c_0$ and $\eta = n - n_0$ are small, when we expand the system (4.15, 4.16) in a Taylor series about (c_0, n_0), we need only keep the first-order terms. Thus,

$$\dot{\xi} = -\delta\xi + \gamma\eta$$
$$\dot{\eta} = -\frac{\omega\delta}{\alpha\gamma}(\alpha - \omega)\xi.$$

We can eliminate ξ in these equations and obtain a second-order differential equation for η:

$$\ddot{\eta} + \delta\dot{\eta} + \frac{\omega\delta}{\alpha}(\alpha - \omega)\eta = 0. \tag{4.17}$$

EXERCISE 15 (i) Verify that this solution satisfies

$$|\eta(t)| \leq Ce^{-\delta t}.$$

(ii) Argue intuitively that the equilibrium must be globally asymptotically stable. That is, even if the initial condition is very far from equilibrium, the resulting solution will tend to the equilibrium.

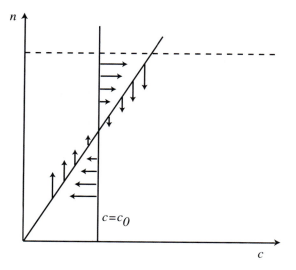

Figure 4.11

So far, we have ignored one thing that can make the equilibrium unattainable and that is the possibility that there is a limit to the extent that the cells can squeeze tightly together. In fact, if v_m is the minimum possible volume for a cell, a physical upper bound on cell density is $\frac{1}{v_m}$. The line $n = \frac{1}{v_m}$ is plotted as a dotted line in Figure 12. In Figure 11, the equilibrium is below $n = \frac{1}{v_m}$, but if the two solid lines had crossed above the dotted line, the equilibrium would not be attainable. The algebraic criterion for this is

$$n_0 = \frac{\delta}{\beta\gamma\omega}(\alpha - \omega) < \frac{1}{v_m}. \tag{4.18}$$

If δ is large (quick inhibitor catabolism) the cells will grow to fill up space.

Notice how little work we had to do to get this criterion for catastrophic growth. Phase-plane analyses are powerful when they can be brought to bear.

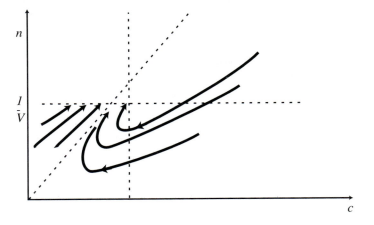

Figure 4.12

We can find an approximate analytic solution to the system (4.15, 4.16) by a reduction of dimension. This solution is commonly called a **quasi-steady state**. It comes about by observing that, since (4.15) describes molecular kinetics while (4.16) is a description of cellular kinetics, the relaxation to an equilibrium associated with the former is much quicker than that of the latter. In this way, the concentration of inhibitor should quickly respond to the current cell density and follow in its wake, so to speak. In fact, according to this approximation, the inhibitor concentration follows the cell density along the line

$$c(t) = \frac{\gamma}{\delta}n(t).$$

The types of trajectories associated with the quasi-steady-state solution are shown in Figure 4.13. One should think of the motion along any trajectory approaching the line as being very fast, while the resulting motion is relatively slow. This asymptotic line is called a *slow manifold*. The cell population dynamics along the slow manifold of the quasi-steady state are given by

$$\dot{n} = \left(\frac{\alpha}{1 + \frac{\beta\gamma}{\delta}n} - \omega \right)n.$$

EXERCISE 16 Solve this differential equation to obtain the formula

$$n^{\frac{1}{\alpha - z\omega}} \left| (\alpha - \omega) - \frac{\omega\beta\gamma}{\delta}n \right|^{(\frac{1}{\alpha - \omega} - \frac{1}{\alpha})} = Ke^t,$$

where K is a constant of integration. Argue that n will approach an equilibrium only if

$$\alpha > 2\omega.$$

What if this condition is not satisfied? Reconcile this with what we derived using the phase plane. Rederive the condition for an attainable equilibrium (4.18).

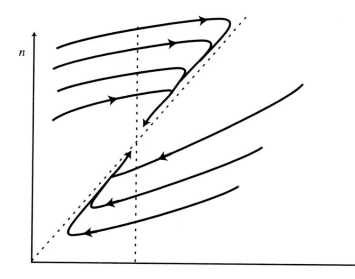

Figure 4.13

We can provide a more precise guide for the applicability of the quasi-steady-state approximation if we introduce nondimensional variables and identify small parameters. To this end, we need characteristic values for the dependent variables n and c and for time. Typical measured values of cell density and chemical concentration can be used, or, perhaps more appropriately, we can use the equilibrium values. For the characteristic time, it is probably best to use the time scale associated with the slowest behavior in our model, that of cell demise: $t_{\text{characteristic}} = \frac{1}{\omega}$.

EXERCISE 17 Introduce the nondimensional variables, θ, σ, τ, as

$$\sigma c_0 = c, \quad \theta n_0 = n, \quad \frac{\tau}{\omega} = t.$$

Show that the resulting equations are

$$\epsilon \frac{d\sigma}{d\tau} = -\sigma + S\theta, \tag{4.19}$$

$$\frac{d\theta}{d\tau} = -\theta + T\frac{\theta}{1 + \beta c_0 \sigma}, \tag{4.20}$$

where $\epsilon = \frac{\omega}{\delta}$, $S = \frac{\gamma n_0}{\delta c_0}$, $T = \frac{\alpha}{\omega}$. The quasi-steady-state solution is found by setting $\epsilon = 0$; thus, the approximation should be okay for $\omega \ll \delta$. Recall the criterion in Exercise 16.

Economic Growth

How can we characterize the growth of an economy? The gross national product (GNP) is a measure of the size of an economy, which we can think of as a big machine with everyone contributing, by way of labor, investments, and purchases, to its grand function.

In the neoclassical growth model ([9]), one considers, in the simplest case, the GNP to be a positive function of labor and investment, and one tracks, over time, the investment per capita. We can think of labor, L, and capital, K, as being the variables input to the big machine, and then output production, Q, is measured by

$$Q = F(K, L).$$

(The dimension of production is rate of accumulation of goods: $[Q] = [K]/T$.) One can imagine an island nation where wheat is grown: the grown wheat can either be consumed directly or a new crop planted (reinvested). Capital is both the goods of production as well as the consumer goods.

What form should F have? F should be positive and depend smoothly on its variables. There should be no output without any input: $F(0,0) = 0$; this is known as "no free lunch." It might be reasonable to suppose that F should be a nondecreasing function of its variables. This assumes that no labor or capital investment is made in a such a contrary way as to neutralize (or worse) the contribution of other labor or capital. There should be a diminishing of returns:

$$\frac{\partial^2 F}{\partial K^2} \le 0, \quad \frac{\partial^2 F}{\partial L^2} \le 0.$$

An important simplifying assumption is *constant returns to scale*. That is, if any unit of production is exactly copied and setup, then production will be doubled. This means that the function is homogeneous of degree 1:

$$F(\lambda K, \lambda L) = \lambda F(K, L). \tag{4.21}$$

This symmetry condition is very valuable because it means that F is essentially a function of a single variable, and it is for this reason that we need only track investment per capita:

$$k = \frac{K}{L}.$$

The functional reduction to this single variable is made by taking $\lambda = \frac{1}{L}$ in (4.21):

$$f(k) = F(k, 1) = F(\frac{K}{L}, 1) = \frac{1}{L}F(K, L). \tag{4.22}$$

Notice that the dimensions of k are $[k] = \dfrac{\text{units of capital}}{\text{units of labor}}$.

Figure 4.14

Based on our assumptions, f should have roughly the form pictured in Figure 14. Notice that the derivatives of F can be written in terms of f by differentiating both sides of (4.22). Thus,

$$\frac{\partial F}{\partial K} = f'(k), \tag{4.23}$$

$$\frac{\partial F}{\partial L} = f(k) - f'(k)k. \tag{4.24}$$

We need a way to relate production to capital growth. The simplest procedure is to suppose that a portion of the output is consumed and the rest invested. Thus, if I is investment, C is consumption, and s is the fraction of production that is invested, then

$$I = sQ,$$
$$C = (1-s)Q.$$

We now equate the rate of capital growth with the investment:

$$\dot{K} = sQ. \tag{4.25}$$

This gives us a differential equation for the per-capita growth rate,

$$\dot{k} = \frac{\dot{K}}{L} - \frac{K}{L}\frac{\dot{L}}{L}$$
$$= s\frac{Q}{L} - k\frac{\dot{L}}{L}.$$

The term $\frac{\dot{L}}{L}$ is the labor force growth rate. A simplifying assumption is exponential growth of the labor force,

$$\frac{\dot{L}}{L} = g,$$

for then

$$\dot{k} = sf(k) - gk.$$

Zero is clearly an equilibrium; is there a positive equilibrium? The graphs of $y = f(k)$ and $y = \frac{g}{s}k$ are plotted in Figure 4.15. There is a nontrivial equilibrium only if the graphs intersect. If the slope g/s is too high, there is no equilibrium.

EXERCISE 18 (i) Find an analytic condition for there to be a nontrivial equilibrium.
(ii) Show (directly from the figure) that the (unique) nontrivial equilibrium is stable.
(iii) Suppose that we allow for exponential depreciation of capital: replace (4.25) with $\dot{K} = sQ - \delta K$. How does this affect the equilibrium?

EXERCISE 19 *Golden Rule* What fraction s of the harvest should be saved? Certainly saving too much would be gloomy, but too little savings might lead to a slump in production and eventual poverty. Suppose that the standard of living is measured by per-capita consumption. Find the savings rate that maximizes the per-capita consumption at equilibrium. Show that this occurs when the rate of profit is balanced by the labor force growth rate.

EXERCISE 20 (i)The simplest example of a production function that satisfies our hypotheses is

$$f(k) = f_0 k^\alpha, \quad 0 < \alpha < 1.$$

Explicitly integrate the corresponding differential equation and find the equilibrium k_e. Also find the optimal savings rate as Exercise 19.
(ii) Do the same with

$$f(k) = \alpha(k^{-\rho} + \beta)^{\frac{-1}{\rho}}.$$

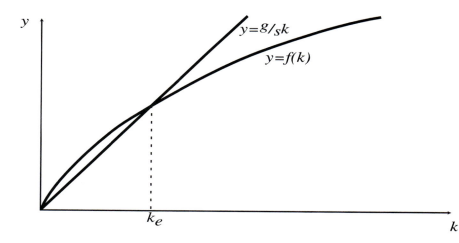

Figure 4.15

The notion that the output of production can be taken in an equivalent fashion to be either the consumer good or the form of capital is primitive, as would be the civilization of our island community. On a neighboring island, where bread is baked using electric ovens, the consumption good is not immediately translated into capital. There are separate production systems for consumer goods (bread) and capital goods (electric power), each requiring its own supply of capital and labor. This leads to two-sector models of economic growth ([75], [76], [28]). The two output functions are given as before,

$$Q_1 = F^1(K_1, L_1), \quad Q_2 = F^2(K_2, L_2),$$

the F^i being neoclassical growth functions as previously described. Q_1 is the output of capital good production and Q_2 is the output of consumer good production. K_i is the amount of capital used in the ith sector and L_i is the size of the labor force employed in the ith sector.

To compare quantities in each sector, we need to know the value, or the price, with which one credits each commodity. Let P_i be the price of one unit of output in sector i. Then the wage rate (the value of labor), being equal in both sectors, can be written as

$$W = P_1 \frac{\partial F^1}{\partial L_1} = P_2 \frac{\partial F^2}{\partial L_2}.$$

Likewise, if we assume that capital depreciation is the same in both sectors, the value of the use of one unit of capital ("rental rate") is written

$$U = P_1 \frac{\partial F^1}{\partial K_1} = P_2 \frac{\partial F^2}{\partial K_2}.$$

The ratio

$$\omega = \frac{W}{U}$$

is independent of price.

EXERCISE 21   (i) As in the one-sector economy, we define $k_i = \frac{K_i}{L_i}$, the capital–labor ratio in the ith sector, and $q_i = f_i(k_i)$. Show that

$$\omega = \frac{f_i(k_i)}{f_i'(k_i)} - k.$$

(ii) Show that ω is a one–one function of k_i so that, inversely, given any ω, the capital–labor ratio k_i in each sector is uniquely determined. Show that $k_i(\omega)$ can be found as in Figure 4.16.

To trace the growth of investment, we need to know how much of the wealth is available at any time to be invested; this is the *gross savings*. Let K be the total available capital and L be the total available labor. The value of the capital is UK, and the value of labor's wealth is $W(L_1 + L_2)$; thus, the gross savings is

$$sUK + \bar{s}W(L_1 + L_2),$$

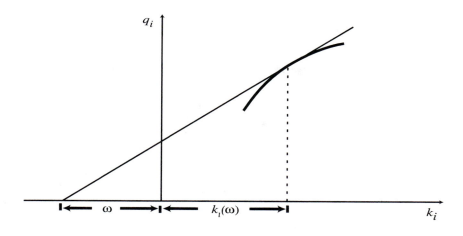

Figure 4.16

where the $0 \leq s, \bar{s} \leq 1$ are the fractions that are saved. Let us assume that labor does not save any income so that $\bar{s} = 0$. Then, equating total gross savings with the gross value of investment, we have

$$P_1 Q_1 = sUK. \qquad (4.26)$$

We can reduce the number of dependent variables if we have another relation. This can be accomplished by assuming *full employment of resources*:

$$K = K_1 + K_2, \qquad (4.27)$$
$$L = L_1 + L_2. \qquad (4.28)$$

EXERCISE 22 (i) If $\rho = L_1/L$, the fraction of total labor employed in the capital goods sector, show that (4.26) and (4.27) can be written as

$$\rho f(k_1) = s f'(k_1)k, \qquad (4.29)$$
$$k = \rho k_1 + (1 - \rho)k_2. \qquad (4.30)$$

(ii) Combine these and use Exercise 21 to express k as a function of ω, $k = k(\omega)$.

The fact that $k = k(\omega)$ is a consequence of our assumption of full employment. Note that this function may not be invertible.
EXERCISE 23 If the production functions are

$$f_i(k_i) = \alpha_i (k_k^{-\rho} + \beta_i)^{\frac{-1}{\rho}},$$

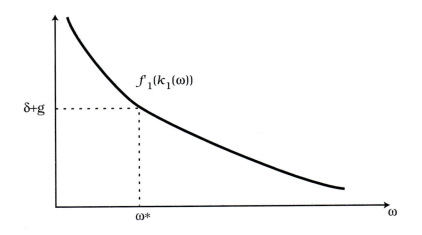

Figure 4.17

show that, for some values of β_i, the resulting $k = k(\omega)$ is not monotone.

The growth laws

$$\dot{L} = gL, \quad \dot{K} = Q_1 - \delta K,$$

can now be combined to form

$$\dot{k} = (f_1'(k_1) - (\delta + g))k.$$

The equilibrium condition

$$f_1'(k_1) = (\delta + g)$$

is satisfied by a unique pair of values (ω^*, k^*), even if $k = k(\omega)$ is not a monotone function. This is because f_1' is a decreasing function of ω (Figure 4.17). If $k = k(\omega)$ is monotone, then the equilibrium is clearly *stable* (Figure 4.18). On the other hand, the equilibrium is *unstable* if $k = k(\omega)$ is not monotone (Figure 4.19). For any $\omega > \omega^*$, we have $\dot{k} < 0$; thus, k will be decreasing. And for any $\omega > \omega^*$, we have $\dot{k} > 0$, and so k will be increasing.

When a trajectory reaches one of the points A or B, a jump must occur. These jumps occur in sequence, and the resulting behavior is cyclic. To see this, suppose that we start with the state on the upper branch; then, as $\omega > \omega^*$, $k(t)$ will be decreasing. When k reaches $\underline{k}$, ω must jump to a stable equilibrium, so the state is now at D, where $\omega < \omega^*$, and $k(t)$ starts to increase, and so on ($ADBC$). Notice that the

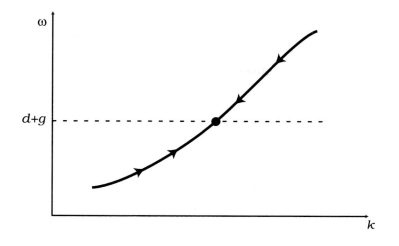

Figure 4.18

middle branch, which contains the equilibrium (ω^*, k^*) ,is inaccessible from the outside. One could conceivably start the evolution on this branch but, once into the cycle, would never see it again.

The per-capita growth rate is continuous, but its time derivative jumps each time it reaches $\underline{k}$ and $\bar{k}$. This is an example of a relaxation oscillation. Note that since ω and $\dot{k}$ suffer jumps so does k_1. This means that the capital–labor ratio in each sector must instantaneously rearrange itself in order to satisfy the constraint of full employment. This may seem unreasonable, but it is a good approximation if the relaxation to full employment is much quicker than the resulting evolution on the curve $k = k(\omega)$. This should remind one of the quasi-steady state seen in the cell growth model,[1] only here the slow manifold is multivalued in the fast variable, ω, and so relaxation oscillations can occur, rather than the monotone behavior seen there.

If we include dynamics for ω, say of the form

$$\dot{\omega} = G(k - k(\omega)),$$

where G is an increasing odd function, then the resulting phase-plane picture is as in Figure 4.20.

We can have oscillation of the variables (k, ω) where both vary smoothly in time. In this case, the oscillation is characterized by a

---

[1] Also look at Problem 11

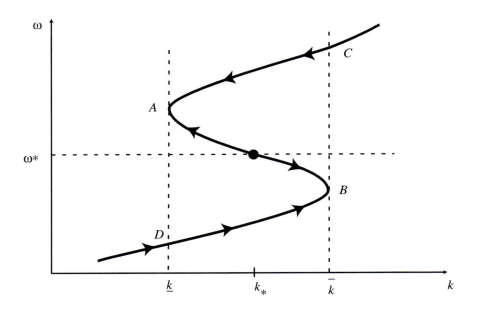

Figure 4.19

closed curve in the phase plane, and all trajectories tend toward this
curve (Figure 4.21). Such a curve is thus the limit of all trajectories
[except the single one (ω^*, k^*)] and is called a *limit cycle*. We will see
more of these in Chapters 5 and 8.

3   Problems and Recommended Reading

1. *Population models*   If we accept the proportional increase law
   (4.4) for population growth, given data at two points, we will
   be able to find the constant of proportionality. The publication
   "Current Population Reports" put out by the Bureau of the Cen-
   sus, lists population by state and year. Pick a state, and pick two
   past years and compute r. Now use the equation to predict the
   population of the latest available year. Does it match? You might
   think that the difficulty will go away if we take into account how
   long people are living, that is, consider separately the birth rate
   and the death rate. Explain why α and ω in (4.14) cannot be
   determined individually. Can you think of better models? You
   might want to look ahead to chapter 7. In [23] you will find a

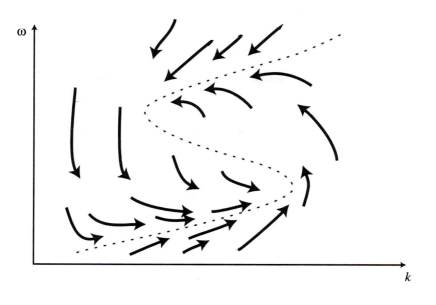

Figure 4.20

discussion of the question of how many people have ever lived, as well as more reference on population modeling.

2. *Relaxation in fluids* Recall that the analogue of the elastic modulus for fluids, the bulk modulus, is given by ρa^2, where a is the sound speed. This gives then the ratio of the stress (pressure) to the strain:

$$p = \rho a^2 \epsilon.$$

Recall that in fluids there are also the effects of viscosity, which gives a stress proportional, not to strain, but to rate of strain. Thus, a more general expression is

$$p = \rho a^2 \epsilon + r \frac{d\epsilon}{dt}.$$

What is the relaxation time for a liquid? When the fluid is compressible, the r that occurs here is not simply the viscosity as previously defined, and furthermore there are other effects such as thermal conduction and molecular relaxation. For details see [40] and [72].

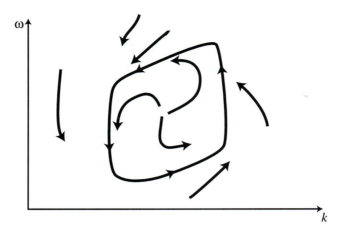

Figure 4.21

3. A city planner notices that everytime a parking garage is built it quickly reaches capacity during certain times of the day. What factors influence the demand for parking? Is there a natural achievable limit to the growth of parking garages?

4. In experimental studies of retrieval from long-term memory ([29]), subjects were asked to list all items that they could think of that belong in a particular category, for example, names of animals or names of girls.

   (a) If we assume that the subject has a finite number of items, M, stored in memory and the rate of recollection is proportional to the possible responses in memory not yet recalled, show that we get a retrieval curve (number versus time) that is exponential relaxation to M.

   (b) It is reported in [29] that they could match such a curve to data when subjects were asked to name all cities in Japan, but when asked to only name cities that are located to the south of cities previously named, the retrieval curve was linear in t. Can you account for this? (See [29] for some early models and [23] for others; also see Chapter 6.)

5. *Wildlife management* Suppose that you are entrusted with the

upkeep of a wildlife refuge. You want to maintain a certain diversity in the ecosystem. You can control the animal population by trapping or killing or transfer, and you can control the vegetation through burning and planting. There are several different species of animals as well as different types of vegetation, and different species of animal may eat different types of flora. To model the interactions of all species, animal and vegetable, do we need a different variable for each species or can we lump some together? The animal populations are made up of different sexes and age groups and array themselves in different spatial patterns.

(a) Let us try a simple model where we lump all vegetation into one variable V and all herbivores into another variable H. How do these variables interact? Let's start with V. In the text, the logistic equation was used as a model for leaf growth that is limited by competition for sunlight. This might be appropriate for a forest in temperate climes, but not for the dry plains of Africa, where the vegetation is never dense and the competition for rainfall is the deciding factor. How should this type of growth be modeled? How is the vegetation growth affected by grazing? The simplest effect would be a decay rate proportional to the number of herbivores. But even with a constant grazing rate, the decay term might be vegetation dependent; for instance, the relative types of flora species might differ according to the absolute amount of vegetation. Investigate V and H by way of the phase plane.

(b) Suppose that we want to model the interaction of two herbivores, each with its own taste and feeding habits. How can you investigate long-term behavior?

(c) What if the herbivore has a predator? What is the simplest model that predicts the long-term growth of the herbivore?

These problems and many more like them are discussed in the book [67].

6. Experiments show that cell tumours grown *in vitro* have a constantly increasing diameter; the diameter growth is linear in t. Can we account for this with a simple model? Model the tumor as a sphere and suppose that growth only occurs in an outer layer of fixed thickness.

7. *Compartment models* The so-called compartment models are very popular in physiology, especially in pharmacokinetics. There are a finite number of *compartments* acting as stores for some qualities, quantified by a mass Q_i. The rate equations are determined by mass balance:

$$\frac{d}{dt}Q_i = \sum R_{ij} - \sum R_{ji},$$

where R_{ij} is the mass-flow rate from compartment j to compartment i. Compartments do not have to be physical containers; in physiology, for example, they do not have to be anatomic: the quantity of a drug in all extracellular fluids could be taken as a compartment. The rates are typically taken to be state proportional, leading to linear equations; but if we allow more general forms it is easy to see that this is a very general modeling scheme, and we will encounter dozens like it in this book.

(a) Start with a simple one-compartment model (Figure 4.22) representing the blood plasma. A drug is injected intravenously and then is eliminated at a rate proportional to the amount present. What is the concentration in the blood as a function of time? IV infusions can be done in different ways. A *bolus injection* is modeled by a finite amount of drug given all at once at $t = 0$ (a delta function). A *short infusion* is constant for a short interval time and then is zero. A *continuous infusion* is constant for all times. What are the differences in the response of the plasma to each of these kinds of injection? Graph them. (See also Problem 12).

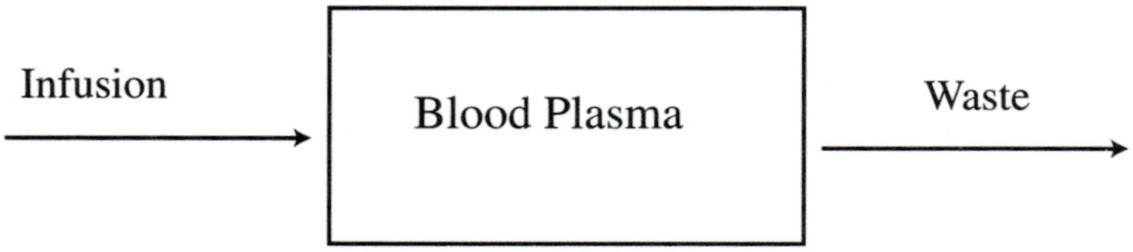

Infusion

Blood Plasma

Waste

Figure 4.22 A simple compartment model.

(b) If drugs are administered orally, they must be absorbed through the gut. This can be modeled by allowing another compartment to represent the gut, and thus drug will flow into the plasma component at a rate proportional to the amount in the gut. This last rate constant depends on many factors, such as the surface area of the gut, the stomache rate flow, and blood flow. Actually, it also depends on the amount of drug present: only a certain amount can be absorbed. Write down the resulting equations and graph the concentration of drug in the blood versus time. Show that the time needed to reach maximum concentration is independent of dose.

(c) In the models considered so far, there has been only a one-way flow of substance. In reality, the drug will diffuse into various tissues and rediffuse from the tissues back into the blood. Model the tissues by one compartment. What are the resulting equations in the case of IV injections? Suppose that blood samples are drawn over 6 hours and the concentrations are as recorded in Table 4.1. What are your estimates of the rate constants?

Table 4.1 Data Obtained after a 150-mg IV Bolus

Time (h)	Concentration (mg/L)
0.5	9.5
1	6.5
2	5.3
3	4.8
4	4.3
5	4.1
6	3.9

(d) Pharmacokinetical models typically involve one central compartment (the blood plasma) that interacts with all the other compartments, and they interact with each other only via the central compartment. What is the general form of the system matrix?

(e) Other types of systems are typically modeled by the so-called train compartment models, where each compartment interacts only with two neighbors and the whole system forms a line of compartments. A famous example of this is the interaction of the Great Lakes system. What is the long-term evolution of pollution in the Great Lakes system? What data do you need to set up a compartment model with each lake a compartment? ([4] has references to some old sources of data.)

(f) These simple compartment models ignore transport properties inside the compartment. Or rather they idealize: we assume that the contents of each model are continuously kept uniform. In fact, lakes may not be able to mix in more than a thin surface region, and, in drug models, different organs belonging to the same compartment will have different absorption rates, and the concentration in even a single organ will not be uniform. How can these nonhomogeneities be accounted for or compensated for in compartment models? (We will study diffusion in Chapter 11. Compartment models will reappear in Chapter 7.)

8. We can model the spread of diseases through compartmental models. Let one compartment represent the infected part of the general population and another compartment represent the part of the population that is susceptible to infection (this might be everyone else).

(a) If the disease is such that recovery from infection does not provide immunity from reinfection and is not fatal, show that this two-compartment model is well defined. Suppose that recovery is exponential relaxation and that the rate of new infections is proportional to the product of the number of infected and the number of susceptible. Write down the appropriate system of equations. (Mass conservation here means a constant population.)

(b) In more complicated systems there are others in the general population besides the infected and the susceptible. For example, there are those who have recovered with immunity (from a viral disease, for example) or have been immunized, and there is the possibility of death from the disease. Show that all these can be accounted for by the introduction of one more class. Note that "mass" can only be conserved in the presence of death by assuming births into the susceptible class. What is the appropriate system of equations?

9. *Chemical reactions* The next several problems discuss the modeling of chemical reactions in terms of concentrations of the reactants.

(i) The simplest reaction is an *endothermic* reaction, which occurs when a substance is heated; absorbing the heat, it undergoes an irreversible transformation into a different substance (or combination of substances). An example is the mixing of steam and hot coal to produce "water gas":

$$H_2O + C \rightarrow CO + H_2.$$

Unless the concentrations are very high, the individual molecules will not interact, and there will be a constant fraction that reacts per unit time. Show that there is a simple exponential relaxation to equilibrium. (Of course, in a reactor the equilibrium is not necessarily the steady state since the products are continually removed.)

(ii) Suppose that in the same reaction there is a small chance that the product could chemically change back to the reactant (a reversible reaction). How does this change the qualitative form of the reaction?

10. *Nonlinear reactions* (i) Consider a bimolecular reaction where substances A and B react irreversibly to form C. Suppose that 1 mole of A reacts with 1 mole of B to form 1 mole of C. This is denoted by

$$A + B \rightarrow C.$$

Suppose that at the beginning there are equal quantities of A and B on hand, say M_0 moles, and there is no product. The reaction will proceed naturally until, at equilibrium, $A = B = 0$, $C = M_0$. Is this an exponential relaxation to equilibrium? To derive a differential equation for C, notice that equal quantities of A and B must physically come together in order to react. It is simple to think that one molecule of A must meet one molecule of B. Argue that the rate of reaction must then depend on the product of the concentrations of A and B, that is,

$$\dot{C} = kAB.$$

The rate constant k can be thought of as the fraction of pairs that are meeting per unit time. Argue that $A(t) + C(t) = M_0$ and likewise for B. Use separation of variables to solve the resulting equation for C, and show that the approach to equilibrium is not exponential but $\sim \frac{1}{t}$. This is an intrinsically nonlinear equilibrium. How would you define a time scale in this case?

(ii) Suppose that in the same reaction the product C can decompose into A and B. This is denoted by

$$A + B \underset{k_{-1}}{\overset{k_1}{\rightleftharpoons}} C.$$

Write down the differential equation for C in this case. Is the approach to the equilibrium C_0 exponential? What is the relaxation time? Compare this equation to the logistic equation.

11. *Enzyme reactions* Enzymes are proteins that act as catalysts for reactions. This means that they only act in an intermediate way, and the final product finds the enzyme molecules unchanged (and looking for some more action). Enzymes are important for most biochemical reactions and are also of importance to industry, especially in food processing. A simple scheme for an enzymatic reaction is that the enzyme reacts with a species called a **substrate** to form a **complex**; a further reaction results in a desired product and free enzyme. Symbolically, this is

$$S + E \underset{k_{-1}}{\overset{k_1}{\rightleftharpoons}} C \overset{k_2}{\longrightarrow} P + E.$$

(a) Write down a system of differential equations for the kinetics of the various concentrations.

(b) At $t = 0$, suppose that $C = P = 0$. Show that

$$E(t) + C(t) = E(0).$$

Eliminate E, leaving a coupled pair of equations for C and S.

(c) Consider the reasonable situation where there is initially much substrate, but little enzyme. (This is clearly believable in the case of cheese production.) Any available enzyme quickly gobbles up (or is gobbled by) the ubiquitous substrate. Thus, the complex concentration quickly attains an equilibrium during a time in which the concentration of substrate hardly changes. Assuming then that C instantaneously attains equilibrium, show that

$$C(t) = \frac{E_0 S(t)}{K_m + S(t)}. \qquad (4.31)$$

This is what the biochemists call the *quasi-steady-state equation*. Draw a picture of the situation in the (S, C)-phase plane. What are the dynamics for S? Discuss more fully the justification for using (4.31). What role do the initial concentrations play? The use and justification of the quasi-steady state equation is discussed more fully in [64] and [45].

12. *System response* The purpose of this exercise is to introduce some notions and terminology that are used when we think of differential equations as input–output systems. That is, the solution of the differential equation is the *output* to the system, which accepts as input a forcing function. The simplest case is when the equation is linear. In this scheme the differential equation

$$\dot{y} + \alpha y = f, \quad \alpha > 0,$$

is represented by

$$f \to \boxed{\tau} \to y,$$

where $\tau = \frac{1}{\alpha}$, and this relaxation time is all that is needed to specify the system. The system could also be characterized by the solution that is obtained with the use of particular forcing functions. The solution to the equation is

$$y(t) = y(0)e^{-\alpha t} + e^{-\alpha t} \int_0^t e^{\alpha s} f(s) \, ds.$$

The first term on the right shows the effect of the initial conditions and is a decaying transient term. The second term is what we will call the **response**, and it is this that is properly the output of the system. Notice that this also may decay in time.

(a) *Impulse response* Suppose that $f = f_0 \delta_0$, where δ_x is the delta function at $t = x$; that is, δ_x has the property that

$$\int_0^\infty f(t)\delta_x \, dt = f(x)$$

for any continuous function f. It can only be defined this way, via its action under the integral sign; it is not a true function. Show that the solution is

$$(y_0 + f_0)e^{-\alpha t},$$

so the response solution is identical in form to the unforced equation with a different initial condition. The effect of the impulse is to simply start at a different place.

(b) *Step response* Suppose that $f = f_0$ for $t > 0$. Show that the output response in this case is

$$\frac{f_0}{\alpha}(1 - e^{-\alpha t}),$$

which is relaxation to the equilibrium $\frac{f_0}{\alpha}$. Notice the reduced size of the output with respect to the input.

(c) *Oscillatory response* Suppose that the input is $f_0 \cos \omega t$. Show that the output response is

$$\frac{f_0}{\sqrt{\alpha^2 + \omega^2}}(\cos(\omega t + \phi) - \frac{\alpha}{\sqrt{\alpha^2 + \omega^2}}e^{-\alpha t}),$$

where ϕ satifies $e^{i\phi} = \frac{\alpha - i\omega}{\sqrt{\alpha^2 + \omega^2}}$ (i.e., $\tan \phi = -\omega/\alpha$). The output is oscillatory, but the phase of the oscillation is shifted. We say that the response *lags* the input. The size of this lag depends on the frequency; in the high-frequency limit, $\omega \to \infty$, the response will be exactly one-quarter out of sync. Show that in the low-frequency limit, $\omega \to 0$, the solution reduces to the step response.

(d) *Transfer function* Force the system with $Ce^{i\omega t}$, and look for a solution of the form $Ae^{i\omega t}$, where A is complex.

Chapter 5

Vibrations

Timing is everything, it has been said. Nobody knows this better than children who like to play on a swing. Timing determines whether their energy will be wasted inefficiently and they remain grounded or whether each pump takes them on a higher arc.

1 Free Vibrations

A single first-order ODE supplies the simple model that is the prototype for relaxation and growth, characterized by an asymptotic approach to an equilibrium steady state. We saw many example of this in Chapter 4. On the other hand, a model that uses a *system* of first-order ODEs can give rise to oscillating behavior, as illustrated by the relaxation oscillations of the two-sector economy in the last section of Chapter 4. In that case, because of the extreme disparity in the size of the time scales, we dealt with the oscillation in a piecewise manner: an instantaneous jump to a curve (the slow manifold) and then a monotone flow on the curve, followed by a jump to another curve, and so on.

In this chapter we will focus on systems that have a smooth oscillation, where energy is continually and smoothly transferred from one form to another. In general, we will use the term *oscillator* to refer to the mathematical model of a phenomenon whose measurement is time periodic. A system may oscillate of its own accord or be directly forced by some external agent or oscillate because some energy has leaked from another system to which it has been coupled. We will start with free mechanical vibrations because they are the easiest to visualize and thereby invested with intuition.

Mechanical Vibrations

Newton's law of inertial motion states that if a particle moves under the influence of force $\mathbf{F}$ then this $\mathbf{F}$ is the time rate of change of the momentum $\mathbf{p} = m\mathbf{v}$. If the mass is constant, then $\mathbf{F}$ is proportional to the acceleration, $\mathbf{a}$:

$$\mathbf{F} = m\mathbf{a}. \tag{5.1}$$

If $\mathbf{r}$ is the vector displacement of the particle with respect to some reference point, then $\mathbf{v} = \dot{\mathbf{r}}$ is the velocity and $\mathbf{a} = \ddot{\mathbf{r}}$ is the acceleration. Thus, (5.1) is a second-order differential equation.

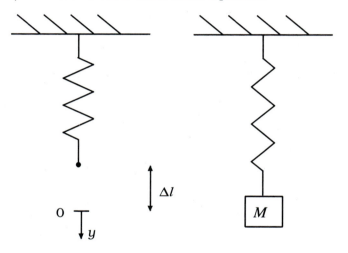

Figure 5.1 Adding a weight to the spring changes its equilibrium position. It is now subjected to a constant gravitational force.

Consider the linear (massless) spring of Chapter 2; equation (2.2) of that chapter is a special case of (5.1). That case is the static situation, which is simply the condition that

$$F_{\text{net}} = 0.$$

Let us now hang the spring from the ceiling overhead. The spring, being massless, will extend to its usual equilibrium as in Figure 2.3. When a point mass is attached to the spring, there is an added contribution of gravity to the forces acting; since this g force is constant, the equilibrium is changed (Figure 5.1, see Exercise 1 in chapter 2), and the equation for the equilibrium is given by the (static) force balance,

$$-k(\Delta l) + mg = 0, \tag{5.2}$$

or

$$\Delta l = mg/k.$$

This relation is the condition

$$U' = 0,$$

where U is the potential of the linear spring–mass system. Let y be the displacement downward of the mass, measured from its equilibrium position, which is found using the force balance condition (5.2). The the equation of motion (5.1) for this one-dimensional system is

$$m\ddot{y} = F_g - F_{\text{spring}} \quad = \quad mg - k(\Delta l + y)$$
$$= \quad -ky.$$

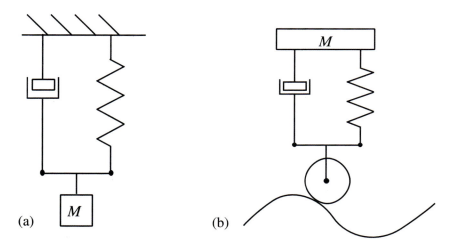

(a)   (b)

Figure 5.2 (a) The addition of a dashpot mechanism supplies velocity proportional damping. (b) A primitive model of an automobile suspension system. Here the spring and damper act in parallel. How do the equations change if they are in series?

If we add a dashpot to our system, we have a primitive model of an automobile suspension system (Figure 5.2). The dashpot is modeled by a viscous damping term that is proportional to velocity,

$$F_{vd} = -\eta v,$$

and the equation of vertical motion is

$$m\ddot{y} + \eta\dot{y} + ky = 0. \tag{5.3}$$

(This holds as long as the car of Figure 5.2 is traveling along a flat road so as not to introduce further accelerations. Later we will develop the proper extension.)

The (mechanical) energy in the spring system is put together from two terms, the **potential energy**,

$$U = \frac{1}{2}ky^2,$$

and the **kinetic energy**,

$$T = \frac{1}{2}mv^2,$$

so that the total energy is

$$E = T + U = \frac{1}{2}m\dot{y}^2 + \frac{1}{2}ky^2. \tag{5.4}$$

If we compute the time rate of change of E:

$$\dot{E} = m\dot{y}\ddot{y} + ky\dot{y} = \dot{y}(m\ddot{y} + ky) = -\eta\dot{y}^2,$$

we find that when $\eta > 0$ the energy is steadily decreasing, while it is conserved if $\eta = 0$.

The energy is a real-valued function of the two variables, the displacement, y, and the velocity, $\dot{y}$. These are the two coordinates of the phase plane.[1] In other words, given a point in the phase plane, a number is computed that is the energy associated with that point:

$$(y, \dot{y}) \rightarrow \text{energy computation} \rightarrow \text{number}$$

In the absence of damping ($\eta = 0$), E is constant along trajectories in the phase plane, so the trajectories in the phase plane must lie on the level curves of E. The quadratic energy (5.4) of the linear oscillator is positive definite. This means that there is one stationary point, corresponding to the 0 level, and all other level curves are ellipses:

$$E_c = \{(y, \dot{y}) : \frac{1}{2}m\dot{y}^2 + \frac{1}{2}ky^2 = c\}.$$

In fact, each level curve, E_c, $c > 0$, is exactly one complete trajectory. Briefly, the reason for this is that, if an ellipse were not a complete trajectory, then it would have to contain at least one stationary point,

---

[1] One could just as well use the momentum in place of the velocity.

where the angular coordinate would rest. A simple change of coordinates makes this transparently false.

EXERCISE 1 (i) Set $k/m = \omega_0^2$, and $x_1 = \omega_0 y$, $x_2 = \dot{y}$. Show that the oscillator equation (5.3), with $\eta = 0$, is equivalent to the system

$$
\begin{aligned}
\dot{x}_1 &= \omega_0 x_2, &(5.5)\\
\dot{x}_2 &= -\omega_0 x_1. &(5.6)
\end{aligned}
$$

Show that the level curves of the energy are circles in the (x_1, x_2)-plane. (ii) Show that in polar coordinates the equivalent system is

$$
\begin{aligned}
\dot{r} &= 0, &(5.7)\\
\dot{\theta} &= -\omega_0. &(5.8)
\end{aligned}
$$

The simple form of the system (5.7) reveals a steady sweeping of the angular coordinate at the **angular frequency**, ω_0, in the (x_1, x_2) - plane. This motion is known as **simple harmonic motion** (SHM). The fundamental **period** of the motion is

$$ T = 2\pi/\omega_0; $$

this is the number of time units it takes to complete one cycle. Figure 5.3 shows the bowl-shaped graph of the energy function and the level curve trajectories.

An important observation to be made concerning this oscillating system is the existence of two poles for the energy function: the total energy is constantly being passed back and forth between them, as water is poured from one glass to another. During an oscillation of the system (5.3), all the energy is kinetic (KE) when the mass passes through equilibrium; when the mass comes to a momentary halt at the end of its reach, all the energy is potential (PE). In this way both KE and PE go through a complete cycle of maximum and minimum, where

$$ KE_{max} = PE_{max} = E, $$

so that

$$ k/m = v_{max}^2/y_{max}^2. $$

This means that, for a harmonic oscillator, if we can measure the maximum kinetic energy and the maximum potential energy, then we get the ratio k/m, which is the angular frequency of oscillation. This simple sharing-of-energy argument can sometimes allow us to compute the frequency of oscillations of a vibrator without having to deal with the differential equation, which in general will be more complicated than (5.3). However, for a highly nonlinear vibrator, this estimate may not be so good.

We have been working with a quadratic energy, E, but the same phase–plane arguments leading to the picture of the trajectories pictured in Figure 5.3, will work whenever the restoring force is obtained from a potential.

EXERCISE 2 *Conservative Dynamics*
(i) Show that if $\mathbf{F} = \nabla U$ then $E = T + U$ is constant along trajectories of the differential equation $m\mathbf{a} = \mathbf{F}$.
(ii) How do the critical points of U relate to those of E?
(iii) *Anharmonic oscillator* If $U(x) = \frac{1}{4}x^4$, show that all trajectories are closed paths.
(iv) *Kepler potential* By studying $U(x) = -\frac{1}{x} + \frac{C}{x^2}$, show that the trajectories of a conservative system need not be closed.

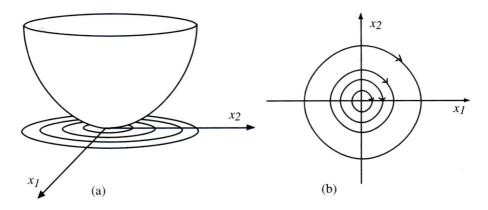

(a)   (b)

Figure 5.3 (a) The graph of the quadratic energy function. (b) The projections of constant energy level curves onto the phase plane are ellipses. Since energy is conserved these are trajectories of the system state.

Other Harmonic Oscillators

In chapter 4, we saw that the charging of a capacitor was entirely analogous to the relaxation of a damped spring. We will now extend this analogy to compare the vibrations of a mass–spring system to an inductive circuit.

Figure 5.4 shows a circuit with three elements, an inductance, a capacitance, and a resistance, in series. We will use the same constitutive laws for the voltage across each element that we used in Chapter 4:

$$V_C = Q/C, \quad V_R = RI, \quad V_L = L\dot{I}.$$

The net change in voltage in a closed cycle is zero.[2] Thus,

$$L\dot{I} + RI + \frac{Q}{C} = 0.$$

Since the current is the same through every element, this is actually a second-order equation for Q, the charge on the capacitor:

$$L\ddot{Q} + R\dot{Q} + \frac{1}{C}Q = 0. \tag{5.9}$$

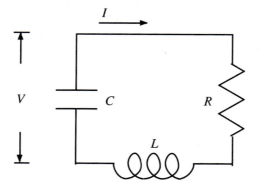

Figure 5.4 Simple linear circuit with three elements.

In practice, the capacitance charge is not directly measured. By differentiating this equation, we obtain an equation for the current in the circuit:

$$L\ddot{I} + R\dot{I} + \frac{1}{C}I = 0.$$

This is the same exact equation as that for the charge. Moreover, this equation has the same form as the equation for the mass-spring-damper mechanical system (5.3).

---

[2]This follows from the fact that the voltage is a potential. See Appendix A.

It is typically convenient to measure the voltage across the capacitor, V_C (think of the membrane voltage in Chapter 1). What is the equation for V_C? From

$$L\dot{I} + RI + V_C = 0,$$

we get

$$\dot{I} = -\frac{R}{L}I - \frac{1}{L}V_C.$$

From

$$V_C = Q/C,$$

we get

$$\dot{V}_C = \frac{1}{C}I.$$

Thus,

$$\begin{aligned}\ddot{V}_C = \frac{1}{C}\dot{I} &= -\frac{R}{LC}I - \frac{1}{LC}V_C \\ &= -\frac{R}{L}\dot{V}_C - \frac{1}{LC}V_C,\end{aligned}$$

or

$$\ddot{V}_C + \frac{R}{L}\dot{V}_C + \frac{1}{LC}V_C = 0. \tag{5.10}$$

Again, we arrive at the same differential equation that we had for I, the current in the circuit, and Q, the charge across the capacitor. In dealing with circuit equations it is best to always assign each variable with care.

The fact that the two equations, (5.9) and (5.3), have the same form impels us to treat the displacement of a mass and the charge of a capacitor as analogous variables. Table 5.1 presents the full details of this analogy between the mechanical and electrical oscillators. It is probable that the most important thing we can learn from this table of analogies is that, knowing already the form of the kinetic energy and the potential energy from mechanics, we can write down the conserved quantities, the energies, in a circuit. The partition of energy between the "kinetic" energy of the inductance and the "potential" energy stored in the capacitance is a reflection of the oscillation of energy in an electromagnetic wave that takes the form of alternating electric and magnetic fields. The fact that we can get these energies quickly without solving Maxwell's equations shows the power of such analogies.

Table 5.1 Analogy between Mechanical and Electrical Oscillators

Mechanical	Electrical
y	Q
v	I
F	V
m	L
η	R
k	C^{-1}
$U = \frac{1}{2}ky^2$	$\frac{1}{2}C^{-1}Q^2$
$T = \frac{1}{2}mv^2$	$\frac{1}{2}LI^2$

By consideration of dimensions in equation (5.10), it is seen that

$$[R/L] = T^{-1}, \quad [1/LC] = T^{-2},$$

so both L/R and $\sqrt{LC}$ have dimensions of time. Each can be associated with a characteristic time of the system. One is the relaxation time and the other is related to the period of oscillation.

All the linear equations that we've seen so far can be put in a unified form:

$$\ddot{x} + 2\lambda\dot{x} + \omega_0^2 x = 0. \tag{5.11}$$

(The factor 2 is there simply to ease the look of some of the formulas.) The parameter ω_0 is the frequency of undamped vibrations, so that the period of such vibrations is

$$T = \frac{2\pi}{\omega_0}.$$

Since ω_0 and 2λ both have dimensions of time^{-1} and the quantity $\omega_0/2\lambda$ is the simplest nondimensional quantity formed from them, this number should characterize the oscillator in some way. It is a straightforward matter to see how this happens because we can explicitly solve equation (5.11). If we expect solutions of the form

$$x = e^{rt},$$

(as we always do with such linear equations), then r must satisfy the characteristic equation

$$r^2 + 2\lambda r + \omega_0^2 = 0,$$

whose solutions are

$$r_+ = -\lambda + \sqrt{\lambda^2 - \omega_0^2}, \tag{5.12}$$

$$r_- = -\lambda - \sqrt{\lambda^2 - \omega_0^2}. \tag{5.13}$$

Clearly, everything depends on the sign of the radical:

If $\lambda > \omega_0 > 0$, then r_+ and r_- are real, and the general solution of (5.11) is a linear combination of decaying exponentials:

$$x(t) = Ae^{r_+t} + Be^{r_-t}.$$

EXERCISE 3 Does the fact that r_+ and r_- are both negative imply that $x(t)$ is necessarily monotonically decreasing?

Figure 5.5 shows the pattern of trajectories formed in the phase plane. This system is called a (stable) node.

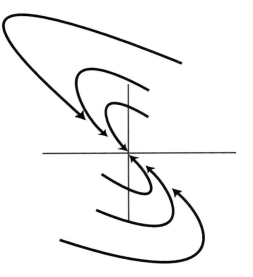

Figure 5.5 Stable node in phase plane. Unstable node has arrows going in the opposite direction.

If $\lambda < \omega_0$, then

$$r_+ = -\lambda + i\omega, \tag{5.14}$$

$$r_- = -\lambda - i\omega, \tag{5.15}$$

where $\omega = \sqrt{\omega_0^2 - \lambda^2}$ is the effective frequency of vibration. Notice that the effect of damping is to lower the frequency. Because

$$|r_+|^2 = |r_-|^2 = \lambda^2 + \omega^2 = \omega_0^2,$$

the roots lie on the same circle in the complex plane of radius ω_0 for any damping λ. The general solution to the equation in this case is

$$
\begin{aligned}
x &= \alpha e^{r_+ t} + \bar{\alpha} e^{r_- t} \\
&= 2Re(\alpha e^{r_+ t}) \\
&= A e^{-\lambda t} \cos(\omega t + \phi).
\end{aligned}
$$

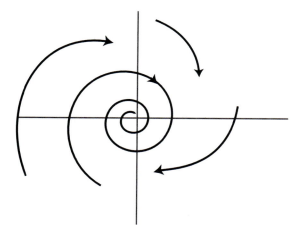

Figure 5.6 Stable focus in phase plane.

The numbers A and ϕ, for which there is the relation

$$\alpha = A e^{i\phi},$$

are determined by initial conditions and are the polar coordinates of a point on the trajectory in the phase plane (Figure 5.6). This family of trajectories in the phase plane is called a (stable) focus. Notice that if the initial velocity satisfies $v_0 = 0$ then

$$\tan \phi = -\lambda/\omega = \frac{-(\lambda/\omega_0)}{\sqrt{1 - (\lambda/\omega_0)^2}}.$$

Figure 5.7 shows the time history of the displacement, $x(t)$, with the parameters A and ϕ labeled. The effects of damping on the harmonic oscillator are,

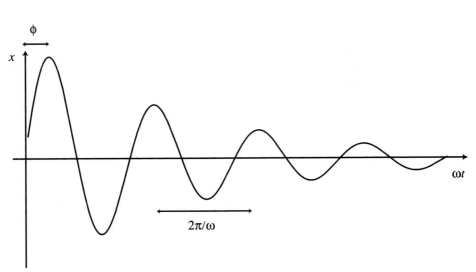

Figure 5.7 Time history of oscillator displacement.

first of all, to lower the frequency so that one oscillation takes longer to complete, and, *second*, during the course of each oscillation there will be a decay in amplitude.

How much has the amplitude decayed in one period of oscillation? The local maxima of displacement are a period T apart,

$$x(t_1), x(t_2), \ldots,$$

where

$$T = 2\pi/\omega,$$

and the $\{t_k\}$ satisfy

$$\omega t_k + \phi = 2k\pi.$$

This means that during one oscillation

$$\frac{x(t_k + T)}{x(t_k)} = e^{-\lambda T} = e^{-2\pi\lambda/\omega},$$

so

$$N_e = \frac{\omega}{\lambda} \frac{1}{2\pi}$$

is the number of oscillations that it takes for the amplitude to decay by a factor of e; in other words, the amplitude is reduced by a factor of e^{-1} in ω/λ radians.

EXERCISE 4 If Ω_e is the number of radians needed for the amplitude to decay by a factor e, show that

$$(\frac{\omega_0}{\lambda})^2 = \Omega_e^2 + 1.$$

In the phase plane, the presence of viscous damping breaks the closed trajectories of the harmonic oscillator. Define the "reduced" energy function

$$E(x, \dot{x}) = \frac{1}{2}\dot{x}^2 + \frac{1}{2}\omega_0^2 x^2,$$

giving the energy per unit mass for the spring oscillator. Computing the rate of energy dissipation along trajectories gives

$$\dot{E} = -2\lambda\dot{x}^2. \tag{5.16}$$

Thus, the energy decrement in one period is

$$\Delta E_T = -2\lambda \int_0^T \dot{x}^2 \, dt. \tag{5.17}$$

EXERCISE 5 Use the complex form of the solution, that is, the form proportional to

$$e^{-\lambda t}e^{i\omega t},$$

to compute ΔE_T and display its dependence on ω_0 and λ/ω_0.

The result of this exercise is that, if

$$\lambda/\omega_0 \ll 1,$$

then the energy decays by a negligible amount in only one period. Now, if we treat the decay in amplitude over one period as negligible, then we can average to get

$$\frac{1}{T}\int_0^T \dot{x}^2 \, dt = \frac{1}{2}\omega^2 A^2;$$

but this is just the total energy, which we think of as the average, E_{av}, in one oscillation of either the potential or kinetic energy. Thus, the right side of (5.17) is

$$E_{av}T,$$

and we have an equation for E_{av} valid for times much longer than T,

$$t \gg T:$$

$$\dot{E}_{av} = -2\lambda E_{av}. \tag{5.18}$$

Here the average is over one cycle, but since the energy changes little over one cycle, we can use this equation for E. We have thus identified

$$\tau_e = \frac{1}{2\lambda}$$

as a relaxation time for the energy.

It must be emphasized that the exponential law (5.18) is, strictly speaking, only valid for the special case of small viscous damping, which, as we will see, is not as common as the simplicity of the analysis would lead us to wish for. However, the procedure that we have carried out here—to wit, derive the equation (5.16), then average over a cycle, treating the system as a nearly harmonic oscillator—can be used in quite general circumstances. One can carry out similar analyses for nonlinear oscillators to get information concerning energy dissipation or to identify possible limit cycles.

In the sense that the energy E represents the amplitude squared of a trajectory in the phase plane, equation (5.18) indicates that the trajectories are logarithmic spirals, tending to $(0,0)$ as $t \to \infty$. The collection of phase-plane trajectories for a second-order system is called a *phase portrait*. The phase portrait of the dissipative oscillator is known as *stable focus* (Figure 5.6).

The dimensionless quantity $\omega/(2\lambda)$ is the number of radians of phase that it takes for the energy to decay by a factor of e. This number is related to the previous dimensionless quantity $\omega_0/(2\lambda)$ (based on the undamped frequency) by

$$\left(\frac{\omega}{2\lambda}\right)^2 + \frac{1}{4} = \left(\frac{\omega_0}{2\lambda}\right)^2.$$

We call the quantity

$$Q = \omega/(2\lambda)$$

the **quality** of the oscillator. Thus systems with high values of Q,

$$Q \gg 1,$$

are models of true vibrations.

EXERCISE 6 (i) Show that the true vibrations are characterized by characteristic roots

$$r_+, \quad r_-$$

that lie in a double wedge in the complex plane.
(ii) For systems that are known to have minimal damping, one looks for a solution of the form $e^{i\omega t}$. Rederive the characteristic equations and show

that the strength of the damping (and energy decay) is determined by the imaginary part of ω.

(iii) Show that reversing the sign of λ,

$$\lambda \to -\lambda,$$

has the same effect as reversing the sign of t,

$$t \to -t.$$

Thus, all phase portraits are the same but with the arrows reversed (Figure 5.8). This portrait is called an *unstable focus*.

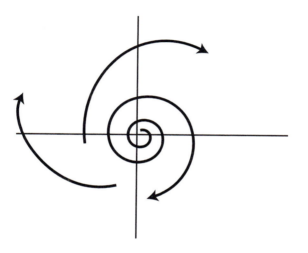

Figure 5.8 Unstable focus in phase plane.

Pendulum

We want to turn now to the analysis of a simple nonlinear oscillator, the pendulum which has all mass concentrated in a bob at the end of an inflexible rod of length l (Figure 5.9); the only outside agent is the downward force of gravity. It is easy to derive the equation of motion for this pendulum by directly accounting for forces and using Newton's law, but we will take a little detour and introduce a method (attributed to Lagrange) that is equivalent to Newton's law. Although not needed in this simple case, it is a useful tool to have at hand, and in more complex circumstances it is the only reasonable way to derive the equations of motion (see the problem section). Suppose that an unmanned mission to Mars is planned: It is necessary to build a robot that can play golf on the rough Martian surface (Figure 5.10). In this case it becomes messy to balance all the torques in Newton's law, but the Lagrangian analysis is straightforward.

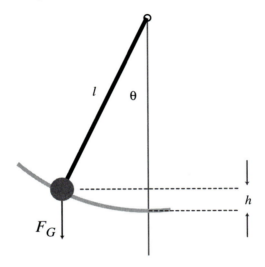

Figure 5.9 Simple pendulum at an angular displacement, θ. The vertical height above equilibrium is h.

This method of Lagrange is an example of the calculus of variations. We will define a functional on the possible trajectories; the actual trajectory, the one that describes the physical motion, is at a *critical point* of the functional.

We illustrate the method in the case of the simple pendulum and start by finding expressions for the kinetic and potential energies. The kinetic energy is

$$T = \frac{1}{2}mv^2,$$

where v is the (rectilinear) velocity of the bob. From Figure 5.10, we see that the potential energy is

$$U(\theta) = mgh = mg(l - l\cos\theta).$$

Notice that

$$\theta = 0$$

is a minimum of U.

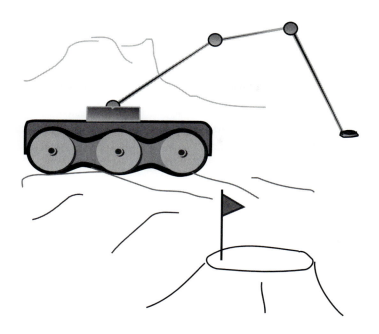

Figure 5.10 Mars golfer.

EXERCISE 7 Using
$$F = -U'(\theta),$$
write down the equation of motion for the pendulum using Newton's law. *(Be careful: What is the acceleration?)*

Since
$$v = l\dot{\theta},$$
the kinetic energy can be written in terms of $\dot{\theta}$:
$$T = \frac{1}{2}ml^2\dot{\theta}^2.$$

We then form a function (similar to the energy function) known as the Lagrangian:
$$\mathcal{L}(\theta, \dot{\theta}) = T - U = \frac{1}{2}mv^2 - mg(l - l\cos\theta).$$

The integral of this function over a time interval $[t_0, t_1]$,
$$L = \int_{t_0}^{t_1} T(\dot{\theta}) - U(\theta)\, dt, \qquad (5.19)$$

is the *action*. It is supposed that the end points of the curve along which the integration is done are fixed,

$$\theta(t_0) = \theta_0, \quad \theta(t_1) = \theta_1, \tag{5.20}$$

but otherwise any smooth curve can be followed. Thus, the action defines a function of phase-plane trajectories. That is, given any hypothesized trajectory in the phase plane,

$$(\theta(t), \dot{\theta}(t)),$$

that satisfies (5.20), we perform the integral (5.19) to obtain a number, the *action* of that trajectory. In contrast to the energy, which is a function of *points* in the phase plane, the action is a function of *trajectories* in the phase plane. (Such a function of functions is commonly called a functional.)

A word of caution here: The conditions (5.20) are not meant to represent input data, but simply reflect that we only care about the *dynamics* of the desired motion. Eventually, the actual trajectory is found by solving an initial–value problem (IVP).

The principle of Lagrange states that the true trajectory, the one that represents the real dynamics of the bob, is the one that furnishes a stationary point for the action functional, L. This means that if we perturb the true trajectory a little bit by considering a *virtual* path that is nearby—imagine, just for a moment, that the bob goes a little faster and then a little slower than it should—the value of L changes insignificantly relative to the size of the perturbation.

More precisely, if each point in the virtual path corresponds to a point on the true path in such a way that we can judge the size of the perturbation by a number $\lambda > 0$ ($\lambda = 0$ corresponds to the true path), then we require that

$$L(\theta^\lambda) - L(\theta^*) = o(\lambda),$$

where θ^* is the true path and θ^λ is the perturbed path.

The astute reader will recognize that this is a way of saying that a certain derivative is zero (Fermat's lemma). But what kind of derivative is this? We seem to be saying that we take the derivative of L with respect to θ and set this derivative equal to zero; but $\theta(\cdot)$ is a function. How are we to take the derivative with respect to a function?

For a clue on how we should proceed, recall that a similar difficulty presents itself when one begins the study of multivariable calculus: how

do we differentiate with respect to a vector? This task is typically reduced to calculating the usual one-dimensional derivative by restricting
the variation of the function to be in one direction; these are the directional derivatives. The derivative of a function f at $\mathbf{w}$ in the direction
of an arbitrary vector $\mathbf{v}$ is

$$\lim_{\lambda \to 0} \frac{f(\mathbf{w} + \lambda \mathbf{v}) - f(\mathbf{w})}{\lambda}. \tag{5.21}$$

The critical points of f are those vectors for which these quotients
vanish for all possible choices of $\mathbf{v}$. If f is a differentiable function,
then by Taylor's theorem, the above limit must be a linear function (of
$\mathbf{v}$), and so there must be a vector $\mathbf{G}$ so that this derivative (5.21) is

$$\mathbf{G} \cdot \mathbf{v}.$$

This is called the *gradient* of f at $\mathbf{w}$. The existence of such a $\mathbf{G}$ (derived
via a generalized Taylor's theorem, for example) for a functional such
as the action is rather complicated; we will simply note that when we
form the directional derivative in a direction of a perturbation η this
derivative must be linear with respect to η:

$$\lim_{\lambda \to 0} \frac{L(\theta^* + \lambda \eta, \dot{\theta}^* + \lambda \dot{\eta}) - L(\theta^*, \dot{\theta}^*)}{\lambda} = \delta L \cdot \eta \tag{5.22}$$

This operator, δL, is called the (first) *variation* of L.

As we don't allow $\theta(t_0)$ and $\theta(t_1)$ to vary, we must choose only
perturbations $\eta(\cdot)$ for which

$$\eta(t_0) = 0 = \eta(t_1).$$

Now, to form the Lagrangian:

$$L(\theta^* + \lambda \eta \quad , \quad \dot{\theta}^* + \lambda \dot{\eta}) \tag{5.23}$$

$$= \int_{t_0}^{t_1} \frac{1}{2} ml^2 (\dot{\theta}^* + \lambda \dot{\eta})^2 + mgl \cos(\theta^* + \lambda \eta) \, dt \tag{5.24}$$

$$= \int_{t_0}^{t_1} \frac{1}{2} ml^2 (\dot{\theta}^* + 2\dot{\theta}^* \dot{\eta} \lambda) + mgl \cos \theta^*$$

$$- \eta \lambda mgl \sin \theta^* + o(\lambda) \, dt. \tag{5.25}$$

Here we used the one-variable Taylor's theorem. The difference quotient (5.22) becomes in the limit

$$\lim_{\lambda \to 0} \frac{L(\theta^* + \lambda \eta, \dot{\theta}^* + \lambda \dot{\eta}) - L(\theta^*, \dot{\theta}^*)}{\lambda}$$

$$= \int_{t_0}^{t_1} ml^2 \dot{\theta}^* \dot{\eta} - mgl \sin \theta^* \eta \, dt. \tag{5.26}$$

This is linear in η, and $\dot{\eta}$ but to put it into the form

$$\delta L \cdot \eta$$

we must integrate by parts. Doing this, we get

$$\delta L \cdot \eta = -\int_{t_0}^{t_1} (ml^2\ddot{\theta} + mgl\sin\theta)\eta \, dt.$$

The integrand is a continuous function of t, and η is arbitrary; therefore, the integrand must be zero.[3]

EXERCISE 8 Prove this last statement.

Now we can catch our breath: the conclusion arrived at by application of the Lagrange principle, that we should have

$$\delta L = 0,$$

is that

$$ml^2\ddot{\theta} + mgl\sin\theta = 0. \tag{5.27}$$

If it were to turn out that the solutions of equation (5.27) did not have continuous second derivatives, then the last step of our derivation would be bogus; but, in fact, the trajectories are smooth, as we will show.

The energy is

$$E = \frac{1}{2}m\dot{\theta}^2 + mg(l - l\cos\theta)$$

and is conserved:

$$\dot{E} = (ml^2\ddot{\theta} + mgl\sin\theta)\dot{\theta} = 0,$$

as it should be, since

$$F = -U'(\theta).$$

(Notice that we could just as well have taken the energy to be

$$E = \frac{1}{2}m\dot{\theta}^2 - l\cos\theta,$$

as this simply corresponds to a shifting of the graph.)

---

[3]The asterisk is no longer necessary.

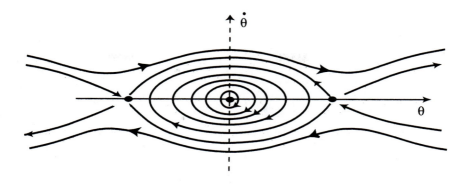

Figure 5.11 Trajectory curves in the phase plane for the simple pendulum. Low-energy curves are closed. The open curves at the top represent counterclockwise spinning of the bob, while the open curves at the bottom correspond to clockwise spinning. (In other words, opposite of the spinning of the physical model.)

The trajectories in the phase plane lie on the level curves of E. These level curves are drawn in Figure 5.11. The critical points,

$$\{(\theta, \dot{\theta}) = (2k\pi, 0)\},$$

are equilibrium points. Notice two different kinds of trajectories: there are some high-energy curves that are not closed curves; for these the angular displacement of the pendulum increases monotonically: this corresponds to the bob spinning around the support (not a true oscillation). Other curves (lower energy) are closed and somewhat elliptical in shape. These correspond to true oscillations. These closed curves and their corresponding bowl-shaped energy graph (near $\theta = 0$) are reminiscent of the harmonic oscillator. In fact, if we linearize about $\theta = 0$ by using

$$\sin \theta = \theta - \frac{\theta^3}{3!} + \cdots,$$

we find that the equation describing small displacements from equilibrium is

$$\ddot{\theta} + \frac{g}{l}\theta = 0.$$

This we recognize as a harmonic oscillator with natural frequency

$$w_0^2 = g/l,$$

that is, with a period

$$T_0 = \frac{2\pi}{w_0} = 2\pi\sqrt{l/g}.$$

For the harmonic oscillator, the period of vibration is independent of the amplitude, but this cannot be true of our pendulum. Look at the energy levels: they flatten out with distance from $(0,0)$ and become almond shaped as they approach those strange curves that join the equilibria at $\theta = \pm\pi, \pm 3\pi, \ldots$

How closely does T_0 approximate the actual period T_p? That is, how big can we take θ_{max} before the linear approximation breaks down?

From dimensional considerations (Chapter 3), we know that

$$T_p = \sqrt{\frac{l}{g}}\, f(\theta_{max}).$$

We want to know more about the function f. We can find an implicit expression for this function. In the course of one-quarter period, the pendulum rises from its bottom position to its maximum height, thus

$$\frac{T}{4} = \int_0^{T/4} dt = \int_0^{\theta_{max}} \frac{d\theta}{\dot{\theta}} = \frac{1}{2}\sqrt{\frac{l}{g}} \int_0^{\theta_{max}} \frac{d\theta}{\sqrt{\sin^2\frac{\theta_{max}}{2} - \sin^2\frac{\theta}{2}}}, \quad (5.28)$$

where we have used energy conservation,

$$\frac{1}{2}ml^2\dot{\theta}^2 = E - mgl(1 - \cos\theta),$$

and the fact that, when $\theta = \theta_{max}$, $\dot{\theta} = 0$, so that

$$E = mgl(1 - \cos\theta_{max}) = 2mgl\sin^2(\frac{\theta_{max}}{2}).$$

The integrand in the expression (5.28) is singular, but if we set

$$K = \sin\frac{\theta_{max}}{2}$$

and introduce a new variable,

$$\sin\phi = \frac{\sin\theta/2}{K},$$

then, as

$$\cos \phi \, d\phi = \frac{\cos \theta/2}{2K} \, d\theta,$$

and

$$\cos \frac{\theta}{2} = \sqrt{1 - K^2 \sin^2 \phi},$$

we have

$$
\begin{aligned}
T_p &= 4\sqrt{l/g} \int_0^{\pi/2} \frac{d\phi}{\sqrt{1 - K^2 \sin^2 \phi}} \\
&= T_0 \frac{2}{\pi} \int_0^{\pi/2} \frac{d\phi}{\sqrt{1 - K^2 \sin^2 \phi}}.
\end{aligned}
\tag{5.29}
$$

This integral, which gives the exact form of the function whose existence was inferred via dimensional analysis, is a complete elliptic integral of the first kind. Given the amplitude θ_{max}, then, with

$$K = \sin^2 \frac{\theta_{max}}{2},$$

we can compute the integral numerically (or look it up in a table), which will give the period to any desired accuracy.

From this integral expression we can calculate the first correction to using T_0 :

$$
\begin{aligned}
T_p &= T_0 \frac{2}{\pi} \int_0^{\pi/2} 1 + \frac{K^2}{2} \sin^2 \phi + \cdots d\phi \\
&= T_0 + T_0 \frac{K^2}{4} + \cdots \\
&\simeq T_0 \left(1 + \frac{\theta_{max}^2}{16} \right).
\end{aligned}
\tag{5.30}
$$

To get an idea of how close we need to be for the linear approximation to give a good approximation, note that if the maximum amplitude of the swing is 30^0 then this last expression (5.30) gives

$$T_p/T_0 \cong 1.0171,$$

while the integral gives $T_p/T_0 \simeq 1.0173$. In any case, this is a factor of less then 2%. A pendulum intended for timekeeping would never be set swinging at such an amplitude. In fact, the clock typically just swings through a few degrees.

EXERCISE 9 A pendulum clock is to be designed. First we need a length of about 1 m to fit in the cabinet. According to linear theory, this will give a period of 1.99 s, just shy of the desired 2 s. It is felt that the lengthening of the period due to the nonlinearity will compensate for this. The clock is made to swing through a total of 5 degrees. How accurate is the clock?

*Equivalent linearization

If a pendulum of length l swings with a maximum amplitude $\theta_{\max}$ and a period T_p, then there is a longer pendulum, of length l', that would oscillate with that same period, T_p, while making oscillations of infinitesimal amplitude[4]; this period is given by

$$T_p = 2\pi \sqrt{l'/g}.$$

Is there some way to compute this l', given l and $\theta_{\max}$, and thereby get the period T_p? In other words can we simultaneously decrease the amplitude and increase the length while the period remains constant? The following method is taken from [8]. The amplitude, as represented by $K = \sin \frac{\theta_m}{2}$, is reduced by the operation

$$S(K) = \frac{1 - \sqrt{1 - K^2}}{1 + \sqrt{1 - K^2}} = \frac{1 - \cos \theta_m/2}{1 + \cos \theta_m/2}.$$

Let $\tau = 2\pi \sqrt{l/g}$ denote the period of a linear pendulum of length l. Let $T_p = T(K, \tau)$ be the period of the nonlinear pendulum with maximum amplitude θ_m and length l. The renormalization operator $R(K, \tau)$ is defined on pairs of numbers, the first of which measures the amplitude of the oscillation, while the second measures the length of the pendulum:

$$R(K, \tau) = (S(K), (S(K + 1))\tau).$$

EXERCISE 10 Show that under the action of the transformation R the amplitude is reduced and the length of the pendulum increases. How is the period affected?

If we define the sequence $\{K_n, \tau_n\}$ by

$$
\begin{aligned}
(K_0, \tau_0) &= (K, \tau), \\
(K_n, \tau_n) &= R(K_{n-1}, \tau_{n-1}),
\end{aligned}
$$

---

[4]Well within the amplitude range, that is, where the linear approximation is sufficient.

then it is easily shown that

$$K_n \to 0$$

and

$$\tau_n \to \tau_\infty = \tau \left(\prod_0^\infty (K_n + 1) \right), \tag{5.31}$$

with a rapid convergence. From the relation

$$\tau_\infty = 2\pi \sqrt{l_\infty / g},$$

we can compute the length of the renormalized pendulum.

If the pendulum is brought to a horizontal position before being released ($\theta_m = \pi/2$), then (5.30) gives a discrepancy of 15% with the linear pendulum; in fact, calculated by use of the elliptic integral (5.29), this error is rather 18%. As the amplitude $\theta_{\max} \to \pi$, the period (5.29) tends towards infinity. A pendulum that is started infinitesimally close to the upper vertical position will make exactly one swing up the other side and will take infinitely long to do this. Of course, most of the time would be spent near the upper equilibrium. This upper equilibrium is clearly unstable. Let's linearize about that point, $\theta = \pi$: since $\sin \theta = -(\theta - \pi) + \cdots$, if $\xi = \theta - \pi$ is to be considered a small displacement, we get

$$\ddot{\xi} - (g/l)\xi = 0.$$

The general solution is a linear combination of a growing exponential and a decaying exponential:

$$\xi(t) = C_1 e^{\sqrt{g/l}\,t} + C_2 e^{-\sqrt{g/l}\,t}.$$

If (C_1, C_2) parametrizes the space of solutions, we see that for all pairs the second component tends to zero, while the first component increases in magnitude without bound. This is clearly an unstable equilibrium. In the phase plane it is a **saddle point.** Locally, saddle points act very much like unstable nodes; however, saddle points are important for their influence on the global phase picture of nonlinear oscillators. This is indicated by the phase picture of the pendulum (Figure 5.11) where trajectories leading *away* from the saddle point separate the two qualitatively different behaviors of the pendulum:

$\to$ the periodic solutions represented by the closed trajectories

$\to$ the run-on solutions represented by the open trajectories

What if a torque is applied to the pendulum? We could hook it up to a geared motor that supplies a constant torque:

$$ml^2\ddot{\theta} + mgl\sin\theta = \tau. \tag{5.32}$$

EXERCISE 11 What is another possible arrangement that will supply a constant torque?

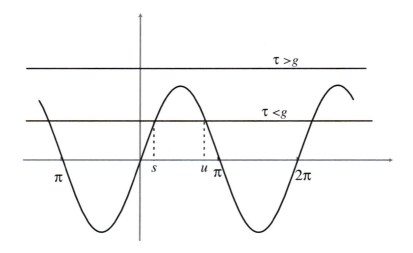

Figure 5.12 The equilibria are found at the intersection of the two curves, where $\sin\theta = \tau/g$. The point marked s is a stable equilibrium and the point marked u is unstable.

From our previous studies in Chapters 2 and 4,, we expect that the main effect of the constant applied torque will be a change in the equilibria: this is explained by Figure 5.12. The equilibria are found at the intersection of the two curves.

The equation for equilibrium is

$$\sin\theta = \tau/g.$$

Thus, for $\tau/g > 1$, there are no equilibria: in this case, we have the spinning behavior. Regardless of initial velocity, the torque supplied is enough to beat out gravity in order to keep it going up each time around.

From Figure 5.12 we see that there are two equilibria between 0 and π, one stable and one unstable. As the parameter τ approaches g

from below, the two equilibria approach and annihilate each other as τ increases through g. This is an example of a limit point instability or fold catastrophe (see Exercise 10 in Chapter 2). Graphical illustrations of this instability are afforded by analogues of the pendulum. These are phenomena that are described by the same equations.

A simple example is the ball rolling on a washboard (Figure 5.13)

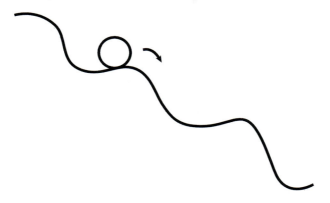

Figure 5.13

The washboard is made of half-circles clamped end to end, and the ball is somehow constrained to the washboard curve. In this case the equation for displacement along the curve is the pendulum equation (5.32), with "torque" proportional to the angle of elevation, and one can see visually the limit point instability (see Problem 2).

A Splash of Reality: Nonlinear Oscillations

The assumption of linear viscous damping for the swinging pendulum is a convenient fiction, for it allows us to compute an exponential energy decay. But from Chapter 3 you know that aerodynamic drag on a body is proportional to the velocity only under certain conditions, specifically for low Reynolds number. For motion in air, this would mean that velocity of motion must be extremely small. For higher velocities, as would be experienced by a pendulum with wide swings, the drag force would be proportional to the square of the velocity; to ensure that the force on the pendulum is always dissipative, we would write it as

$$F_d \propto |v|v.$$

Then the energy dissipation is computed as

$$\dot{E} = -\lambda v^3, \tag{5.33}$$

and since average energy is not proportional to v^3 in any sense, we do not get exponential relaxation. However, the expression (5.33) is enough to ensure the decay of the energy and the asymptotic stability of the equilibrium. Let us examine this carefully because the same type of stability argument can be used in other circumstances.

The expression (5.33) says that energy is never increasing, in fact it is always decreasing except when the velocity is zero. Except for the times when the velocity is zero, the trajectory in the phase plane is passing down the gradient of E, from high levels to low. This decrease will be monotone, unless the velocity is zero on some nontrivial time interval. But this cannot happen because the velocity is zero only at

(i) equilibria

(ii) local extrema of the potential energy.

In case (ii), $v = 0$ can hold only for an instant before the energy begins to shift back to the kinetic form. Thus,

$$E \longrightarrow 0$$

at a monotone rate. In fact, it is faster than exponential for large velocities; but when

$$v \approx 0,$$

the appropriate damping law is

$$F_d \propto v,$$

and so we have exponential approach to equilibrium.

As the pendulum shows, the potential energy may have a non-quadratic form, and thus the equations will be nonlinear. The elastic restoring force of a spring will, for large enough displacements, cease to follow Hooke's law (usually it will be stiffer, but for some plastic materials it will be more compliant). A "hard spring" is one whose stiffness increases with large displacement, while a "soft spring" has the opposite behavior. We can model the different behaviors with an extra quartic term in the potential:

$$\text{Hard spring}: \quad U_{hd}(x) = \tfrac{1}{2}kx^2 + \tfrac{1}{4}\bar{k}x^4,$$
$$\text{Soft spring}: \quad U_{sf}(x) = \tfrac{1}{2}kx^2 - \tfrac{1}{4}\bar{k}x^4.$$

EXERCISE 12 (i) Show that for a hard spring (with U_{hd}) all motions are periodic. *(Hint:* use facts about energy levels.)
(ii) Show that, for a soft spring, motions are periodic if the initial conditions are small.

Notice that the pendulum has a "soft" restoring force.

If the spring is made of a plasticlike material that softens as it is stretched, but stiffens as it is compressed, then the restoring force is not symmetric, and we must add a cubic or other odd term to the potential. This type of potential also characterizes the interaction between two atoms (see the problems).

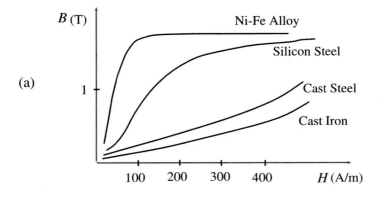

Figure 5.14 (a) Magnetization curves for inductor cores made from different materials. Magnetic induction is measured in teslas, and the applied field is measured in amperes per meter.

The components of electrical circuits, more often than not, also satisfy nonlinear constitutive laws. The relation

$$V_L \propto \frac{dI}{dt},$$

which we have used, is approximately true for the case of a coil in a vacuum, where the flux is proportional to the current,

$$\phi \propto i,$$

but in a working coil with a saturable core, the curve of flux versus current is more like the nonlinear curves of Figure 5.14. We can match this curve to a polynomial,

$$ni = a_1\phi + a_3\phi^3 + \cdots,$$

where n is the number of turns in the coil. It is convenient to take the magnetic flux as the state variable, and the resulting equation is the same equation as for the hard spring with symmetric restoring force.

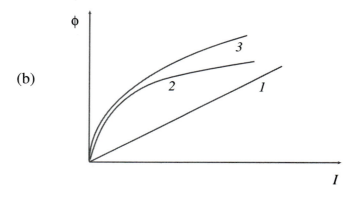

Figure 5.14 (b) In a vacuum, an inductance coil would have magnetic flux proportional to current (1), but with an iron core, it can have the characteristic of curve (2). A cubic curve (3) is shown for comparison.

In many circuit components, the relation between voltage and current is not one of strict proportionality. Figure 5.15 shows the I–V curve for a typical diode. Notice that for some values of the voltage, the "resistance" is negative. This same type of relationship holds between the plate current, i_p, and the grid voltage, v, in the vacuum-tube oscillator of Figure 5.16. In the problems, you will investigate the circuit in detail, especially the effect of the mutual inductance, M, which acts as a control parameter. But for now, let's assume that the relation between plate current and grid voltage is

$$i_p = f(v) = v - v^3.$$

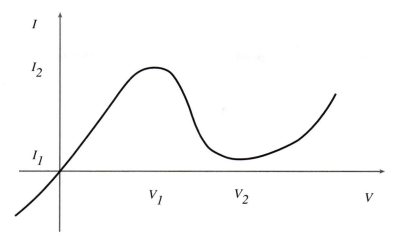

Figure 5.15

The equation for the current through the LC oscillator is

$$L\frac{di}{dt} + \frac{1}{C}\int i - i_p = 0.$$

This current is coupled to the grid voltage via the inductance M:

$$v = M\frac{di}{dt}.$$

Differentiating this equation and substituting the former, we get

$$\frac{dv}{dt} = \frac{M}{LC}(v - v^3 - i),$$

which, after another differentiation, gives

$$\frac{d^2v}{dt^2} = \frac{M}{LC}(\dot{v} - 3v^2\dot{v} - \frac{1}{M}v).$$

By introducing the new time variable,

$$\tau = \omega_0 t, \qquad \omega_0 = \frac{1}{\sqrt{LC}},$$

we get

$$\frac{d^2v}{d\tau^2} + \alpha(3v^2 - 1)\frac{dv}{d\tau} + v = 0,$$

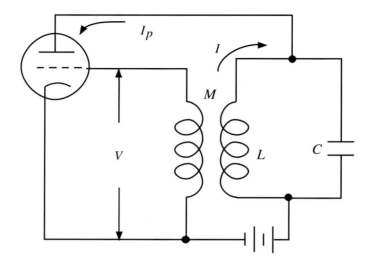

Figure 5.16 Vacuum-tube oscillator circuit.

where

$$\alpha = M\omega_0.$$

If we think of M as a control parameter, then this equation tells us that it controls the magnitude of the damping term; but for any fixed M, the damping term is negative for

$$v < \frac{1}{\sqrt{3}}$$

and positive for

$$v > \frac{1}{\sqrt{3}}.$$

Thus, the phase plane equilibrium,

$$(v = 0, \dot{v} = 0),$$

is an unstable critical point; however the oscillator is globally stable since the energy,

$$E(v, \dot{v}) = \frac{1}{2}(v^2 + \dot{v}^2),$$

satisfies

$$\dot{E} = \alpha(1 - 3v^2)(\dot{v})^2, \tag{5.34}$$

which is negative for v sufficiently large. This argument is logically unsatisfactory, for we can't rule out the posssibility that $\dot{v}$ becomes large while v remains moderate, wavering. Some detailed phase-plane analysis is needed to show that this never occurs and that the oscillator is indeed stable (see problems), but we can confidently use this energy argument if we assume that α is a very small number,

$$\alpha \ll 1.$$

For $\alpha = 0$, the trajectory in the phase plane is a circle

$$x_1 = v = A \cos t, \tag{5.35}$$
$$x_2 = \dot{v} = -A \sin t. \tag{5.36}$$

So for $\alpha > 0$ the trajectory will be close to a circle, so we can evaluate the energy decrement in one cycle, as we did for the linear oscillator:

$$\Delta E = \int_0^T \dot{E} \, dt.$$

Integrating along the circle (5.35, 5.36), keeping terms of first order in α, we get

$$\Delta E = \int_0^{2\pi} \dot{E}(A \cos t, -A \sin t) \, dt + O(\alpha^2) \tag{5.37}$$
$$= \alpha \int_0^{2\pi} (1 - 3A^2 \cos^2 t) A^2 \sin^2 t \, dt + O(\alpha^2) \tag{5.38}$$
$$= \alpha \pi A^2 (1 - \frac{3}{4}A^2) + \cdots + O(\alpha^2), \tag{5.39}$$

where we have used (5.34). The first term on the right side of the last expression is positive for

$$A < \frac{2}{\sqrt{3}}$$

and negative for

$$A > \frac{2}{\sqrt{3}}.$$

At a single amplitude

$$A = \frac{2}{\sqrt{3}},$$

we have

$$\Delta E = 0,$$

to first order. This shows that for α small there is a distinguished amplitude of oscillation (near $A = 2/\sqrt{3}$), which is independent of the initial state of the oscillator. It is a limiting solution, because for larger amplitudes the energy is decreasing and therefore so is the amplitude, while for smaller amplitudes the energy is increasing and therefore also the amplitude. This type of oscillation is known as a *limit cycle*. In the phase plane, the cycle is close to the circle

$$v^2 + \dot{v}^2 = 4/3.$$

This type of solution behavior is to be contrasted with the case $\alpha = 0$, which is the harmonic oscillator, where there are oscillations of arbitrary amplitude, the amplitude of oscillation depending on the initial conditions.

Other sources of nonlinearity arise in the damping of the oscillator. Viscous damping has been assumed to be linear, but in many circumstances a closer fit to data show that damping is proportional to a higher power of velocity,

$$\text{damping force} \propto v^\alpha, \quad \alpha > 1.$$

Another source of nonlinearity is *friction*. The so-called *dry friction* consists of two parts: a *sticking friction*, which is a minimum force that must be applied in order for motion to begin, and a *dynamic friction*, which is often independent of velocity, although it could be an increasing, or even decreasing, function.

2   Forced Vibrations

If an oscillator is subjected to a constant force, such as gravity, we know that the behavior is not qualitatively changed, but the equilibrium is shifted. (Of course, nonlinear vibrators are subjected to bifurcations, if the force is strong enough, that may change the nature of the equilibria.) The response of an oscillator to a time-varying force is more interesting, and the possible kinds of behavior, especially in the nonlinear case, are astounding.

The response of an oscillator to a steady periodic force is worthy of study for a number of reasons: it is the simplest time-dependent forcing, and it is easy to supply such a forcing to a wide range of physical (and biological) oscillators.

Linear Response

Is it possible to construct a speed-bump configuration on a road so that the effect on a car will depend on the car's speed in such a way that the eccentricities in the road surface will be transparent to a car going well below a certain speed limit, but will become apparent as the speed approaches the limit? If the car is to be modeled as a mechanical oscillator with spring (suspension) and damper (shock absorber), then the variations in the road act as a forcing function to the oscillator. The whole system is to act as a filter so that slow speeds are "filtered out."

The general solution of a nonhomogeneous linear system such as

$$\ddot{y} + 2\lambda\dot{y} + \omega_0^2 y = C(t),$$

is, as you probably know, given by the variation of parameters formula. This involves a splitting of the solution into two pieces

$$y = y_h + y_p,$$

where the initial conditions are taken care of by y_h. This is not necessarily the same thing as the decomposition into the transient and steady-state parts,

$$y = y_t + y_{ss}, \tag{5.40}$$

since y_p will in general contain a part that decays in time. We now turn our attention to this latter decomposition, (5.40), and we will look first at the steady-state solution.

Consider a simple case: the undamped harmonic oscillator that is harmonically forced. We take a forcing function of the form

$$C(t) = Ce^{i\omega t},$$

where C is a positive real number. Of course, the actual forcing amplitude as a function of time should be taken as the real part of this complex function. We look for a solution to the equation

$$\ddot{y} + \omega_0^2 y = Ce^{i\omega t}$$

in the form

$$y = Ae^{i\omega t}. \tag{5.41}$$

In general, A will be a complex number with real and imaginary parts,

$$A = A_1 + iA_2,$$

which reflects the fact that the equation has a two parameter family of solutions. Differentiating the function (5.41) gives

$$\dot{y} = iw A e^{iwt}$$

and

$$\ddot{y} = -w^2 A e^{iwt},$$

so, plugging into the equation, we get

$$(w_0^2 - w^2)A = C. \tag{5.42}$$

All the quantities in (5.42) are real. However, A must change sign from negative to positive as w increases through the value w_0. That is,

$$\lim_{w \uparrow w_0} A(w) = \infty$$

and

$$\lim_{w \downarrow w_0} A(w) = -\infty.$$

It is simplest to consider the complex number A in its polar form

$$A = |A|e^{i\theta},$$

so that

$$|A| = \frac{C}{|w_0^2 - w^2|} \tag{5.43}$$

and

$$e^{i\theta} = \begin{cases} 1 & \text{if } w < w_0 \\ -1 & \text{if } w > w_0 \end{cases}$$

In the first case, for frequencies below w_0, we can choose $\theta = 0$, while for the second case θ can be any odd multiple of π. It turns out to be most convenient to have $\theta = -\pi$ so that

$$\theta = \begin{cases} 0 & \text{if } w < w_0 \\ -\pi & \text{if } w > w_0 \end{cases}. \tag{5.44}$$

The amplitude $|A|$ and phase θ are plotted in Figure 5.17. The important thing to notice is the singularity that occurs in both graphs.

This type of behavior is called **resonance**.

Notice that (5.43) says that the output amplitude is obtained from the input amplitude through multiplication by the rational function

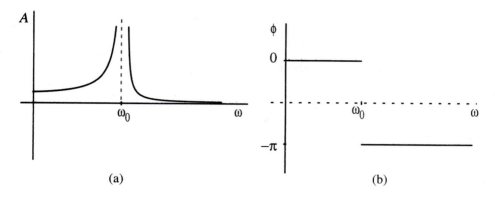

Figure 5.17 Resonance curves with no damping. (a) Amplification factor.
(b) Phase lag.

$(\omega_0^2 - \omega^2)^{-1}$. A function used in this way is called a *transfer function*
or *response function.*

   Of course, A is not defined for $\omega = \omega_0$ because a solution of the
form $Ae^{i\omega_0 t}$ does not exist in that case. A solution is found having the
form $Ate^{i\omega_0 t}$, whose real part supplies the physical displacement:

$$at \cos(\omega_0 t + \phi);$$

these are oscillations with a growing amplitude envelope at (Figure
5.18).

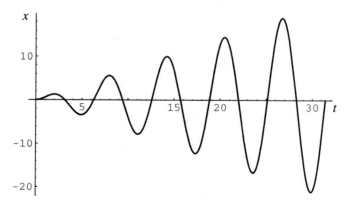

Figure 5.18 Displacement versus time for a system at resonance $\omega = \omega_0$.

Although there will be dissipation in a physical system that will

temper the mathematical singularity of resonance, it is a real phenomenon. A familiar example of resonance is the opera singer who
breaks a glass when she hits a certain note (singing glasses, rubbing
rims). To understand how spectacular this really is, consider the forces
involved in sound propagation. The singer's sound level is certainly
less than 85 dB, which corresponds to an intensity of

$$5 \cdot 10^{-4} \text{ W/m}^2$$

or a pressure amplitude of

$$\Delta p \approx 1 \text{ N/m}^2.$$

Atmospheric pressure is about 10^5 N/m^2. Those glasses are subject to
larger forces in the dishwasher.

EXERCISE 13 (i) Solve the initial–value problem for the harmonic oscillator. There are actually two frequencies that compete here. (In other words,
y_t is not really transient.)
(ii) *Beats* If $\omega_0 \approx \omega$, show that the solution is a carrier frequency of $\omega_1 + \omega_2$,
modulated by the difference frequency $|\omega_1 - \omega_2|$.

Let's now consider the response of the car suspension–shock absorber system (Figure 5.2) to a periodically varying road surface. We
will assume that the car tire is always touching the road. Then there
are two possibilities for the types of speed bumps: (i) low-curvature
bumps where the tires touch all parts of the road (Figure 5.19a), (ii)
high-curvature bumps where the tire is too big to fit into the grooves
(Figure 5.19b). We will consider (i) here and leave (ii) for the problems
at the end of the chapter.

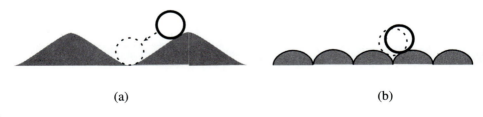

(a)   (b)

Figure 5.19

he profile of the road is given by

$$\tilde{y} = S \cos \kappa x. \tag{5.45}$$

The distance between successive crests is $\frac{2\pi}{\kappa}$: there are $(\frac{2\pi}{\kappa})^{-1}$ bumps per unit length (a spatial frequency).[5] If the speed of the car is v, then, as a function of time,

$$\tilde{y}(t) = S \cos \kappa v t,$$

so $\omega = \kappa v$ is the temporal (angular) frequency as felt by the car.

Let y be the displacement of the mass (car) from equilibrium; this equilibrium belongs to an inertial frame, so the acceleration of the mass is $\ddot{y}$. The spring and damper are acting in parallel: the spring force is proportional to the extension of the spring, which is $y - \tilde{y}$, and the damping force is proportional to the velocity of this extension. Newton's law gives

$$m\ddot{y} = -k(y - \tilde{y}) - \eta \frac{d(y - \tilde{y})}{dt}$$

or

$$
\begin{aligned}
m\ddot{y} + \eta\dot{y} + ky \;&=\; kS \cos \kappa v t - \eta S \kappa v \sin \kappa v t & (5.46) \\
&=\; S \cos(\kappa v t + \phi), & (5.47)
\end{aligned}
$$

where

$$\tan \phi = \frac{\eta}{k} \kappa v.$$

This can be written as

$$\ddot{y} + 2\lambda\dot{y} + \omega_0^2 y = \mathrm{Re}(Ce^{i\omega t}), \tag{5.48}$$

where $|C| = S/m$. Since we are concerned with the steady-state solution, we will set the phase of the forcing function so that C can be taken to be a positive real number. We will look for a solution of the form

$$y = Ae^{i\omega t},$$

where A is a complex number. This function satisfies (5.48) if

$$[(\omega_0^2 - \omega^2) + i2\lambda\omega]A = C. \tag{5.49}$$

Solving for A gives

$$A = \frac{C}{(\omega_0^2 - \omega^2) + i2\lambda\omega}. \tag{5.50}$$

The ratio of input to output, A/C, is often called a *transfer function*, since it "transfers" input to output. In a mechanical system, it is

---

[5] Such a quantity is called a wavenumber (see Chapter 9).

sometimes called the *compliance*. It is recognized by its characteristic complex-rational form (5.50), but it can easily be separated into its real and imaginary parts:

$$A_1 + iA_2 = \frac{C(\omega_0^2 - \omega^2)}{(\omega_0^2 - \omega^2)^2 + 4\lambda^2\omega^2} - i\frac{C2\lambda\omega}{(\omega_0^2 - \omega^2)^2 + 4\lambda^2\omega^2}. \tag{5.51}$$

Furthermore, the modulus of the amplitude is

$$|A| = \frac{C}{\sqrt{(\omega_0^2 - \omega^2)^2 + 4\lambda^2\omega^2}}, \tag{5.52}$$

and the phase, δ, is given by

$$\tan\delta = \frac{-2\lambda\omega}{\omega_0^2 - \omega^2}. \tag{5.53}$$

These functions are graphed in Figure 5.20. The main effect of the damping is to remove the singularity seen in Figure 5.17. The phase, which is the phase difference between the output and the input, is a smooth function that decreases from 0 to $-\pi$. Let us make note of the special limiting cases:

Low–frequency response: as $\omega \to 0$,

$\delta \to 0$; this means that the response is in phase with the forcing.

$|A| \to \frac{C}{\omega_0^2} = \frac{S}{k}$; this expresses a balance between injected force and magnitude of the restoring force, $|A|k$. At very low frequencies, the driver is concerned mainly with pushing against the spring, and not in overcoming inertia. The low–frequency response is *stiffness limited*.

High–frequency response: as $\omega \to \infty$,

$\delta \to -\pi$, so response is exactly out of phase with the driver.

$|A| \to 0$, fadeout of response.

The maximum of $|A|$, A_{max}, is not $|A(\omega_0)|$, but obtains at a lower frequency, ω_*. How close is ω_* to ω_0? Let's reintroduce the dimensionless number $Q = \frac{\omega_0}{2\lambda}$. In terms of Q,

$$|A| = \frac{C}{((\omega_0^2 - \omega^2)^2 + (\frac{\omega\omega_0}{Q})^2)^{1/2}} \tag{5.54}$$

$$= \frac{C}{\omega_0^2} \frac{\omega_0/\omega}{((\frac{\omega_0}{\omega} - \frac{\omega}{\omega_0})^2 + \frac{1}{Q^2})^{1/2}}, \tag{5.55}$$

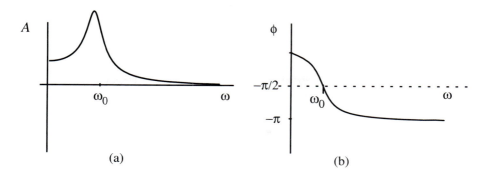

Figure 5.20 Resonance curves for system with damping. Here, $Q = 3$.

and

$$\tan \delta = [Q((\frac{\omega_0}{\omega} - \frac{\omega}{\omega_0})]^{-1}.$$

EXERCISE 14 Show that

$$\omega_* = \omega_0(1 - \frac{1}{2Q^2})^{1/2},$$

$$A_{\max} = \frac{A(\omega_0)}{(1 - \frac{1}{4Q^2})^{1/2}};$$

and thus, when

$$Q \gg 1,$$

$$\frac{\omega_*}{\omega_0} \approx 1 - \frac{1}{4Q^2}.$$

Clearly, for high-Q systems, we can identify the resonance frequency with the frequency of maximum response. This turns out to simplify much of the analysis.

EXERCISE 15 Show that a force that oscillates at the resonant frequency produces an amplitude that is Q times as big as a nonoscillating force of the same magnitude.

EXERCISE 16 What happens when $Q \leq \frac{1}{\sqrt{2}}$?

Figure 5.21 displays the graphs of $|A|$ for various values of the parameter Q. Higher Q means a greater peak at resonance, but it also means a steeper ascent to the peak. In fact, the width of the resonance

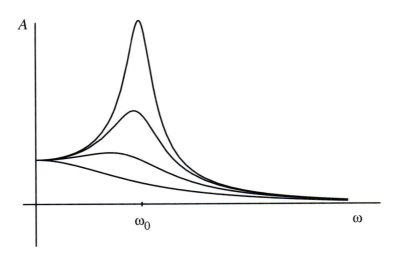

Figure 5.21 Resonance curves for the values of $Q = 0.5, 1.0, 2.0, 4.0$.

peak is proportional to $\frac{1}{Q}$. The quickest way to see this is to note that the function $\arctan(\lambda x)$ has most of its change in the symmetric interval

$$|x| < \lambda^{-1}.$$

EXERCISE 17 (i) What is a road washboard configuration (represented by the parameters S, κ) designed to be effective at 55 mph (at that speed the vibration amplitude should be at least twice the low–frequency amplitude, S/k) for a sports car with $m = 300$ kg, $k = 10^4$ N/m, $\eta = 1000$ Ns/m.
(ii) What is the response of a heavier car ($m = 600$ kg, $k = 10^4$ N/m, $\eta = 2000$ Ns/m)?
(iii) Is it possible to design a configuration that is effective [according to the criterion in (i)] for both these cars?
(iv) If a car accelerates fast enough, it can push through the "peak-response" portion of speeds and reach a low-response region at high speeds. What really happens in that case? Why does the model break down? (What we really want is a low-pass filter: filters are studied in the problem section to this chapter and in Chapter 9.)

At $\omega = \omega_0$, the amplitude response is 90° out of phase with the input. Meanwhile, in comparing the displacement, $y = Ae^{i\omega t}$ with the velocity $\dot{y} = i\omega Ae^{i\omega t}$, we see that the one is 90° out of phase with the other. The reason is that, since

$$i = e^{i\pi/2},$$

multiplication by i means a phase change of $\pi/2$. Thus, at resonance, the velocity is in phase with the input. This leads us to consider a different transfer function, one that transfers the input to the velocity. This is the **admittance**,

$$Y = G + iS,$$

whose real and imaginary parts are known as the conductance and the susceptance, respectively. In fact, it is customary to take the reciprocal of *that* transfer function; this is the **impedance**:

$$Z(\omega) = \frac{1}{Y} = \frac{Ce^{i\omega t}}{\dot{x}(t)} = \frac{C}{iA\omega}.$$

This can be written in real and imaginary form, using (5.50):

$$R + iX = 2\lambda - i\frac{(\omega_0^2 - \omega^2)}{\omega}. \tag{5.56}$$

This simple form is another reason why this transfer function finds much use. Notice that the real part of Z is just the "resistance," whence the name impedance. (The imaginary part is called the *reactance*. The names *resistance, reactance, conductance, and susceptance* all come from electrical circuit theory.) By writing (5.56) in the following way,

$$Z = 2\lambda + i\omega + \frac{\omega_0^2}{i\omega},$$

we see the complete dynamics involved in recovering the input from the velocity. There is a proportional term, 2λ, a differentiation, $i\omega$, and an integration $\frac{\omega_0^2}{i\omega}$. [Compare equation (5.48)]. The real part determines the behavior at resonance (notice that at $\omega = \omega_0$ and $Z = 2\lambda$, which is purely resistive), while the imaginary part determines the high– and low–frequency behavior.

For low frequencies,

$$Z \approx \frac{\omega_0^2}{i\omega};$$

so to recover input from velocity, integrate and multiply by ω_0^2 (this expresses the force balance $mg = kl$). This is the so-called *quasi-static limit*.

At the high-frequency end,

$$Z \approx i\omega;$$

so the input acts to overcome inertia (as we have already noted).

The reciprocal of Z, the admittance,

$$|Y| = |Z|^{-1} = \frac{1}{\omega_0} \frac{1}{((\frac{\omega_0}{\omega} - \frac{\omega}{\omega_0})^2 + \frac{1}{Q^2})^{1/2}},$$

has a maximum at precisely $\omega = \omega_0$, which is yet another reason to commend the use of this particular transfer function. The admittance or impedance tells us how much and in what manner the system resists being forced.

The Energy Cycle and the Power Absorption Curve

Closely related in spirit to these ideas is how much power and energy must be input to an oscillator to keep it vibrating. In a mechanical oscillator the work done by a force, F, in effecting a displacement dx is $dW = F\,dx$. Power is the rate at which this work is done, or

$$P = \frac{dW}{dt} = F\frac{dx}{dt} = Fv.$$

EXERCISE 18 Show that, in the linear oscillator without damping ($\lambda = 0$), there is no net power input over the course of one cycle. Interpret the cycle physically.

In the presence of dissipation, some net power must be lost over the course of each cycle. The average power absorbed by the oscillator can be found through some simple algebra if we use the complex representation:

$$F = \text{Re}(Ce^{i\omega t}), \qquad v = \text{Re}(i\omega Ae^{i\omega t}).$$

EXERCISE 19 (i) Show that for any two complex numbers z and w,

$$\text{Re}(z)\text{Re}(w) = \frac{1}{4}[zw + \bar{z}\bar{w} + z\bar{w} + \bar{z}w].$$

(ii) Apply (i) to F and v and integrate over one cycle to find the average power input per cycle.

From this exercise, you know that the net power input per cycle is

$$P_{\text{av}} = \frac{\omega C}{2}\text{Re}(iA) \tag{5.57}$$

$$= -\frac{\omega C}{2}\text{Im}(A), \tag{5.58}$$

so, using (5.51),

$$P_{\text{av}} = -\frac{\omega C}{2} \frac{-2\lambda\omega C}{(\omega_0^2 - \omega^2)^2 + 4\lambda^2\omega^2} \tag{5.59}$$

$$= \frac{C^2}{2\omega_0 Q} \frac{1}{(\frac{\omega_0}{\omega} - \frac{\omega}{\omega_0})^2 + \frac{1}{Q^2}}. \tag{5.60}$$

The maximum power absorbed occurs at exact resonance, $\omega = \omega_0$, and

$$P_{\text{max}} = \frac{C^2 Q}{2\omega_0}.$$

EXERCISE 20 Show that the frequencies at which $P_{\text{av}}(\omega)$ is $\frac{1}{2}P_{\text{max}}$ are

$$\omega_0 \pm \Delta\omega,$$

where

$$\Delta\omega \approx \frac{\omega_0}{2Q}.$$

The width of the PAC is proportional to $\frac{1}{Q}$; that is, the bandwidth is

$$\text{bandwidth} = 2\Delta\omega \approx \frac{\omega_0}{Q} = 2\lambda.$$

Notice that the dimensionless number Q has the two important interpretations:

1. It determines the rate of energy decay:

$$\frac{\Delta E}{E} \approx -2\lambda\Delta t,$$

so

$$\left(\frac{\Delta E}{E}\right)_{\text{cycle}} \approx -\frac{2\pi}{Q}.$$

2. A "sharp resonance" means

$$\frac{\text{bandwidth}}{\omega_0} \ll 1$$

or

$$Q \gg 1.$$

But, besides these more or less obvious facts, there is a more subtle way in which the importance of the parameter Q can be presented. Notice that the form of the PAC shows that more power is absorbed for frequencies near resonance. This means that for a fixed value of the *quality*, Q, we must supply more power to get a particular amplitude when forcing at frequencies near resonance. However, for a fixed forcing frequency the required power decreases inversely as Q increases, roughly,

$$\text{power requirement } \propto \frac{1}{Q}.$$

This is simply another way of looking at interpretation 2 above. In a recasting of interpretation 1, we are given a relaxation time constant. Indeed, if we take the liberty to write

$$\frac{dE}{dt} = -2\lambda E = -\frac{\omega_0}{Q}E,$$

we see that the exponential relaxation time for energy decay is

$$\tau = Q/\omega_0.$$

But this will also be the characteristic time constant for all relaxation, in particular that which determines the extent of the transient phenomena. If a resonant force is applied suddenly to a quiescent system, then the amplitude will grow gradually over many cycles to the resonant amplitude of

$$Q\frac{C}{\omega_0^2}.$$

The oscillating response will be multiplied by an envelope of the form

$$Q\frac{C}{\omega_0^2}(1 - e^{-\omega_0 t/Q}) = K(1 - e^{-t/\tau}),$$

as in Chapter 4. Thus, although interpretation 1 shows that high-Q resonators, having high relaxation times, are true oscillators in the sense that dissipation need not be accounted for, this same factor appears as determining a long, slow approach to the ultimate steady state. The high-Q oscillation takes longer to build up to steam, since it is absorbing more power.

Paradoxically, then, the resonant phenomena are studied by ignoring transients and considering the steady state; yet it is precisely those high-Q systems for which resonance is the most dramatic that need the longest amount of time for transients to decay. (Beats can be observed in these systems; see the problems.) High-Q systems, as resonators, are designed for those situations that must be taken to be steady-state and are less useful for situations where transients are important. This is related to the famous uncertainty relation (Chapters 1 and 10).

In real life, of course, all signals and forces are transient. This makes the use of resonators very difficult for transduction when the time intervals over which stimuli act are very short.

EXERCISE 21 Let a resonant force be applied to a damped oscillator for N cycles. How close is the approach to A_{max}? How long does the signal "ring" after the force is turned off?

*General Resonance

The concept of resonance is suitable to describe the response of any linear system with viscous dissipation, even if the actual dynamics seem far removed from the simple harmonic oscillator. The essence is that there is some control parameter, such as ω above, and that the response of the system is determined by two other parameters that identify (i) where the response is greatest and (ii) the bandwidth of the response (in the simple harmonic oscillator, these parameters are ω_0 and Q, respectively). There is an input–output relation given by a transfer function. We've seen examples of such transfer functions: impedance, compliance, susceptibility, elastic modulus, and so on. The fundamental structure of the transfer function is best seen when damping is slight. In that case, for $\omega \approx \omega_0$, we have

$$\frac{\omega_0}{\omega} - \frac{\omega}{\omega_0} \approx \frac{2(\omega_0 - \omega)}{\omega_0},$$

and so

$$P_{av} \propto \frac{1}{(\omega_0 - \omega)^2 + \lambda^2}.$$

This type of rational function seems to pop up whenever the term resonance is used.

An example that is rather different from the mass–spring system is nuclear magnetic resonance (now called MRI, presumably in order to avoid use of the N word). Here we think of protons as little spinners with a defined axis or, equivalently, as little bar magnets. There are only two possible symmetry directions, let's say up and down. While in the presence of an ambient magnetic field, incident radiation of the right frequency can make the protons flip their axes from one direction to the opposite.When a large number of the protons in one place do this, his results in a changing magnetic field, which induces a voltage. The voltage response to varying frequency of the radiation is measured. Actually, it turns out to be equivalent to fix the radiation frequency and vary the level of the magnetic field; in that case, the voltage response function has the form

$$V \propto \frac{1}{(B - B_0)^2 + \Delta^2},$$

where B_0 identifies the resonant field strength, and Δ is the bandwidth. This relation explains the use of the term resonance in that context.

Nonlinear Response

If we compare the experimental response of a nonlinear system to a periodic input, expecting the type of behavior seen in the linear oscillator, the results can be surprising. Recall that when a linear system is subjected to a periodic input, f, with frequency ω the steady-state output is

$$x_{ss} = H(\omega)f;$$

after the effects of the initial condition have died away (at an exponential rate), the resulting signal is proportional to the input. If the input is a sinusoid of frequency ω, the output is a sinusoid of frequency ω. Although, in general, the output will be amplified and phase shifted, these effects will be strictly determined by ω, via the resonance relations (5.52, 5.53). The form of the A versus ω curve is qualitatively the same, regardless of the size of the input amplitude.

We have seen that the assumption of linearity is untenable in many systems. If the induction of an RLC circuit is due chiefly to an iron core coil, the model equation is not linear. This is also true of the mechanical system if the spring is "hard" or "soft". If the amplitudes are big enough or if stops are use to control the travel of the spring or pendulum, any of the physical systems will display nonlinearity. Let us record what happens when we experiment with one of these systems.

Suppose that the input is a small-amplitude sinusoid of frequency ω (Figure 5.22). After transients have died away, we observe that the output, although periodic with frequency ω, does not have a sinus waveform (Figure 5.23).

If we increase the strength of the input signal, we may notice that the output is changed not only in strength but in shape of waveform as

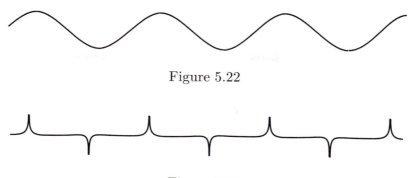

Figure 5.22

Figure 5.23

well. Another effect we can find by experimenting is that for different initial conditions the "steady-state" response can be different. If we carry out a sequence of experiments, fixing C and x_0 while varying the input frequency, with the intention of discovering the "resonance curve" analogous to (Figure 5.20), we find that with increase in the forcing frequency the curve does indeed resemble the linear response curve— but only up to a certain frequency, ω_u, at which point the response suddenly drops in strength; further increase in the frequency leads to a more gradual decline in the response (Figure 5.24).

When we start with a high forcing frequency and slowly decrease the frequency, we find that the process is not reversible: as the frequency is lowered through ω_u, the response does not suddenly show a discontinuous change. Such behavior happens at a lower frequency; this is *hysteresis* in the response curve, reminiscent of the experiments with the Euler strut in Chapter 2. If $|C|$ is made small enough, the hysteresis disappears; for $|C|$ large, the effect is correspondingly more pronounced: it appears as though we have come across a cusp catastrophe (see Chapter 2).

There are many other oddities to be discovered with nonlinear systems. For instance, by changing the forcing amplitude and/or the initial conditions, we may find an output that looks like Figure 5.25, where the fundamental frequency is not the same as the forcing frequency, ω, but is $\frac{1}{3}\omega$. This is an example of a *subharmonic* solution. Nonlinear systems exhibit subharmonics of order 2, 3, ..., as well as superharmonics, where the frequency is a multiple of the driving frequency. There may even be rational resonances.

And all of this is the response to a single sinusoid.

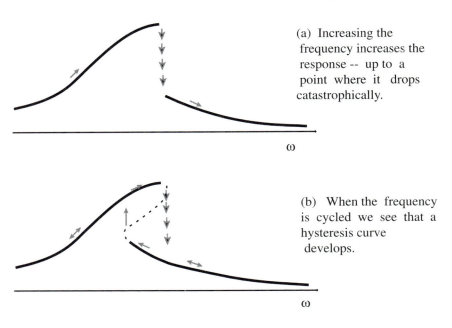

(a) Increasing the frequency increases the response -- up to a point where it drops catastrophically.

(b) When the frequency is cycled we see that a hysteresis curve develops.

Figure 5.24

If the periodic input to a *linear* system is not a sinus, but compounded of many frequencies, the superposition property allows an easy resolution: the output is a linear combination of the responses to the individual components. In a nonlinear system, superposition is not valid, and the output is not made up of only the frequencies that appear in the input (subharmonics are a case in point), but other combinations will appear. Go ahead and put the function

$$f = A \sin \omega_1 t + B \sin \omega_2 t$$

through a squaring element:

$$f \rightarrow \boxed{x^2} \rightarrow g.$$

Besides the terms of frequencies ω_1 and ω_2, there will be combination frequencies $\omega_1 + \omega_2$ and $|\omega_1 - \omega_2|$.

EXERCISE 22 What are the new combination frequencies that arise with a cubic nonlinearity? Do you see the third-order subharmonic?

There are many other strange things that can happen in nonlinear systems, as we will see in Chapter 8. Of course, not all these things

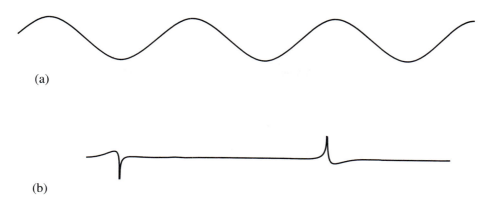

(a)

(b)

Figure 5.25 When a sinusoid (a) is input to a nonlinear oscillator it can result in a subharmonic response (b).

necessarily occur in every system. In general, it is an involved matter to model a nonlinear physical system. One must rely on qualitative hunches, and it is best to concentrate on one aspect of the behavior at a time. To take a specific example, experiments with a mass–spring system give the frequency-response curve of Figure 5.24. We will show that this can be accounted for by supposing a cubic spring law. Instead of a quadratic spring potential

$$\frac{1}{2}x^2,$$

we have instead the potential of a "hard" spring:

$$\frac{1}{2}\left(x^2 + \frac{\beta}{2}x^4\right).$$

If the forcing function is a sinusoid of frequency ω, the equation of motion for the displacement is

$$\ddot{x} + 2\lambda\dot{x} + x + \beta x^3 = C\cos\omega t. \tag{5.61}$$

We will work under the assumption that β and C are small:

$$\beta \ll 1, \qquad C \ll 1.$$

Since the nonlinearity is small, we may expect that there is a solution close to the solution of the harmonic oscillator, in other words, we try a solution of the form

$$x = A\cos\omega t + B\sin\omega t. \tag{5.62}$$

(Although this represents the general solution of the linear harmonic oscillator, we have to expect, based on our discussion of the experiments with nonlinear oscillators, that this is a very special form of solution for a nonlinear oscillator.) Differentiating this solution, we have

$$\dot{x} = -\omega A \sin \omega t + \omega B \cos \omega t \tag{5.63}$$

and

$$\ddot{x} = -\omega^2 A \cos \omega t - \omega^2 B \sin \omega t. \tag{5.64}$$

The cubic term gives

$$x^3 = A^3 \cos^3 \omega t + 3A^2 B \cos^2 \omega t \sin \omega t + 3AB^2 \cos \omega t \sin^2 \omega t + B^3 \sin^3 \omega t. \tag{5.65}$$

Temporarily, let us work under the assumption that $\lambda = 0$; in that case, we can assume that $B = 0$. Using the trigonometric identity

$$\cos^3 x = \frac{3}{4} \cos x + \frac{1}{4} \cos 3x,$$

and upon substitution of (5.62), (5.64), and (5.65) into the equation of motion, we get

$$\left(A - \omega^2 A + \frac{3}{4}\beta A^3 - C\right)\cos \omega t + \frac{\beta}{4}A^3 \cos 3\omega t = 0;$$

so the equation that must be satisfied for the amplitude A is

$$A - \omega^2 A + \frac{3}{4}\beta A^3 = C,$$

giving a relation between amplitude and frequency, which reduces to (5.42) when $\beta = 0$. In contrast to the linear case, where a finite forcing amplitude, C, resulted in a finite solution amplitude, A, for every frequency except for the resonant frequency, $\omega = 1$ (the singular curve is then the straight line $\omega = 1$), the nonlinear response singular curve is a parabola in the (ω^2, A)-plane (Figure 5.26). In other words, the resonance frequency is not simply determined by the system, but depends on the amplitude of oscillation as well; the singular curve is given by

$$\omega^2 = 1 + \frac{3}{4}\beta A^2. \tag{5.66}$$

This equation can be thought of as an expression of the balance between average kinetic energy, as given by

$$KE_{\text{av}} = \omega^2 A^2,$$

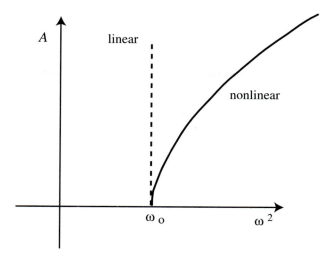

Figure 5.26 Position of critical resonance curve in the (A, ω^2)-plane. In the linear system, resonance occurs at ω_0, independently of amplitude. The frequency of a *nonlinear* response depends on amplitude. For the nonlinearity of (63), this is a parabola in the (A, ω^2)-plane.

and average potential energy given by

$$PE_{\text{av}} = A^2 + \frac{3}{4}\beta A^4.$$

Since β is small, we can expand to first order to get the small–amplitude resonant frequency

$$\omega_0 = \sqrt{1 + \frac{3}{4}\beta A^2} = 1 + \frac{3}{8}\beta A^2 + \cdots. \tag{5.67}$$

Now, in order to extend this to the (lightly) damped case, we will make a move similar to what we had found useful for the pendulum: we are going to replace the actual potential by a quadratic potential that supplies the natural frequency ω_0. For each amplitude of vibration, we must choose a different harmonic oscillator with which to match frequencies. But we know what the frequency response of a harmonic oscillator looks like: it is (for light damping) the classical resonance relation,

$$A^2 = \frac{C^2}{(\Delta\omega)^2 + (2\lambda)^2}.$$

We substitute (5.67) for the resonant frequency and get the resonance relation:

$$A^2 = \frac{C^2}{(\omega - 1 - \frac{3}{8}\beta A^2)^2 + (2\lambda)^2} = \frac{(C/2\lambda)^2}{(\frac{\omega - 1 - \frac{3}{8}\beta A^2}{2\lambda})^2 + 1}.$$

Multiplying out,

$$(\frac{\omega - 1}{2\lambda})^2 + (\frac{3}{4}\frac{\beta A^2}{2\lambda})^2 - \frac{3}{4}\frac{\beta A^2(\omega - 1)}{(2\lambda)^2} + 1 = (\frac{C}{2\lambda A})^2,$$

and setting $X = \frac{\omega - 1}{2\lambda}$, we have

$$X^2 - \frac{3}{4}\frac{\beta A^2}{2\lambda}X + (\frac{3}{8}\frac{\beta A^2}{(2\lambda)^2})^2 = (\frac{C}{2\lambda A})^2 - 1,$$

which, upon completing the square, gives

$$X = (\frac{3}{8}\frac{\beta C^2}{\lambda^3})(\frac{2\lambda A}{C})^2 \pm \sqrt{(\frac{C}{2\lambda A})^2 - 1}.$$

This should be compared to the analogous expression for the linear harmonic oscillator,

$$\omega - \omega_0 = \pm 2\lambda\sqrt{(\frac{C}{2\lambda A})^2 - 1},$$

where the two branches are the two sides of the resonance peak. The nonlinear resonance shows the same fundamental shape, except that both sides of the resonance peak are tilted toward the X-axis by a sloping that depends on β (Figure 5.27) which results in the form of the resonance curve that was observed in experiment.

A phase–plane analysis (see chapter 8) can complement a frequency response curve (Figure 5.27) and help to shed light on the jumping and hysteresis. It is easy to see that in the absence of forcing there are three equilibrium points: the origin is a saddle point and there are two stable foci (Figure 5.28). It is reasonable to suppose that for an initial condition near a stable equilibrium and for small periodic forcing, there will be a periodic solution that remains close to an equilibrium of the unforced system. If we increase the strength of the forcing and try to follow what happens, we find that, globally, there are two asymptotically periodic solutions, a large–amplitude solution and a small–amplitude solution and, as we vary the frequency, the basin of attraction of the

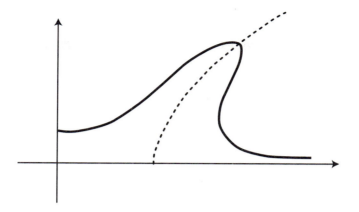

Figure 5.27 Nonlinear frequency response in the presence of damping. The dotted curve is the same parabola as in the previous figure.

solutions changes so that at one frequency we may tend to one solution and then switch to the other as we pass through a certain frequency.

This short introduction to nonlinear systems suffices to give the impression that one must use great care in extending the analysis of linear systems. However, physical (and biological, sociological, economic, ...) systems do exhibit nonlinear behavior and so one must be prepared to deal with them. Further adventures for the reader will be found in the following problem section and in Chapter 8.

3 Problems and Recommended Reading

1. Using the variational method of Lagrange, derive equations of motion for the double-jointed pendulum (Figure 5.29) and the Mars golfer (Figure 5.10). For further details on Lagrange's method and other variational methods, consult a classical mechanics text.

2. Show that the equilibrium position of the ball in Figure 5.13 undergoes a fold catastrophe (limit point instability) as the angle of limit varies.

3. Find the forcing function for high-curvature speed bumps (Figure 5.19b).

Figure 5.28

4. Show that the phenomenon of beats is highly pronounced in slightly damped oscillators.

5. Is it possible by vibrating the support of a pendulum to make the upper position stable and lower position unstable?

6. How can you oscillate the plates of a parallel-plate capacitor in order to increase the amplitude of oscillations in a circuit?

7. Explicitly contrast the resonance in the two circuits of Figure 5.30.

8. Investigate the stability of the circuit in Figure 5.31, where the nonlinear diode has the nonlinear I–V curve of Figure 5.15. (This is done with a phase plane analysis.)

9. Model plasma oscillations as simple harmonic motion. (You will have to suppose some simple form of the spatial oscillation in order to calculate the restoring force. See Figure 3.6).

10. Model the diatomic molecule of Figure 5.32 via a linear restoring force. What is the natural frequency of vibrations?

11. In ionic molecules, the potential corresponding to the interaction force is often taken to be of the form

$$V(r) = -\frac{e^2}{4\pi\epsilon_0 r} + \frac{\beta}{r^n},$$

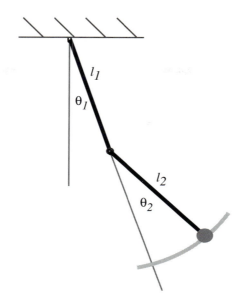

Figure 5.29

where n is around 10. The parameter β can be found in terms of the bond length, the equilibrium value of r. Find the dependence of the frequency of small oscillations on the bond length.

12. Consider the Euler strut of Chapter 2. What is the frequency of small vibrations about the vertical position for $m < m_c$? What happens to the frequency as $m \to m_c$? Plot frequency versus m.

13. Complete the analysis of the circuit of Figure 5.16, identifying all of the behavioral changes associated with varying the parameter M.

14. (i) A classical example of a relaxation oscillator is pictured in Figure 5.33. Suppose that the block sitting on the conveyor belt is subject to dry friction. Show that when the belt moves at a constant speed (or torque – is there a difference? – and the sticking friction isn't too great) the position of the block undergoes a discontinuous oscillation. (ii) A related example is given by the classical clock mechanism. How does a constant gravitational force acting on a damped oscillator lead to a perfectly periodic oscilation?

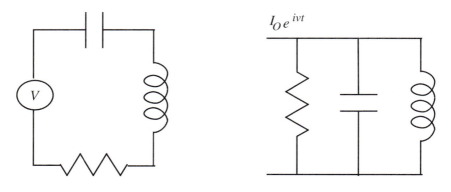

Figure 5.30

15. Josephson junctions are weak links, or gaps, between two pieces of superconducting material. To some extent the interaction of these two materials across the gap can be characterized by the phase-difference of macroscopic quantum wave functions (see [19]):

$$\phi = \phi_1 - \phi_2.$$

If a voltage, V, is applied across the gap, then $\dot{\phi} = 2eV/\hbar$, and there is a tunneling current, $j_t = j_0 \sin \phi$. Suppose that a current j is applied across the gap. Contrast the equations governing two contrasting situations: the SIS junction, where the response to the applied current is that a charge builds up across the insulating layer at a rate $C\dot{V} = j - j_t$, and the SNS junction, where a normal resistive current flows, $j - j_t = V/R$.

16. Here are some further examples of transfer functions.

   (a) *Perfect electric conductors* Let the velocity of charge carriers in a conducting material be $\mathbf{v}$, let q be the electric charge of each, and let n be the density of these carriers. Then the electric current (density) is

$$\mathbf{J} = nq\mathbf{v}.$$

   In the presence of an electric field of intensity $\mathbf{E}$, a charge carrier experiences an acceleration,

$$\mathbf{a} = \dot{\mathbf{v}} = \frac{q}{m}\mathbf{E}.$$

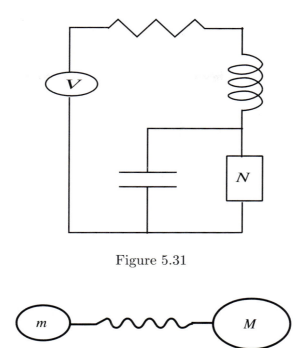

Figure 5.31

Figure 5.32

Meanwhile, the carriers are scattered in such a way that they experience an exponential relaxation as given by a relaxation time, τ. Write down a first-order equation for the velocity. Treat the situation as a one-dimensional input–output system,

$$J = \sigma E,$$

where the transfer function, σ, is known as the conductivity. Show that the conductivity is a rational function of input velocity that is purely imaginary in the limit of zero dissipation ($\tau \to 0$).

(b) *Electric susceptibility* A simple model of the polarization (electric dipole moment) of a crystal is arrived at by assuming that it is a one-dimensional row of atoms where electrons are attached to an atom by springs:

$$m\ddot{x} + 2\lambda\dot{x} + kx = qE.$$

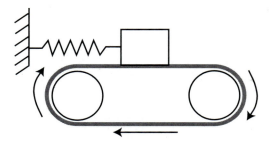

Figure 5.33

Suppose that the electric field acts on all masses in the same way so that the polarization can be defined as

$$\mathbf{P} = P\mathbf{i} = nqx\mathbf{i}.$$

Write down an equation for P, and get an expression for the electric susceptibility, χ, which is defined by

$$P = \epsilon_0 \chi E,$$

where ϵ_0 is the dielectric constant. What are the parameters appearing here that can be measured? Which of them can be inferred from experiment?

Chapter 6

Random Thinking

Professor Shoda has gone to the rathskeller for lunch three times this week and every single time his order has pulled number 53. Is that an amazing coincidence? Professor Shoda advertised for a lab assistant one week ago. The first two days he got 6 applicants, none of whom he felt was qualified. He hasn't received any applicants since then. What can explain that?

1 Probabilities

What do we mean when we state the chances of a particular thing happening? During the evening news the weatherman says that there is an 80% chance of rain tomorrow. He is making a prediction concerning the uncertain future based, presumably, on consideration of the certain past (as certain as the accuracy of our observations). Among the possible scenarios for tomorrow are those with rain and those without. The weatherman has decided to weight the former case as being four times more likely than the latter.

In discussing probabilities, such weights are *normalized* so that each possible outcome is weighted as a *fraction of unity*, and the sum of the weights over all possible outcomes is 1. We are used to dealing with such fractions of unity in many contexts; we call them *frequencies*, or *rates*, or *percentages*. It is often useful in modeling to think of rates as probabilities.

The weatherman splits the set of all possible outcomes into two kinds of outcomes; we call these *events*: the event of rain and the event of no-rain. And he assigns a weight, or *probability*, of 0.8 to the first event and, since the events are exclusive, a probability of 0.2 to the second. The assignment of these weights is an inductive procedure and, if the weatherman is operating rationally, is based on a study of the meteorological conditions that prevail up to the point at which these weights are assigned. This assignment is the weatherman's *model*. Of course, we won't be able to test this model, since we can only sample tomorrow once. These probabilities are *conditional* upon prior experience, but are expressed as probabilities, not certainties, to reflect the lack of sufficiency in data. If the weatherman had no information to work with, if he didn't know the season, yesterday's weather, or even astronomical positions, then he could not justify any assignment but equal probabilities to the events of rain and no-rain.

If I flip a coin, what is the chance that it will land heads up? Most people would answer that the chance is 50 : 50 or that the probability is $1/2$ (some would insist on more information). In fact, without further knowledge of the physical state of the coin, knowing only that, like any coin, it has two sides, either of which can turn up, we are forced to assign equal probabilities to each possibility. Bernoulli recognized this in 1713 and phrased it as the *principle of insufficient reason*:

> In the absence of prior knowledge we must assume that all events have equal probabilities.

This has formed, part and parcel with methods of counting, the basis for a practical calculation of probabilities: assign probabilities, as fractions of unity, to possible outcomes using whatever information is available, and otherwise equably partition the set of outcomes. Notice that this must be done in such a way that the probability of an outcome A *or* an outcome B occurring is not to exceed the sum of the probabilities assigned to each event.

The prior information that we use to assign probabilities can take many forms. As an example, suppose that five of the six sides of a cube are marked with H and the remaining side with T. There are two possible outcomes: either H lands up or T lands up. Are we then to assign each outcome a probability of $1/2$? Of course not. Our prior information here consists in *knowing* that a cube has six sides. The six events, each corresponding to a particular side landing up, form the finest partition we can conceive of the set of outcomes; a better assignment of probabilities is to assign $1/6$ to each of these outcomes.

EXERCISE 1 Look at the hemispherical coin of Figure 6.1. There are only two outcomes: either T lands up or H lands up. Justify or criticize the equiprobable assignment $P(H) = 1/2 = P(T)$.

Notice that typical physical considerations such as symmetry will often play a role in the assignment of probabilities.

Figure 6.1

The principle of insufficient reason expresses a notion of *randomness*: the degree of randomness corresponds to the lack of prejudice that we bring to our model. The least random state of affairs is the one where we are so sure of the outcome that we assign to this particular outcome a weight (probability) of unity; we say that the outcome is *determined* or *certain* to occur. On the other hand, the most random situation is the one that uses the least amount of prior knowledge: this is the equiprobable assignment. Suppose that there are N mutually exclusive outcomes. This forms a partition of the set of possible outcomes, and this equiprobable assignment is

$$p_i = p, \quad i = 1, \ldots, N.$$

But since one of these possible outcomes must, in fact, be the one that occurs, we must have

$$1 = \sum p_i = Np.$$

Thus, the probability, p, of any particular occurence is $1/N$.

In this way, the principle of insufficient reason can be used to calculate probabilities, at least when the number of possible outcomes is finite.

The roll of a six-sided die is, in accepted terminology, a *random experiment*, and the possible outcomes are assigned probabilities. What probability do we assign to the roll of a 5? If we can do that, how do we calculate the probability of rolling either a 5 or a 6? What is the probability that the number we roll is an odd prime? Without more information, we must assign equal probabilities to possible physical outcomes. This expresses the notion that the die is *fair*. The first thing to do is to enumerate the physically possible outcomes. Physically, the cube must land with exactly one of its faces up. The set of physical outcomes can be labeled $\{1, 2, \ldots, 6\}$ and, by the physical symmetry of the cube, each of these outcomes has probability $1/6$. The answer to our first question follows immediately: $p(5) = 1/6$. This was obvious, because the event "roll a 5" is a *simple* event; it corresponds to a physical outcome. To answer the second question, which is not simple in the sense used here, let us reason as follows: out of the six possible outcomes, the event {5 or 6} is made from two of the six possibilities; thus, $p(5 \text{ or } 6) = 1/3$. Note that

$$p(5) + p(6) = 1/3.$$

As for the third question, we note that the set {odd, prime} = {3, 5} and thus, as for the second question, the probability is 1/3. However, notice that here we have

$$P(\text{odd}) = P\{1, 3, 5\} = 1/2,$$

and

$$P(\text{prime}) = P\{2, 3, 5\} = 1/2,$$

and so

$$P(\text{odd}) + P(\text{prime}) = 1/2 + 1/2 = 1.$$

On the other hand,

$$P(\text{odd or prime}) = P\{1, 2, 3, 5\} = 2/3.$$

Thus, unlike in the previous case,

$$P(\text{odd or prime}) \neq P(\text{odd}) + P(\text{prime}).$$

The reason for this is that on the right side the *simple* events {3} and {5} were counted twice. We can discount for such occurences by noting that

$$\begin{aligned}
P(\text{odd or prime}) &= P(\text{odd}) + P(\text{prime}) - P(\text{odd and prime}) \\
&= 1/2 + 1/2 - 1/3 = 2/3.
\end{aligned}$$

This is indicated by the Venn diagram of Figure 6.2.

These machinations should give us enough confidence to be able to say what we mean by a probability assignment in general circumstances. Simply put, a **probability** is a real-valued function, taking values between 0 and 1, that is defined on the set of outcomes of a random experiment, and satisfies three properties. The universe of all outcomes is the space of events or, as it is usually termed, the *sample space S*. The properties that the probability function must satisfy are[1]

---

[1] Properly speaking, the function p should be referred to as a *probability measure*, and its value on any subset of S is a *probability*.

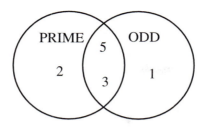

Figure 6.2

i. $p(\emptyset) = 0$,

ii. $p(\mathcal{S}) = 1$,

iii. $p(A \cup B) = p(A) + p(B) - p(A \cap B)$,

where A and B are subsets of $\mathcal{S}$. We call A and B *events*. The properties i and ii together simply say that something *must* occur, and property iii is best explained at this point by reference to the last example and Figure 6.2.

We will say that $\mathcal{S}$ is partitioned into simple events, $\{s_k\}$, if

$$\mathcal{S} = \cup s_k, \quad \text{and} \quad s_k \cap s_j = \emptyset, \quad \text{for all } k, j$$

Using properties, i, ii, iii, we can find the probability of any event if we have already assigned probabilities to the simple events, $p(s_k) = p_k$. We must use *a priori* information to assign these probabilities; if no information is forthcoming, we have no choice but to assign equal probabilities. In that case, if $\mathcal{S} = \{s_1, \ldots, s_N\}$, then $p(s_k) = 1/N$. In this random situation, the calculation of the probability of an event, $E \subset \mathcal{S}$, reduces to counting the number of elements in E:

$$P(E) = \frac{|E|}{|\mathcal{S}|} = \frac{|E|}{N}. \tag{6.1}$$

EXERCISE 2 (i) Show that this is a proper way to compute probalities; that is , verify properties i, ii, iii.
(ii) What if $\mathcal{S}$ is infinite? Can we define, in that case, a *random* situation?

2   The Law of Averages

Let us return to the example of the coin toss. Without prior information, we must assign a weight of 1/2 to the event that heads occurs,

$$P(H) = 1/2.$$

Now, that may not be a reflection of the true state of affairs. The coin
may be loaded. However, we can accept this as a hypothesis. In other
words, we propose a mathematical model of a fair coin; this means we
assign equal probabilities to the two possible outcomes. Proceeding
from there, we can derive, in a rigorous manner, certain statements
about our model of the coin; these can suggest experiments that we
can perform on our coin that will test the validity of our model. In
broad strokes, this is what the study of statistics is all about.

How can we test our hypothesis that the coin is fair? Our feeling
may be that if the coin is fair then we should expect that the probability
of getting heads half of the time will increase as we make more and more
tosses. But this is far from true, as we will see below. Can we expect,
on average, to see heads about half of the time? What do we mean
by "on average"? Perhaps we mean that if we toss the coin a large
number of times, N, then as N gets bigger the number of heads that
appear should approach $N/2$. We could simulate this by the following
experiment. Toss N_1 coins and count the number of heads, say there
are H_1; toss N_2 coins, $N_2 > N_1$, and count the number of heads, H_2;
and so on. At each step measure the error

$$|H_N - N/2| = \Delta_N.$$

EXERCISE 3   Try this experiment with your coin, keeping records. Plot
Δ_N versus N. Is there a way to save time? What do you conclude? Are you
suspicious of your coin?

The result of this exercise tells us nothing, because, if the coin is
truly fair, that is, if $P(H) = 1/2$, it does not follow that $\Delta_N \to 0$. But,
you say, hold on! This is an absolute error. Shouldn't we be tracking
the *relative error*,

$$\frac{|H_N - N/2|}{N}?$$

Does the hypothesis that $P(H) = 1/2$ imply that the limit of this
expression is zero and thus supply a statistical test for the fairness of

any particular coin? This version of the "law of averages" turns out to be inaccurate, but we can salvage something that resembles it closely.

Let us start by computing the probability of tossing a certain number, say k, of heads, in N tosses of the coin. Let us call this probability $P_N(k)$. We can compute this number by the counting method developed in the previous section. To that end, we must identify the simple events, the finest partition of the sample space. The sample space is the set of all sequences of length N made from the symbols $\{H, T\}$. A simple event is a *distinct* N-long string of Hs and Ts. How many of these are there? We must choose N symbols all together; there are two ways of choosing the first, two ways of choosing the second, and so forth. Thus, the sample space has 2^N elements, and each of these elements is assigned a probability of 2^{-N}. We know that

$$P_N(k) = \frac{\text{number of ways to toss } k \text{ heads}}{2^N} \tag{6.2}$$

How many ways are there to pick k heads?

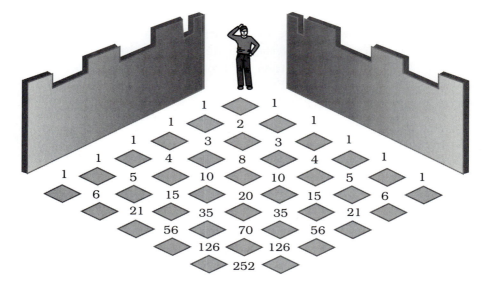

Figure 6.3 Pascal's triangle.

A method for counting this number can be found in a famous construction of Pascal, who was thinking of the following problem. The blocks of a city are laid out as in Figure 3. Starting at the top and moving steadily downward, never backing up, by how many different

paths can one arrive at a particular intersection? Finding ourselves at
a particular intersection, we must have arrived there directly from one
of the two intersections directly above it. Thus, the number of ways
of arriving at our present intersection is the total number of ways of
reaching *either* of the two intersections that are directly above. In this
way, we generate *Pascal's triangle*, where each number in the triangle
is the sum of the two numbers directly above it (Figure 6.3).

Suppose that we number the (horizontal) rows of intersections,
starting at the solitary intersection at the top, which is labeled 0. The
next row, which has two intersections, is labeled 1. And so on. Let n
denote the label of a row. Given a row, n, number the intersections
in each row, starting at the left with 0. Call this label k. Then the
pair (n, k) serves to identify a particular intersection. These pairs are
usually written in the form

$$\binom{n}{k}$$

and are called the *binomial coefficients*.

Let us use Pascal's triangle to compute the number of ways of getting k heads in N coin tosses. We will negotiate a passage through the
streets of the city, starting at the top, and at each intersection we flip
a coin. If we toss "heads," then we proceed to the next intersection
on our left (facing down in Figure 6.3); if "tails," then we go to our
right. Thus, at the start, at $\binom{0}{0}$, if "heads," we go to $\binom{1}{1}$; if "tails," to
$\binom{1}{0}$. At each intersection we repeat: starting at $\binom{n}{k}$, we go to $\binom{n+1}{k}$ if
"tails," and $\binom{n+1}{k+1}$ if "heads." It is clear that the bottom index, k, is
the running count of heads. So the number of ways of getting k heads
in N coin tosses is $\binom{N}{k}$. Thus, from (6.2),

$$P_N(k) = \binom{N}{k} 2^{-N}. \tag{6.3}$$

EXERCISE 4 (i) In a set with n elements, how that the number of subsets
with k elements is $\binom{n}{k}$. We say that there are $\binom{n}{k}$ ways of choosing k elements
from n elements. The symbol $\binom{n}{k}$ is usually read as " n choose k." (*Hint:* To
create a subset with k elements, walk through the n elements one at a time,
picking and choosing.)
(ii) How many subsets are there of a set with n elements?
(iii) Show that
$$\binom{N}{0} + \binom{N}{1} + \cdots + \binom{N}{N} = 2^N.$$

(iv) Show that

$$\sum_{k=0}^{N} P_N(k) = 1.$$

The probability of getting heads exactly half of the time is

$$P_N\left(\frac{N}{2}\right) = \binom{N}{N/2} 2^{-N}.$$

From Pascal's triangle, we know that the first term grows as $N \to \infty$, while the second term decays. Which term dominates? We can compute any binomial coefficient recursively via Pascal's triangle, but to make a quantitative estimate of the limit as $N \to \infty$, we need a different way of computing the binomial coefficients.

Recall that the number of permutations of n objects is

$$n! = n(n-1)(n-2)\ldots 2 \cdot 1. \tag{6.4}$$

Any given permutation of the n objects is given by specifying the particular *ordering*. What are the different ways to order n objects? For the *first* element, any of the n elements can be picked, but once the first element is known there are only $(n-1)$ choices for the *second*, and so on, giving expression (6.4).

To compute the binomial coefficient, we start by asking; what is the number of different ways that a group of k items can be selected from n items if we care about the *order* in which they are selected? This number is

$$n(n-1)\ldots(n-k+1) = \frac{n!}{(n-k)!}.$$

And now, if we no longer care about the order, we have to discount for the number of ways of permuting the k elements. Thus, the number of different ways of choosing k items from n items is found:

$$\binom{n}{k} = \frac{n!}{k!(n-k)!}.$$

EXERCISE 5 (i) How many different words (regardless of meaning) can be made from the letters BOOB?
(ii) In how many different ways can we paint a picket fence if 4 of the slats are to be red, 4 are to be white, and 5 are to be blue?
(iii) Derive this more general (multinomial) coefficient: the number of distinct

permutations of n objects of which k_1 are alike, k_2 are alike, ..., k_m are alike, is

$$\frac{n!}{k_1!k_2!\ldots k_m!}.$$

We now have a nonrecursive method for computing the probability (6.3), and we see that the probability of heads half the time is

$$P_N(N/2) = \binom{N}{N/2}2^{-N} = \frac{N!}{(\frac{N}{2})!(\frac{N}{2})!2^N}.$$

We want to evaluate this as $N \to \infty$. This is easy if we make use of Sterling's asymptotic approximation for the factorial:

$$n! \approx e^{-n}n^n\sqrt{2\pi n} \quad \text{(for large } n\text{)}.$$

Then

$$\begin{aligned} P_N(N/2) &\approx \frac{e^{-N}N^N\sqrt{2\pi N}}{2^N e^{-N}N^N 2^{-N}\pi N} \\ &= \sqrt{\frac{2}{\pi}}\frac{1}{\sqrt{N}}. \end{aligned}$$

It turns out, then, that

$$P_N(N/2) \to 0,$$

as $N \to \infty$. The probability of getting heads exactly half the time actually goes to zero as the number of tosses increases.

EXERCISE 6 Find the probability that, at some time in our series of coin tosses, the number of heads equals the number of tails.

EXERCISE 7 Suppose that we fix $\epsilon > 0$. Can we expect that as N increases the number of heads tossed is within ϵ of $N/2$?

We are now in a position to analyze our intuitive conjecture that the "relative error" in the coin tossing is small in the limit:

$$\left|\frac{H_N - N/2}{N}\right| \to 0, \quad N \to \infty.$$

As mentioned earlier, this is inaccurate, and for a very simple reason: there are obvious sequences of coin tosses that do not satisfy this. There is, for example, the sequence $HHHHHHHH\ldots$ which consists only of heads. We certainly don't expect to ever see this sequence; it is too unlikely. We should only have to worry about "reasonable" sequences. To use this insight, we will have to modify our notion of the limit in order to sift through the sequences. Although the limit cannot be true of *all* possible sequences, it can be true of most of them (after we throw out the unreasonable ones, the ones like $HHHHHHH\ldots$). The *fraction* of all sequences that satisfy the limit should increase to 1 as N gets large. More precisely, given any tolerance $\epsilon > 0$, the *proportion* of all coin-toss sequences satisfying

$$|\frac{H_N - N/2}{N}| < \epsilon$$

should approach 1 as N increases.

But we know that a proportion of occurrences is a *probability*.

Since each particular sequence is a *simple* event, or outcome, in our sample space consisting of all coin tosses of length N, the probability of any event is the sum of the probabilities of all the simple events making up that event. Each simple event has probability 2^{-N}, and the event of interest to us is that where the proportion of heads differs from $1/2$ by less than ϵ. We want that

$$P(|\frac{H_N}{N} - \frac{1}{2}| > \epsilon) \to 0, \quad N \to \infty,$$

where

$$1 - P(|\frac{H_N}{N} - \frac{1}{2}| > \epsilon) = \sum_{\frac{1}{2}-\epsilon<\frac{k}{N}<\epsilon+\frac{1}{2}} P_N(k) = \sum_{\frac{1}{2}-\epsilon<\frac{k}{N}<\epsilon+\frac{1}{2}} \binom{N}{k} 2^{-N}.$$

Bernoulli's proof of this *weak law of large numbers*, as this version of a law of averages is called, was based on a direct computation of the above probabilities. It follows directly from the following modern inequality, which is a special case of the Chebyshev inequality (see Problem 1):

$$P(|\frac{H_N}{N} - \frac{1}{2}| \geq \epsilon) \leq \frac{1}{4N\epsilon^2}.$$

As N increases, the proportion of cases that don't constitute our event goes to zero. In this way, unreasonable sequences such as $HHHHHH\ldots$ have zero probability of occurring.

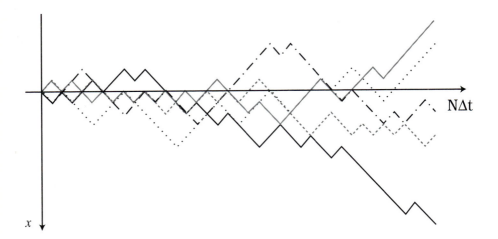

Figure 6.4 Space-time plot of four coin-tossing experiments. These are random walks in one dimension. Here, $N = 40$.

Drunkard's Walk

Suppose that a drunkard (who happens to live in a one-dimensional world, which to us is the x-axis) sets out from a bar that is located at $x = 0$ and, at regular intervals of time, $\Delta t, 2\Delta t, 3\Delta t, \ldots$, takes a step (of size Δx) to the right or to the left, the direction to be chosen at random. Thus, if at time $t = m\Delta t$, his position is x, at time $t + \Delta t = (m + 1)\Delta t$ it is either $x - \Delta x$ or $x + \Delta x$. Since we have no reason for bias, each of the two possibilities can occur with probability $\frac{1}{2}$. Does the drunkard wander away from the bar or is he doomed to continually return to the bar?

EXERCISE 8 Simulate this drunkard's progress: draw a line and number it with the integers; place a pawn at 0 and flip a (fair) coin to effect each step. If heads show, move right; if tails, step left. Every so often, after you have followed this procedure for many steps, record your position and start all over. Figure 6.4 shows the results [in the (x, t)-plane] of several such simulations. Try to answer the following questions. What is the probability that, after N steps, the pawn is back at the origin? Does this probability increase or decrease as N gets large? Do you expect the pawn to somehow stay near the origin?

EXERCISE 9 Generalize the drunkard's walk to the situation where there is a built-in bias as to which direction is taken at each step (one can imagine that the bar is on a hill). Model this by supposing that steps to the left occur with probability p and those to the right with probability $(1-p)$. How is the built-in bias at each step reflected in the accumulation of steps? Where do you expect the drunkard to be after N steps?

EXERCISE 10 Let us mine another nugget in this vein. Consider a one-dimensional arrangements of N magnetic dipoles. (Just imagine a row of little iron bar magnets, painted red on one end). These dipoles can point up or point down at random, but can have no intermediate orientation (Figure 6.5). (This is a crude model for the atomic spins in a magnetic material.) Let us say that the magnetic moment, μ_i, of the ith bar is $\mu_i = +1$ if it points up and is $\mu_i = -1$ if it points down. The total magnetic moment is defined to be

$$M = \sum_{i=1}^{N} \mu_i.$$

Argue that this situation is analogous to the drunkard's walk. What is the variable in the drunkard's walk that is analogous to M? What is the analogous variable for the case of N coin tosses? (These are examples of what are called *random variables*. They are variables that depend on the outcome of random experiments.) What is the probability distribution of M?

EXERCISE 11 Exercise 10 can be generalized to where we include the effect of an applied magnetic field. Suppose that the magnetic field is pointing up with flux density B, and suppose that this changes the probability of being up to $p = \frac{1}{2}(1 + \tanh B)$. Argue that this is analogous to the situation in exercise 2.

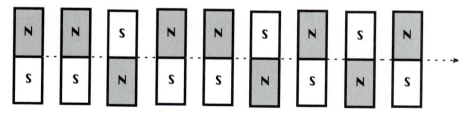

Figure 6.5

3   Counting on Probabilities

The situation that we have referred to as completely random consists of a collection of N simple events making up a set $\mathcal{S}$ that we call a sample space. Each simple event is to be weighted on an equal basis with the others. To calculate the probability of a compound event A, we simply count the number of simple events that make up the event A and divide that number by N. That is,

$$P(A) = \frac{|A|}{N},$$

or

$$P(A) = \frac{\text{number of outcomes in } A \cap \mathcal{S}}{\text{number of outcomes in } \mathcal{S}} = \frac{|A \cap \mathcal{S}|}{\mathcal{S}}. \qquad (6.5)$$

This is the classical (18th century) definition of probability, and it is certainly acceptable as long as we concern ourselves with sample spaces that are finite. The question is, whether we can modify this definition for use when the sample space is more complicated. First, we see that there is no way to assign equal probabilities to all the simple events and have these probabilities sum to 1. But what if $\mathcal{S}$ is infinite but bounded, say an interval on the real number line; then we would replace the cardinality of the set by the appropriate measure, such as length, and we identify "events" with subsets of the interval. Then the definition gives the probability as the *relative size* of intervals; it is *normalized*.

EXERCISE 12   (i) Suppose that a point, x_0, is chosen at random from the interval $(0, 5)$. What is the probability that $3 < x_0 < 4$? What is the probability that $3 \leq x_0 \leq 4$? What is the probability that $x = 2.5$?
(ii) A person is expected to arrive at a random time within the next hour. What is the probability that she arrives in the next 15 minutes?
(iii) A couple has agreed to meet on the Eiffel tower between 17:00 and 18:00. If each arrives at a random time in the hour and stays for 20 minutes, what is the probability that they meet? (*Hint:* the sample space is a certain two-dimensional interval.)

Unfortunately, the problems that arise when trying to extend the classical definition of probability to infinite situations is not simply one of normalization. The paradox that is suggested by the following exercise is due to Bertrand.

EXERCISE 13   (i) Two points are chosen at random on a circle. What is the probability that the chord of the circle passing through these two points is greater than the radius?
(ii) A point is picked at random on a given diameter. What is the probability that the chord through the given point that is perpendicular to the diameter is longer than the radius?
(iii) A point is chosen at random inside the circle. What is the probability that the chord that is bisected by this point is longer than the radius?

The modern, or axiomatic, approach to probability theory, due to Kolmogorov, is to first pick out those properties that a probability assignment should have.

EXERCISE 14 Using the classical definition of probability, (6.5), show that
1) $P(\emptyset) = 0$
2) $P(\mathcal{S}) = 1$
3) $P(A \cup B) = P(A) + P(B) - P(A \cap B), \quad A, B \subset \mathcal{S}$

EXERCISE 15 (i) Compute $P(A \cup B \cup C)$.
(ii) If A^c denotes the complement of a set, show that $P(A^c) = 1 - P(A)$.

Nowadays, one defines a **probability** to be a function defined on subsets of a sample space $\mathcal{S}$ that satisfies 1),2),3) of Exercise 14. In fact, we replace 3) with the stronger condition:
   3′) If $E_1, \ldots, E_n, \ldots$ are subsets of $\mathcal{S}$ that are mutually disjoint, that is, $E_i \cap E_j = \emptyset$, for all i, j, then

$$P(\cup_{k=1}^{\infty} E_k) = \sum_{k=1}^{\infty} P(E_k).$$

Aside on Entropy and Information

The reader is perhaps asking whence comes the function P. After all, if we are going to model a phenomenon using probabilistic methods, where do we find our probabilities? The axiomatic approach to probability assumes that your assignment of probabilities is already in place; it is up to the modeler to make this decision, and the classical definition of probability is as good a place as any to look for inspiration. We can modify the basic idea behind the classical approach to arrive at the relative-frequency approach: Suppose that an experiment is repeatedly and uniformly carried out. For each event, let $N(E)$ be the number of times E occurs in the first N repetitions. Then define

$$P(E) = \lim_{N \to \infty} \frac{N(E)}{N}.$$

That is, $P(E)$ is the *limiting* frequency of E. Of course, the problem (for the pure mathematician) is that this limit may not exist in any acceptable sense, and then it must be assumed. There is a consistency between this approach and the axiomatic approach, which is made precise by the (various) laws of large numbers. If an experiment is repeated over and over again, then, with probability 1, the fraction of the cases when E occurs will be $P(E)$.

We see that we should define probabilities as fractions, according to the classical definition, unless circumstances conspire to allow *more information*. This is the idea of maximizing the *entropy*.

To see this, imagine an experiment with M possible outcomes performed N times. Let N_k be the number of times that the kth event occurs;

$$P_k = \frac{N_k}{N}$$

is the *frequency*, and

$$\{P_k\}, \quad k = 0, 1, \ldots$$

is the *frequency distribution*. In any particular series of N experiments there are M^N possible outcomes. Each outcome will give rise to a distribution, $\{P_k\}$. It is possible that different outcomes will generate identical distributions. We would like to pick out the distribution that is generated by the greatest number of outcomes. The number of outcomes that generate any particular distribution $\{P_k = \frac{N_k}{N}\}$ is

$$W = \frac{N!}{N_1! N_2! \ldots N_M!}.$$

EXERCISE 16   Use Sterling's formula to show that

$$\frac{\ln W}{N} \underset{N \to \infty}{\approx} \sum_{k=1}^{M} P_k \ln P_k.$$

Notice that the number on the right in the formula is independent of N; it is known as the *entropy* of the distribution. We can sum up the above discussion by saying that one should choose the distribution with maximum entropy, and this statement remains true if we are given prior information as long as we work with the constraints.

EXERCISE 17   (i) Consider a coin toss. Show directly that the assumption of a fair coin gives the distribution with maximum entropy.
(ii) We are given the information that a certain six-sided die is weighted so that the side with 6 lands up 1/5 of the time. What is the distribution that maximizes the entropy and satisfies this constraint?
(iii) Suppose that we are told that for a particular six-sided die the even numbers show up 2/3 of the time. What is the maximum entropy distribution?

The notion of maximum entropy makes more precise the rule of insufficient reason. The reader is possibly familiar with the term entropy in connection with the second law of thermodynamics. There is a connection. It suffices here to say that the states of a system tend to be those with the largest entropy because they occur in the greatest number of ways.

Conditional Probabilities

Sometimes events are conditional upon another event happening. For example, if two fair coins are tossed, we ask what is the probability of two heads appearing? This is one possibility out of four, so we assign the probability of 1/4. Now suppose we *knew* somehow that at least one head appeared. This eliminates the outcome where two tails appear and reduces the list of eligible outcomes to 3, and so the probability (given the additional information) is 1/3. The conditional probability of an event A given the certainty of the event B is denoted $P(A|B)$. Note that if

$$P(A) < P(B) < 1$$

and

$$A \subset B,$$

then

$$P(A|B) > P(A).$$

On the other hand, if A is incompatible with B, then

$$P(A|B) = 0,$$

regardless of the value of $P(A)$. If $P(B) > 0$, we can define the **conditional probability** as

$$P(A|B) = \frac{P(A \cap B)}{P(B)}.$$

EXERCISE 18 (i) Show that this definition is consistent with the classical notion of probability.
(ii) Let $P(\cdot|B)$ be the function on sets defined from P by fixing B. Show that this is a probability, that is, show that conditions 1),2), and 3') hold.
(iii) Show that

$$P(A) = P(A|B)P(B) + P(A|B^c)(1 - P(B)).$$

(iv) Partition $\mathcal{S}$ into mutually exclusive events, F_i; show that

$$P(E) = \sum P(E|F_i)P(F_i).$$

It is sometimes convenient to use trees to compute conditional prob-abilities. Let us illustrate their use in a slightly unorthodox way in resolving the Prisoner's paradox: Two of three political prisoners are to be released on election day. Prisoner 1 can learn from the jailer the name of a prisoner (not himself) who will be released. He decides not to take advantage of this information because he thinks it will decrease his chances of being released. He reasons that without any knowledge his chance of being released is 2/3; but if he knows the name of the other prisoner, then his chance is now 1/2. What is the flaw in his computation? Let

$$R_k, \quad k = 1, 2, 3,$$

be the event that prisoner k is released and

$$A_j, \quad j = 2, 3,$$

the event that the name of prisoner j is announced to prisoner 1. We can form the tree (Figure 6.6) and compute the probabilities by count-ing the appropriate paths through the tree. Thus,

$$P(R_1|A_k) = \frac{P(R_1 \cap A_k)}{P(A_k)} = \frac{1/3}{\frac{1}{3} + \frac{1}{3}\frac{1}{2}} = \frac{2}{3}.$$

EXERCISE 19 I roll a (fair) die and then toss a (fair) coin as many times as the number of pips on the die. Then I tell you that no heads appeared among my coin tosses. What would you guess that I rolled on the die? (*Note that you are computing a conditional distribution,* $P(k|no\ heads), k = 1, \ldots, 6.$)

Events are said to be independent of each other or, simply, **inde-pendent** if conditioning is irrelevant. That is, if

$$P(A|B) = P(A), \qquad P(B|A) = P(B).$$

EXERCISE 20 Show that if A and B are independent then
$$P(A \cap B) = P(A)P(B).$$

We implicitly used the idea of independence when computing the probabilities for a sequence of coin tossing. That is to say that each coin toss is independent of the others amounts to saying that for any sequence, say $HTHHH$, we have

$$P(HTHHH) = P(H)P(T)\ldots P(H).$$

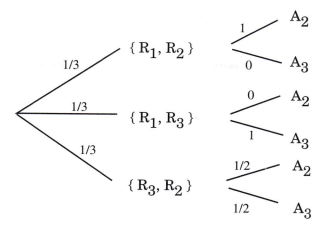

Figure 6.6 Tree diagram for figuring probabilities.

4 Random Variables

A series of coin tosses is an *experiment*, and each particular toss is a *trial* of the experiment.

A series of such trials performed under identical conditions (the probability distributions don't change from trial to trial), and where the outcome of each trial is independent of the others, is termed **Bernoulli trials**. If each experiment has only two possible outcomes, this series is called *binomial trials*. In this latter case we label the outcomes S for success and F for failure, although the usual connotations of these words may not be appropriate in some experiments.

EXERCISE 21 Let $P_N(k)$ denote the probability of k successes in N binomial trials. Assume that the distribution on each toss is probability of success, $P(S) = p$, and probability of failure, $P(F) = q = 1 - p$. Show that

$$P_N(k) = \binom{N}{k} p^k q^{N-k}.$$

Explain why this is called a **binomial distribution**. (*Hint* : $p+q=1$.) This reduces to the formula (6.3) when $p = q = \frac{1}{2}$.

EXERCISE 22 (i) In the scenario of the drunkard-on-a-hill (Exercise 9), what is the probability that after N steps the drunkard is k steps to the right? (ii) For the magnetic spins in the presence of an applied field (Exercise 11), what is the probability that k spins are up? (iii) Place N points at random in the interval $(-T/2, T/2)$. Let $(a, b) \subset$

$(-T/2, T/2)$ be a subinterval. Let $P(k)$ be the probability that k of the N points lie in (a, b). Find the distribution.

When $p = q = \frac{1}{2}$, the binomial distribution is symmetric as illustrated by Figure 6.7, The case where $p < q$ results in a skewed distribution as shown in Figure 6.8. As $p \to 0$, this skewness becomes more apparent; if N is very large and p is very small, then the histogram becomes smoother looking and has a decidedly exponential-looking tail. This observation leads to a very useful approximation to the binomial distribution when p is small and N is large. If $p \ll 1$ and $k \ll N$ so that $Np \sim 1$, and $kp \ll 1$ so that

$$\binom{N}{k} \approx \frac{N^k}{k!},$$

and

$$q = 1 - p \approx e^{-p}$$

so that

$$q^{N-k} \approx e^{-Np},$$

then

$$\binom{N}{k} p^k q^{N-k} \approx \frac{(Np)^k}{k!} e^{-Np}.$$

If, when $N \to \infty$ and $p \to 0$, we have $Np \to \alpha$, then, asymptotically,

$$\binom{N}{k} p^k q^{N-k} \approx \frac{\alpha^k}{k!} e^{-\alpha}.$$

The function (of k) on the right side is known as the **Poisson distribution** with parameter α (Figure 6.9). This distribution arises frequently in the modeling of rare events, such as the number of insurance claims in a given period, the number of defects in a manufactured article, or the number of α-particle emissions during a certain time. We will be seeing more of this distribution.

EXERCISE 23 (i) (Refer to Exercise 13) Suppose that we consider a constant density (but sparse) distribution of N points on the real line, that is, $N \to \infty$, and $T \to \infty$, but $\frac{N}{T} \to \lambda$, where λ represents the constant density of points. For a fixed interval (a, b), show that the probability of finding k points in (a, b) is a Poisson distribution. (Note that both the probabilities in this and the previous exercise depend only on the length of the interval.)
(ii) Here is another way of looking at the same thing. Suppose that an event occurs in any time period Δt at a constant rate λ, that is, with a probability

$\lambda \Delta t$. How will the number of events occurring during any time interval be distributed? (*Hint* : Partition the time interval into subintervals of length Δt, which will then be sent to zero.)

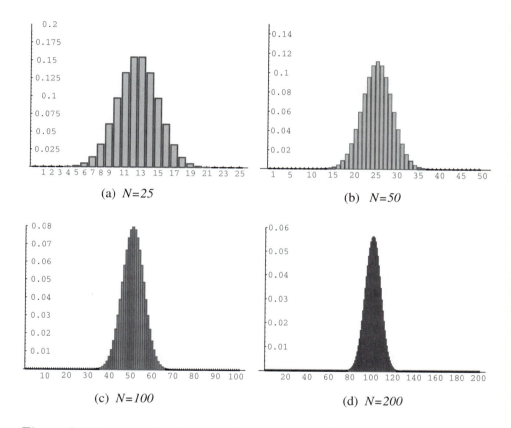

(a) *N=25*

(b) *N=50*

(c) *N=100*

(d) *N=200*

Figure 6.7 Binomial distribution with $p = q = 1/2$. Notice that the vertical axis is re-scaled in each case.

If the value of a variable depends on the outcome of a random experiment or a series of random trials, such as Bernoulli Trials, we call the variable a **random variable**. More formally, a random variable (RV) is a real-valued function defined on a sample space.

We have already seen examples of random variables. In a coin toss, the possible outcomes are H or T. We can define a random variable on this space of outcomes by

$$X(H) = 1, \quad X(T) = 0.$$

Or, on the same space, we can define a different random variable Y by

$$Y(H) = 1, \quad Y(T) = -1.$$

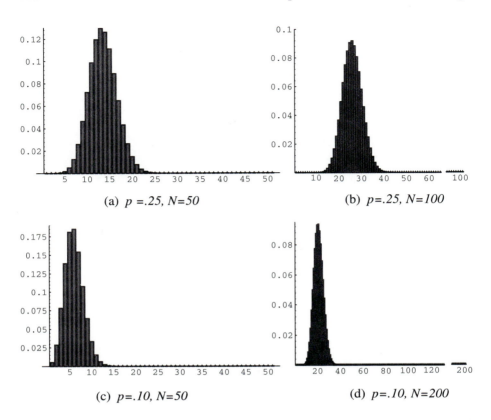

Figure 6.8 Binomial distribution. Notice the re-scaling of the vertical axis in each case.

We implicitly used X when we counted heads in the sequence of coin tosses and implicitly used Y when we measured progress to the right in the random walk or counted up-spins and down-spins in the magnet.

As another example, suppose that S_N denotes the number of successes in a series of N binomial trials; then S_N can take on the values $0, 1, \ldots, N$, and it does so with the probabilities $\{P_N(k) = P(S_N = k)\}$. Notice that

$$S_N = X_1 + X_2 + \cdots + X_N,$$

where X_k is the random variable X, defined in the previous paragraph, applied to the kth trial.

EXERCISE 24   Express R_N and M_N in terms of Y when the random variable R_N is the net distance to the right in the random walk, and the random variable M_N is the net magnetization in the row of magnets.

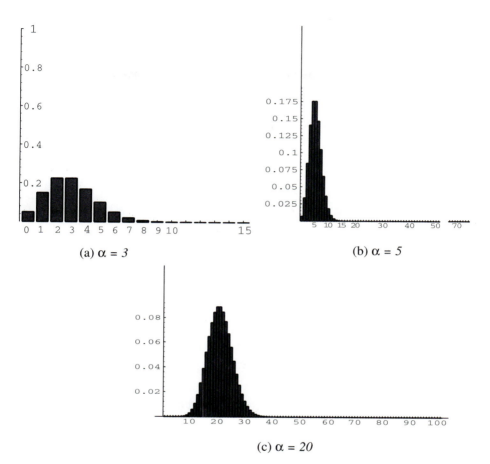

(a) $\alpha = 3$

(b) $\alpha = 5$

(c) $\alpha = 20$

Figure 6.9 Poisson distribution. Compare to the binomial distribution.

Notice that a random variable can be defined before we know anything about how to compute probabilities in the sample space. With the probability structure we get the probability distribution of the random variable. In general, if a random variable X has possible values in a discrete set $\{x_1, x_2, \ldots, x_n, \ldots\}$, we can define its *probability distribution*,

$$P(x_k) = P(X = x_k).$$

Some would prefer that we call this the *probability mass function* so that you won't confound it with the similarly named. Examples of such probability mass functions are the binomial distribution and the Poisson distribution.

Given any function $g : R \to R$, we can define a new RV, $Y = g(X)$. For example, if X takes on values $\{x_1, x_2, x_3\}$, then the RV $Z = aX + b$ takes on values $\{ax_1 + b_1, ax_2 + b_2, ax_3 + b_3\}$, and since the function relating them is one–one, Z has the same distribution as X.

EXERCISE 25 X is a RV that takes on values $\{x_n\}$ with a distribution $p_k = P(X = x_k)$. What is the range of the RV $Y = X^2$, and what is its distribution (probability mass function)?

Two random variables X and Y are said to be **independent** if the events $\{x \in A\}$ and $\{Y \in B\}$ are independent. In other words,

$$P\{X \in A, Y \in B\} = P\{X \in A\}P\{X \in B\}. \qquad (6.6)$$

The distribution defined for both X and Y on the left of (6.6) is called the *joint probability function* of X and Y. If X and Y are independent, then it is the product of their respective probability distributions. Such a probability function can be defined for any number of RVs. Consider a sequence of N Bernoulli trials. Suppose that each experiment can result in one of M possible outcomes, and the probabilities of these happening are $p_1, \ldots, p_M$. That is, $\{p_1, \ldots, p_M\}$ is the probability distribution of each trial. Let Q_k equal the number of experiments whose result is the kth outcome. Then

$$P\{Q_1 = N_1, \ldots, Q_M = N_M\} = \frac{N!}{N_1! \ldots N_M!} p_1^{N_1} p_2^{N_2} \ldots p_M^{N_M},$$

where $\sum N_k = N$. This is known as a *multinomial distribution*. Of course, the Q_k are not independent.

EXERCISE 26 Show that the binomial distribution is a joint distribution for appropriate RVs.

EXERCISE 27 Consider a random scattering of points in an interval of the real axis, $(-T/2, T/2)$. Let (a_1, b_1) and (a_2, b_2) be two disjoint subintervals of $(-T/2, T/2)$. Let X_1 be the RV that counts the points in (a_1, b_1) and X_2 be the RV that counts the points in (a_2, b_2). Show that X_1, X_2 are *not* independent. Now let $N \to \infty$, $T \to \infty$, and $\frac{N}{T} \to \lambda$ so that the counting RVs become Poisson distributed. Show that now they *are* independent.

We can also define a conditional mass function if a RV is restricted in its values. For example, a coin is tossed twice and the RV H_2 is the number of heads. We know that H_2 has the distribution

$$P\{H_2 = 0\} = \frac{1}{4}, \quad P\{H_2 = 1\} = \frac{1}{2}, \quad P\{H_2 = 2\} = \frac{1}{4}.$$

Suppose that we are given to know that there is at least one head; then the distribution becomes

$$P\{H_2 = 0|H_2 \geq 1\} = 0, P\{H_2 = 1|H_2 \geq 1\} = \frac{2}{3}, P\{H_2 = 2|H_2 \geq 1\} = \frac{1}{3}.$$

More generally, the conditional mass function of a RV X given that $X \in B$ is $P\{X = x_k | X \in B\}$.

More important is the notion of a conditional probability mass function of X given information about some other RV. This is

$$P(X = x|Y = y) = \frac{P(X = x, Y = y)}{P(Y = y)}.$$

EXERCISE 28 If X and Y are independent RVs, show that the conditional probability mass function is the same as the unconditional probability mass function.

EXERCISE 29 Show that

$$P(X = x) = \sum_y P(X = x|Y = y)P(Y = y). \tag{6.7}$$

EXERCISE 30 A radioactive nucleus is emitting a random stream of alpha particles. It is assumed that the particles can be counted according to a Poisson distribution with parameter λt, where t is time (thought of as the length of an interval). However, our Geiger counter has only a limited sensitivity and only picks up the more energetic particles. Let's say that a particle has a probability p of being recorded. What is the distribution of the random variable that represents the number of particles actually registered in an interval of length t?

The **mean** of RV X is

$$\bar{X} = E(X) = \sum_{x \in S} X(x)P(X = X(x)).$$

Thus, if the range of X is $\{x_1, x_2, \ldots, x_N\}$, then

$$\bar{X} = E(X) = \sum_k x_k P(X = x_k).$$

The mean is also called the *expectation* or the *expected value*. Of course, since it is a weighted average, the mean is not necessarily a value in the range of X.

EXERCISE 31   Give an example of a random variable whose mean or expectation is not one of the possible values of X.

EXERCISE 32   Is the following betting game fair? A has a chance of $999/1000$ of winning against his adversary B; accordingly he antes up \$999 as opposed to $B's$ \$1.

EXERCISE 33   Find the mean of the RV that is Poisson distributed with parameter α.

An important property of the mean is its linearity,

$$E(X + Y) = E(X) + E(Y).$$

Let's put this to use in computing the mean of a series of trials. Let a sequence of experiments be performed; to be specific, a series of binomial trials where each trial results in success or failure and is described by the RV X, where $X(S) = 1$ and $X(F) = 0$. This generates a sequence of RV. Then

$$S_N = \sum_{k=1}^{N} X_k$$

is the total number of successes, and the mean or expectation of S_N is

$$\overline{S_N} = E(S_N) = \sum_{k=1}^{N} E(X_k) = \sum_{k=1}^{N} \overline{X_k},$$

the sum of the means. Notice that this is true even if the experiments are neither independent nor identically distributed.

EXERCISE 34 In the examples of random walk and magnetic spin, compute the mean position after N steps and the mean magnetic moment for N dipoles. If there is a definite bias in each step (trial), how is this reflected in the sum?

If X is a RV, then so is X^k; its mean, $E(X^k)$, is called the kth *moment* of X. The first moment is the usual mean. Some indication of how a RV spreads out about its mean can be had by looking at the shifted RV, $X - E(X)$. The second moment of $X - E(X)$,

$$
\begin{aligned}
\sigma^2(X) &= E((X - E(X))^2) \\
&= \sum_{x \in S} [X(x) - E(X)]^2 P(X = x),
\end{aligned}
$$

is called the **variance** of X. The square root, σ, is called the **standard deviation**. Either the variance or the standard deviation can be referred to as the *dispersion*.

EXERCISE 35 Find the variance of a RV X that has a Poisson distribution with parameter α.

EXERCISE 36 (i) Show that $\sigma^2(X) = E(X^2) - E(X)^2$.
(ii) Show that $\sigma^2(aX + b) = a^2\sigma^2(X)$.
(iii) Is the variance linear: $\sigma^2(X + Y) = \sigma^2(X) + \sigma^2(Y)$?

If X and Y are RVs, we can define their **covariance**,

$$\mathrm{cov}(X, Y) = \sigma^2(X + Y) - \sigma^2(X) - \sigma^2(Y).$$

Having done the last exercise, you already know that, in general, the covariance is nontrivial. However, if X and Y are independent RVs, then

$$\mathrm{cov}(X, Y) = 0.$$

This is equivalent to the statement that if X,Y, are independent, then

$$E(XY) = E(X)E(Y). \tag{6.8}$$

EXERCISE 37 Show that

$$cov(X, Y) = E(X - E(X))E(Y - E(Y)).$$

Let's calculate the dispersion about the mean in a generalized random walk. We will allow a variable step length. The step length at the ith stage, either to the right (positive) or to the left (negative), will be given by the value of the random variable Λ_i. We assume, however, that the probability mass function of each of these RVs is the same, regardless of i; this is to say that they are *identically distributed*. In addition, the RV Λ_i is independent of the RV Λ_j, $i \neq j$. Let L be the RV that represents the net length after N steps; then

$$L = \sum_{i=1}^{N} \Lambda_i.$$

The variance of L, $\sigma^2(L)$, is the second moment of the RV

$$
\begin{aligned}
L - E(L) &= L - \overline{\sum \Lambda_i} \\
&= L - \sum \overline{\Lambda_i} = L - N\overline{\Lambda}.
\end{aligned}
$$

We used the fact that the Λ_i are identically distributed. If they are not, then

$$L - E(L) = \sum (\Lambda_i - \overline{\Lambda_i}).$$

The dispersion of L, valid in the more general case, is

$$\sigma^2(L) = \overline{\left(\sum \Lambda_i - \overline{\Lambda_i}\right)^2}$$
$$= \overline{\sum (\Lambda_i - \overline{\Lambda_i})^2} + \overline{\sum_{i \neq j} (\Lambda_i - \overline{\Lambda_i})(\Lambda_j - \overline{\Lambda_j})},$$

and the second term is zero because the Λ_i are independent [see (6.8)]. Thus,

$$\sigma^2(L) = \sum \overline{(\Lambda_i - \overline{\Lambda_i})^2}. \tag{6.9}$$

But when we assume that the Λ_i *are* identically distributed,

$$\sigma^2(L) = N\sigma^2(\Lambda)$$

or, in terms of the standard deviation,

$$\sigma_L = \sqrt{N}\sigma_\Lambda.$$

The expression (6.9) is quite general; notice that we didn't assume anything about the Λ_i in its derivation, other than independence. For the last formulas, identical distribution was assumed.

In the even more special case (that reduces to binomial trials), where the distribution is

$$p(\Lambda_i = l) = p, \quad P(\Lambda_i = -l) = q = 1 - p,$$

then $\overline{\Lambda} = l(p - q)$ and

$$\sigma_\Lambda^2 = \overline{(\Lambda - \overline{\Lambda})^2} = (l - l(p-q))^2 p + (-l - l(p-q))^2 q$$
$$= l^2(p(2 - q)^2 + q(2 - p)^2)$$
$$= 4l^2 pq.$$

So, in this very special case,

$$\sigma_L = 2l\sqrt{pqN}.$$

Let's do a little scaling here. If we write $L = L^* l$, then L^* expresses the net position in units of l, and

$$\overline{L^*} = N(p - q), \quad \sigma_{L^*} = 2\sqrt{Npq}.$$

Now we'll introduce the passage of time. Suppose that jumping from one step to the next is accomplished in a time Δt. (You can think of Δt as the time to "relax" at each step before instantaneously hopping to the next, or the time it takes to go from one step to the next.) Thus, after N steps, a time of

$$t = N\Delta t$$

has elapsed. Thus, we have the relation

$$\sigma_{L^*} = 2(\frac{pq}{\Delta t})^{1/2}t^{1/2}.$$

This says that the spread in position scales as the $1/2$ - power of time. This identifies the random walk as a diffusion process (cf. chapter 11).

EXERCISE 38 (i) Consider again the chain of magnetic dipoles (two-state systems are sometimes called spin-$\frac{1}{2}$ particles), where moment has probability p of being μ_0 (up) and probability $q = 1 - p$ of being $-\mu_0$ (down). Find the mean and variance of the net moment.
(ii) Systems with three possible states are referred to as spin-1 particles. Suppose that we have a line of such particles. Suppose that the moment of each particle has the probability distribution $P(\mu_0) = p, P(-\mu_0) = p, P(0) = q = 1 - 2p$. Find the mean and variance of the net magnetic moment.

If X and Y are RVs we can define the conditional mean of X given Y:

$$E(X|Y = y) = \sum xP\{X = x|Y = y\}.$$

EXERCISE 39 Show that this satisfies all the usual properties of a mean.

Notice that the conditional mean, as above, is a function of y, unless X is independent of Y, in which case it is the constant

$$E(X|Y = y) = E(X).$$

By letting Y range over all its values, out of the conditional means comes a RV

$$E(X|Y).$$

In that case, we can recover $E(X)$ by averaging over Y:

$$\begin{aligned} E(X) &= E(E(X|Y)) \\ &= \sum E(X|Y = y)P(Y = y) \end{aligned}$$

This *conditioning* procedure is sometimes useful in computing means or even probabilities (see problems).

Continuous Random Variables

Some random variables do not have a discrete range of values. In a previous example (Exercise 14), we considered a random scattering of events in time and showed that this gave rise to a Poisson distribution for the number of events in a given interval. One could term the Poisson distribution a *counting distribution*, which indicates that its nature is discrete. What if we are interested in the actual times of the events or the waiting time for the next event? We can define the RV T, which gives the actual points in time, and the RV W, which gives the waiting time between any two events. Each of these RVs has a continous range of values. As we pointed out in the introduction to this chapter, it is fruitless to define $P(X = a)$ to be anything but zero for such RVs. Instead, we must replace the notion of a counting probability with a suitable (continuous) measure.

For any RV, we can define the **cumulative distribution function (CDF)**:

$$F(b) = P(X \leq b).$$

For a discrete RV, like the ones we have been studying,

$$F(a) = P(X \leq a) = \sum_{x_k \leq a} P(X = x_k),$$

and

$$P(X = a) = F(a) - P(X < a),$$

so

$$P(X \leq a) \neq P(X < a),$$

for any a, unless $P(X = a) = 0$.

We call a RV *continuous* if the opposite is true *for all a*:

$$F(a) = P(X < a). \tag{6.10}$$

This, of course, leaves open the possibility for the RV to be neither continuous nor discrete, or rather, and more accurately, to have features of both types. There may exist some points for which (6.10) holds and other points for which it doesn't. We'll call the latter type of point *atoms*, because they are like point masses.

EXERCISE 40 Show the following properties of the CDF when X is any RV, continuous or discrete.
(i) $a < b \Longrightarrow F(a) \leq F(b)$.
(ii) $\lim_{x \to \infty} F(x) = 1$.

(iii) $\lim_{x \to -\infty} F(x) = 0$.

(iv) F is continuous from the right,

$$\lim_{x \downarrow x_0} F(x) = F(x_0+) = F(x_0).$$

(v) F is discontinuous at atoms.

A continuous RV is best described through the use of its **probability density function** (PDF), f, where

$$P(X \in A) = \int_A f(x)dx. \tag{6.11}$$

If $A = [a, b]$, then

$$P(a \le X \le b) = \int_a^b f(x)dx,$$

and

$$F(a) = P(X \le a) = P(X < a) = \int_{-\infty}^a f(x)\,dx.$$

Thus,

$$F'(x) = f(x).$$

The differential dF gives the probability increment:

$$dF = f(x)\,dx = P(x \le X \le x + dx). \tag{6.12}$$

This shows that if F is not differentiable in the usual sense then f doesn't exist as a function (in the usual sense). This is the case, for example, with a RV that is neither discrete nor continuous (has atoms).

EXERCISE 41 A continuous RV has a *Cauchy distribution* (with parameter θ) if it has the probability density function

$$f(x) = \frac{1}{\pi} \frac{1}{1 + (x - \theta)^2}.$$

What is the cumulative distribution function, $F(x)$?

Independence of continuous random variables can be characterized in the same way as for discrete random variables. Instead of a joint probability mass function, we define a *joint distribution function*,

$$F(a, b) = P\{X \le a, Y \le b\},$$

and a *joint density function*, $f(x, y)$, so that

$$F(a, b) = \int_{-\infty}^{a} \int_{-\infty}^{b} f(x, y)\, dx dy.$$

Notice that the (marginal) distributions of X or Y can be found by integrating away the other RV:

$$f_X(x) = \int_{-\infty}^{\infty} f(x, y)\, dy,$$

where the notation is obvious. The *conditional probability density function of X given $Y = y$* can be defined for all values of y with $f_Y(y) > 0$ by

$$f_{X|Y}(x|y) = \frac{f(x, y)}{f_Y(y)}.$$

EXERCISE 42 Explain why $f_{X|Y}(x|y)\, dx dy$ represents the probability that X is between x and $x + dx$ given that Y is between y and $y + dy$.

Using the last two formulas, we have the appropriate analogy to (6.7):

$$f_X(x) = \int_{-\infty}^{\infty} f_{X|Y}(x|y) f_Y(y)\, dy,$$

which is often useful.

EXERCISE 43 (i) Show that independence of two continuous RVs X and Y, reduces to the condition $f(x, y) = f_X(x) f_Y(y)$.
(ii) Show that the density function of the *sum* of two independent RVs is the convolution of the densities:

$$\begin{aligned} f_{X+Y}(z) &= (f_X * f_Y)(z) \\ &= \int_{-\infty}^{\infty} f_X(\zeta) f_Y(z - \zeta)\, d\zeta. \end{aligned}$$

It is natural to want to define the mean or expectation of a continuous RV as

$$E(X) = \int_{-\infty}^{\infty} x f(x)\, dx, \qquad (6.13)$$

but we have to be careful here. The density, f, may not exist, and even if it does, the improper integral in (6.13) may not converge.

EXERCISE 44 Show that the RV with the Cauchy distribution has no mean.

To cover all bases, we can define the mean in terms of the Stieltjes integral,

$$E(X) = \int_{-\infty}^{\infty} x \, dF,$$

which by its very form, in combination with (6.12), generalizes directly and naturally the definition of mean that we gave to a discrete variable.

If $Y = g(X)$ is a RV defined in terms of X, then it has its own mean, $E(Y)$, defined in the usual way. In fact, as one would hope,

$$E(Y) = E(g(X)) = \int_{-\infty}^{\infty} g(x) \, dF. \tag{6.14}$$

The reader might want to prove this.

The conditional expectation of X given Y is found from the conditional density, if it exists:

$$E(X|Y) = \int_{-\infty}^{\infty} x f_{X|Y}(x|y) \, dx \tag{6.15}$$

$$= \frac{\int_{-\infty}^{\infty} x f(x,y) \, dx}{\int_{-\infty}^{\infty} f(x,y) \, dx} \tag{6.16}$$

Notice that this is a function on the sample space of Y; that is, it is a RV in its own right.

Time between Random Events

Suppose that it has been observed, over a long time, that California earthquakes are Poisson distributed at a rate of two per week. How long should one wait until the next quake? We know that, in any time interval, say $[0, t)$, the number of earthquakes in that interval is a RV, E, with the distribution

$$P(E = k) = \frac{2^k}{k!} t^k e^{-2t},$$

where the time unit is taken to be one week. The waiting time until the next quake is a continuous RV, W. It has the distribution function (CDF)

$$F(t) = P(W \le t) = 1 - P(W > t) \tag{6.17}$$

$$= 1 - P(E = 0) \tag{6.18}$$

$$= 1 - e^{-2t}, \tag{6.19}$$

and so the density function (PDF) is

$$f(t) = F'(t) = 2e^{-2t}. \tag{6.20}$$

How long is the expected wait? This will be given by the mean of W:

$$\int_0^\infty 2te^{-2t} dt = 2(\frac{1}{4}) = \frac{1}{2}.$$

That we should expect to wait $1/2$ week is not surprising.

More generally, if any sequence of events is counted according to a Poisson distribution with parameter α, then the expected wait between events is $1/\alpha$. This is easy to remember if one keeps in mind that α is the mean "frequency" of events, so that $1/\alpha$ should be the mean "period" between events. Notice that, although we had asked about the waiting time between any events, we actually calculated the distribution of the waiting time until the *first* event. The reader should look back at the derivation of the Poisson distribution from a *random* distribution of points to see that the process has no point of reference in time and is thus *memoryless*. (The reader who is bothered by the fact that here we consider intervals whose left end point is always zero can go back and rederive the Poisson distribution starting from such intervals.)

An RV whose PDF has the form of (6.20) is said to be *exponentially distributed*. The **exponential distribution** (with parameter λ) is characterized by the PDF:

$$f(x) = \lambda e^{-\lambda x}. \tag{6.21}$$

The exponential distribution is associated with waiting time between random events, but it can be shown that the memoryless assumption alone is enough to ensure exponential distribution. Let W be a RV (with CDF F) that represents a waiting time. W is **memoryless** if the probability that the wait has a certain duration is independent of starting time, but depends only on its duration. In other words,

$$P(W > t + \Delta t | W > t) = P(W > \Delta t). \tag{6.22}$$

This equation can be written in terms of the distribution function of W:

$$\frac{1 - F(t + \Delta t)}{1 - F(t)} = 1 - F(\Delta t).$$

This can be rearranged to give

$$F(t + \Delta t) - F(\Delta t) = F(t) - F(t)F(\Delta t).$$

Expanding $F(t + \Delta t)$ about t and $F(\Delta t)$ about 0 gives

$$F(\Delta t) + (F'(0)F(t) - F'(t))\Delta t)\Delta t = o(\Delta t).$$

Dividing by Δt and taking the limit $\Delta t \to 0$, we see that F satisfies the linear differential equation

$$F'(t) + F'(0)F(t) = F'(0)$$

with the solution

$$F(t) = 1 - e^{-F'(0)t},$$

which is the distribution function of an exponential RV with parameter $\alpha = F'(0) = f(0)$.

EXERCISE 45 Let the RV T be the time of arrival of some event and suppose that

$$P(T < s | \text{one event in } (0, t)) = \frac{s}{t}.$$

Show that T is exponentially distributed.

This exercise and Problem 13 illuminate the important relationship between the Poisson distribution and the exponential distribution. We will investigate this more deeply in chapter 7, but it is worthwhile to review here what we have found. The Poisson distribution and the exponential distribution are both associated with "random", "rare" events. If points are sampled on an interval of the real number line in an equiprobable way, that is, the probability of a point coming from some particular subinterval is proportional to the size of that subinterval (see Exercise 13), we have a case of Bernoulli trials. In the limit of an infinite-sized sample space with a constant density of points, the limiting distribution will be Poisson. The lengths between consecutive points follows the exponential distribution. Conversely, if the lengths between consecutive points are exponentially distributed, then the count of points follows the Poisson distribution (see Problem 13).

5　*Simulation of Random Variables

Random variables must sometimes be generated to be used in computations. For example, if we wish to estimate a certain phenomenological

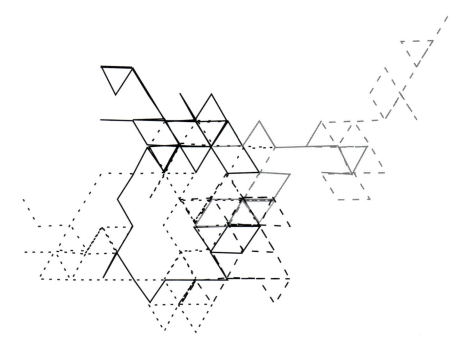

Figure 6.10 Five random walks on a hexagonal grid. These were done by hand with the roll of a die deciding the direction of travel.

parameter μ by use of the central limit theorem (see problems), we can sample a random variable, ξ, that has μ as its mean. We do that many times, generating the sequence of samples $\{\xi_1, \ldots, \xi_N\}$ and then make the estimate

$$\mu \approx \frac{1}{N} \sum \xi_k.$$

We need a way to to generate the sample $\{\xi_1, \ldots, \xi_N\}$. We could do this through physical sampling. For example, the uniform distribution $\{\frac{1}{37}, \ldots, \frac{1}{37}\}$ can be sampled with a roulette wheel; other analogue methods (dice) may suggest themselves (Figure 6.10). Sampling from a uniform (continuous) or equiprobable (discrete) distribution is called *random number generation*. Tables can be generated by using roulette wheels or analogue noise in electrical circuits. Nowadays, with computations on digital computers being common, it is convenient for any computer programmer to be able to call up a sampling from a uniform or equiprobable distribution. Of course, since a computer's number set is finite, such a program will, in fact, generate a periodic sequence of numbers; but number sequences can be generated with very long periods and act as random numbers. They are properly called pseudo-random numbers. Most computer systems and many hand-held calculators now make available a random number generator (RNG), so we will not give here either a table of random numbers nor discuss such programs (see [66]).

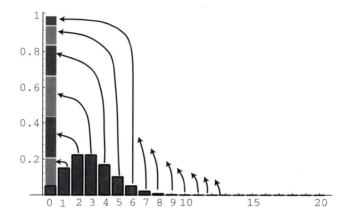

Figure 6.11 The Poisson distribution illustrates the simulation of a discrete distribution.

The output of a RNG is a sampling from the uniform distribution on $[0, 1]$, that is, a random number $0 < \gamma < 1$. A uniform distribution on $[a, b]$ can be simulated by using

$$(b - a)\gamma + a$$

as the random number. To simulate a nonequiprobable discrete distribution $\{p_k\}$, we can split the interval $\{[0, 1]\}$ into the appropriate number of pieces, the size of each piece being proportional to the corresponding probability (Figure 6.11)

To simulate a nonuniform continuous distribution of a RV X with $dF = f(x)\, dx$, we let $Y = F(X)$ [if $f(x) > 0$], and we sample Y from a uniform distribution. Since

$$F(x) = P(X \le x) = P(F(X) \le F(x)),$$

just solve for ξ:

$$\gamma = F(\xi). \tag{6.23}$$

[If $f(x) \geq 0$ only, then note that we can still get this to work, since $P(F^{-1}(X) \geq x) = P(y \geq F(x))$.]

EXERCISE 46 Show that an exponential distribution with parameter α can be sampled from

$$\xi = -\frac{1}{\alpha} \ln \gamma.$$

Sometimes (6.23) is hard to solve. Another method, based on the idea of sampling from proportional areas in the plane can be used: suppose that $f : [a, b] \to [0, M]$; then

1. generate two random numbers, γ_1, γ_2, and then:
   a uniform RV on $[a, b] : \delta_1 = a + \gamma_1(b - a)$
   and a uniform RV on $[0, M] : \delta_2 = \gamma_2 M$.

2. If (δ_1, δ_2) lies under the graph of f,
   then $\xi = \delta_1$,
   if not,
   then back.

Notice that this gives a probabilistic method of estimating the area under a curve: simply keep track of the number of times that you accept a number in the algorithm given above. That is, the area under the curve is to the area of the circumscribed rectangle as the proportion of "successes" during the running of the algorithm (Figure 6.12). This is the "Monte Carlo" method of integration and has lent its name to a more comprehensive notion of simulation (see Chapter 7).

EXERCISE 47 Suppose that a RV has the distribution $F(x) = \sum c_k F_k(x)$. Explain how we can simulate this RV by simply choosing according to the components the correct proportion of times.

Any of these methods can be adapted to multivariable distributions.

6   Problems and Recommended Reading

1. Derive Markov's inequality: if X is a RV, $X \geq 0$,

$$P(X \geq \alpha) \leq \frac{1}{\alpha} E(X).$$

Use this to get Chebyshev's inequality:

$$P(|X - \bar{X}| \geq k) \leq \frac{\sigma^2}{k^2}.$$

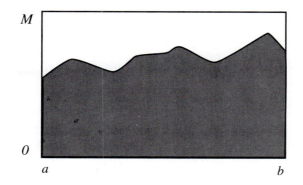

Figure 6.12 The Monte-Carlo method for computing the definite integral. Imagine throwing darts at the rectangle while blind-folded. The relative number of "hits" in the colored region is the area.

Derive the form that we used in the text

$$P(|\frac{H_N}{N} - \frac{1}{2}| \geq \epsilon) \leq \frac{1}{4N\epsilon^2}.$$

The weak law of large numbers could be restated in the following form:

$$\lim_{N\to\infty} 2^{-N} \sum_{k/n-\epsilon\leq k/n\leq k/n+\epsilon} \binom{N}{k} = 1.$$

2. Can you count the number of different ways to place n particles in m boxes?

   (a) (*Maxwell–Boltzmann statistics*) Particles are distinguished and any number can be put in one box.

   (b) (*Bose–Einstein statistics*) Particles are not distinguished and any number can be put in one box.

   (c) (*Fermi–Dirac statistics*) Particles are not distinguished and no more than one can be placed in a box.

3. (a) The probability of an event is 0.005. What is the probability that this event will occur exactly 50 time during 10,000 trials? Obtain the answer using both the binomial distribution and an appropriate approximate distribution.

   (b) In an experiment, we are monitoring the α-particles coming off a 2.45 gram specimen. It seems that the average is 3.33

per second. What is the probability that no more than one will be measured in the next 5 seconds?

4. Here are some problems that exemplify the computation of expectations by conditioning.

    (a) In a primitive pinball game, the ball, on its way down, will enter one of three chutes at random. Going through the left chute will earn 10,000 points and the ball will be thrown back up, going through the middle will earn 1000 points and the ball will drain; and taking the right chute will earn 100,000 points and the ball will be recycled. What is the expected number of points in a game?

    (b) On the average, 30 insurance policies are sold every day in an office. If the average policy is $10,000, how much is the expected total amount insured per day?

5. Do you buy an annual parking ticket at a reduced price and risk losing it or having it stolen, or do you buy the monthly and pay a small amount more?

6. (*Genetic inheritance*) Many traits are inherited. Each person inherits 23 chromosomes from each parent. There are 22 matching pairs and 2 sex chromosomes. These chromosomes are made up of *genes*. The same gene is located at the same place on each of the chromosomes in a pair. Each gene has different manifestations, or *alleles*. Since everyone has two copies of each gene, there are always two alleles for every gene. A *genotype* is a particular combination of alleles for a given gene. A *phenotype* is the observable trait that is due to the genotype. In the simplest form of genetics (Mendelian), a single gene determines a single trait, and one allele is dominant and the other recessive, meaning that it is simply the presence of or absence of the dominant allele that determines the phenotype. A clear example of such a trait is Huntington's disease, where the allele that causes the disease (phenotype) is dominant. Eye color, in many cases, also follows this simple model, brown eyes being dominant, although often a mixed genotype will produce a mixed phenotype and so not follow this simplest model. Suppose that a particular gene has two alleles, A and a, with A dominant, so that there are three genotypes:

$$AA, Aa, aa.$$

Suppose that the proportions of the different genotypes in the population are

$$p_{AA}, \quad p_{Aa}, \quad p_{aa},$$

and so represent the probabilities of receiving a particular genotype. Define

$$p = \frac{p_{AA} + p_{Aa}}{2} = 1 - q.$$

Show that the probabilities of the three genotypes in the next generation are

$$p^2, \quad 2pq, \quad q^2 \tag{6.24}$$

What are the probabilities of the various phenotypes? As a matter of fact, (6.24) gives the proportions in all succeeding generations, regardless of the original distribution. This is an example of a stationary Markov chain.

7. (*More Mendelian Genetics*)

   (a) The father of a child is unknown. Find the probabilities of the genotype child when the mother is known to be AA (and also for: Aa, aa). Find the probabilities of the mother given that the child is AA (and also for: Aa, aa).

   (b) Suppose that eye color follows this simple genetic model; suppose that the two alleles are brown and blue, with brown dominant. If a man has one brown-eyed parent and one blue-eyed parent, what is the probability that his first child is blue-eyed? What is the same probability if it is known that his wife has one brown-eyed parent and one blue-eyed parent?

8. Human blood is classified into four types depending on the type of agglutinogen (protein) on the outer layer of the wall of the corpuscle. The gene that determines the inheritance of the agglutinogen has three alleles, which we will call α, β, ω; so there are six different genotypes that determine the four phenotypes, or blood groups, in the following way

Phenotype	Genotype
A	$\alpha\alpha$ or $\alpha\omega$
B	$\beta\beta$ or $\beta\omega$
O	$\omega\omega$
AB	$\alpha\beta$

It is seen that the α and β are both dominant over ω, but that neither is dominant over the other. This is an example of *incomplete dominance*. Suppose that the proportions of α, β, ω in the gene pool are p, q, r, respectively. Find the probabilities for the blood groups in the succeeding generation.

9. Inheritance of skin color is more complicated, since it is not determined by a single gene and dominance is incomplete. Assuming the next simplest model of two genes each with two alleles, determine the number of possible shades of skin color. Set up a probabilistic model to determine the proportion of different shades that appear in succeeding generations.

10. One of Mendel's findings was that genes of two different characteristics are transmitted from parents to offspring independently. How could this have been deduced? Sometimes a trait is only apparent if a certain combination of alleles from different genes is present. Suppose that a person is a "problem drinker" only if he is both a heavy drinker and a frequent drinker. Suppose that these last two traits are associated with genes, G_1, which has the alleles $\{L, H\}$, and G_2, which has the alleles $\{I, F\}$. Given the proportion of genotypes in a population, what is the probability that someone is a problem drinker?

11. *Gamma distribution* Show that the sum of n independent RVs that are all exponentially distributed with parameter λ has the density

$$f(x) = \lambda e^{-\lambda x} \frac{(\lambda x)^{n-1}}{(n-1)!}.$$

This is called a gamma distribution with parameters (n, λ).

12. Suppose that the RV W_N is the waiting time until the Nth event in a random stream. How is W_N distributed?

13. The Poisson distribution can be derived by conditioning on exponential waiting times:

$$P(n \text{ events in}(0, t))$$

$$= \int_{y \leq t} P(n\text{th event in }(y, y + dy))$$

$$\cdot P(\text{wait until next event} \geq t - y)\, dy.$$

Verify that this does give the Poisson distribution.

14. A light bulb has a certain probability of failing in a certain period of time; once it fails, it is replaced by a new one. Let L be the lifetime of the current bulb. Explain how to find the distribution and mean of L. Is L a continuous or discrete RV? Do it both ways.

15. *Mixing* A sense that randomness embodies is that of being *well-mixed*. A shuffled deck of cards is more "random" than a new one. The degree of mixing of a system should be expressed by the distribution of sampling from that system. Think of tokens in a raffle lottery or chocolate chips in a cookie dough. This idea is also useful in designing ways to count large populations, such as cell counts. Explain how to obtain the Poisson distribution with parameter α by considering binomial trials of randomly placed things into a number of bins that cover a region of space [these could be squares (two-dimensional) or cubes (three-dimensional)] and letting the number of bins get arbitrarily large while the density, α, is fixed. Propose a procedure for quality control in a cookie baking factory.

16. *Probability generating functions* Let X be a RV with the the discrete values $\{n_1, n_2, \ldots\}$ and the probability distribution

$$P(X = n_j) = p_j.$$

The generating function (associated to this distribution) is

$$G(s) = \sum_{j}^{\infty} p_j s^j.$$

(*Note:* Sometimes the variable is changed: $s = e^t$; as a function of t it is usually referred to as the moment-generating function; see the following.)

(a) Show that the series always converges for $|s| \leq 1$.

(b) Show that G determines the p_j. (*Hint:* Taylor series about $s = 0$.)

(c) Show that without loss of generality we can assume that X takes on only nonnegative integer values.

(d) Show that the probability generating function (for X) is just the mean (or expectation) of the RV,

$$Y = s^X.$$

(e) Show that $\overline{X} = G'(1)$. Find a formula for the variance of X, involving G and its derivatives evaluated at a point.

(f) Let $X_1, X_2, \cdots, X_n$ be an independent RV taking noninteger values. Show that the generating function for the RV $Y = X_1 + ... + X_n$ is the product of the generating functions of the X_i.

(g) Using the fact in 16d define the generating function for a continuous RV that takes on nonnegative real values. (You should notice that this is the Laplace transform of the density of X. If you are not familiar with the Laplace transform, acquaint yourself now.) Verify all the above properties for this case.

(h) What is the mean of the gamma distribution?

(i) Show that for the Poisson distribution with parameter α we have $G(s) = e^\alpha (s - 1)$.

(j) Suppose that the number of insurance claims in a fixed time period is Poisson distributed with parameter α. Suppose that the amount of the claims are RVs, X_i, that are independent and follow the same distribution. Find the expected total amount in terms of the mean of the X_i. (*Hint:* you will need to condition over the number of claims.)

17. *Normal approximation* In Bernoulli trials, the number of successes is a RV S with the binomial distribution

$$P_N(k) = \binom{N}{k} p^k q^{N-k}.$$

Show that for N very large (assume that $Np \approx Npq \gg 1$) we have

$$P_N(k) \approx \frac{1}{\sqrt{2\pi Npq}} e^{-(k-Np)^2/2Npq}.$$

(*Hint:* Write $\ln P(k)$ as a Taylor series about its mean, Np. Show that the error in this approximation is of the form $C/\sqrt{N}$ and is fourth-order when $p = \frac{1}{2}$.)

18. A continuous RV with the density function $\frac{1}{\sqrt{2\pi}} e^{-x^2/2}$ is called a (unitary) *normal distribution*. Show that the mean is 0 and the variance is 1. For this reason, the unitary normal distribution is denoted by $N(0,1)$. In general, a RV X is said be $N(\mu, \sigma^2)$ if the

RV $\frac{X-\mu}{\sigma}$ is $N(0,1)$. What is the density function of a $N(\mu, \sigma^2)$ RV?

19. Show that if X_1 and X_2 are independent RV with distributions $N(\mu_1, \sigma_1^2)$ and $N(\mu_2, \sigma_2^2)$ then $X_1 + X_2$ has the distribution $N(\mu_1 + \mu_2, \sigma_1^2 + \sigma_2^2)$. (*Hint:* use generating functions.)

20. Derive the following relation valid for a $N(\mu, \sigma^2)$ RV:

$$P(|X - \mu| \leq \eta\sigma) = \sqrt{\frac{2}{\pi}} \int_0^{\eta} e^{-x^2/2} \, dx.$$

21. *Central limit and statistical inference* Recall that the RV that has the binomial distribution is the sum of RVs, X_i, that are 1 if success and 0 if failure. Problem **17** says that this sum is approximately $N(Np, Npq)$. This is very general: if X_i are independent RVs with a common mean μ and variance σ^2 and satisfy some uniformity condition (they could share a distribution function), then the RV $S_N = X_1 + \cdots + X_N$ satisfies

$$\lim_{N \to \infty} P(\frac{S_N - N\mu}{\sigma\sqrt{N}} \leq a) = \int_{-\infty}^{a} \frac{1}{\sqrt{2\pi}} e^{-x^2/2} \, dx.$$

It is not possible, in general, to estimate the error in the approximation obtained by stopping the limit at some high value of N. In practical statistical applications this is simply ignored, and the following version is used:

$$P(a\sigma\sqrt{N} < S_N - \mu N < b\sigma\sqrt{N}) \approx \int_a^b \frac{1}{\sqrt{2\pi}} e^{-x^2/2} \, dx.$$

(a) Suppose that a physical quantity (perhaps dipole moment with respect to a given axis) is being measured a number of times and generates a sequence of measurements, $x_1, \ldots, x_n$. Suppose we know that the value of this quantity is restricted to the interval $-1 \leq x \leq 1$, but we have *no other prior information*. In this case, we will use the mean of the measurements $\frac{1}{n}(x_1 + \cdots + x_n)$ as an estimate for x. How good is this? That is, what is the probability, P, that the relative error in this estimate is less than η? Think carefully about what assumptions are being made about the measurement process. (Tables involving the normal distribution are widely available.)

(b) Typically, in statistical inference one fixes the above probability in advance. Thus, P is called a *significance level* or *confidence level* and is given as a percentage. The relative error, η, is referred to as the *margin of error*. How many measurements of our physical quantity must we make to ensure that we are within 5% at a 95% confidence level?

(c) Come up with statistical method to determine whether a gambling die is loaded. You will need a model of a "fair" die. The simpler this model, the easier will be the statistical method, but, of course, you must justify your model.

22. *Litigate or settle?* Surveys have shown that well over 95% of legal claims are settled before going to trial. An even smaller fraction of litigated disputes is ever appealed. Are the litigated disputes representative of all claims, or do they somehow form a small class? This is pertinent to social scientists, who often study cases in appellate courts to determine developments in the interpretations of laws. A model of this settling process is attempted in [55].

(a) What are the variables and factors involved? According to the simple model presented here, they are

   i. Direct costs: Lawyer and court fees.

   ii. Expected payoffs: these are costs, of course, for the defendent.

   iii. Prior information needed to estimate the likelihood of success.

   Consider a claim to be "litigated" if a verdict is rendered; otherwise, it is a settlement. Assume the expected payoff to be characterized by a single number, the *judgment*; this is what the defendant pays to the plaintiff in the case of a "liable" verdict. If the verdict is "not liable," then no money changes hands. It is assumed that there is a deterministic decision standard operating. That is, given the facts of the case, the fault of the defendant will be determined *according to the particular jury or judge*. The facts of the case include everything except the amount of money involved, even such factors as the racial prejudice of the judge or jury (see [55]). Note that although the decision function is deterministic it is not known; each rendered verdict constitutes a "measurement." Thus, if Y represents the fault of the defendant and if there is Y^* such that $Y < Y^* \Rightarrow$ "not liable" and $Y > Y^* \Rightarrow$ "liable," then the true value, Y', of fault must be estimated by both parties to the dispute in order to predict the likelihood of liability should a trial take place. Suppose that each forms an independent, unbiased estimate: the plaintiff's error, ϵ_p, and the defendent's error, ϵ_d, are independent RVs with zero mean and identical variance σ^2. Call these estimates Y_p and Y_d. Define $P_p = P(Y' \geq 0|Y_p)$ and $P_d = P(Y' \geq 0|Y_d)$. What information is needed to "know" how ϵ_p and ϵ_d are distributed?

(b) Describe the stakes of the dispute in terms of the expected judgment J. The direct costs are C_p and C_d, the litigation costs for plaintiff and defendant, respectively, and the settlements costs S_p and S_d. Let A be the plaintiffs minimum settlement demand. The plaintiff will decide on A by subtracting from the expected jury award the amount saved by not going to trial. The defendant will set the value of B by adding to the expected payout the extra costs that a trial would incur. Show that a sufficient condition for litigation is

$$P_p - P_d > \frac{(C_p + C_d) - (S_p + S_d)}{J} = D.$$

Explain why D should satisfy $0 < D < 1$. Show graphically the result that litigated disputes are not a random sampling of all disputes. What characterizes claims that go to trial?

(c) Show that σ small implies that the probability of plaintiff victory is close to 50%.

(d) There are several assumptions that could be relaxed. What if we allowed partial-compensation verdicts? What if the stakes are not equal for both parties?

23. *Flash cards* A commonly recommended method to learn, say the cases of noun declensions is to label cards with the cases and and then to associate the correct case with the form of a given noun. After one pass through, the cards are shuffled. This is continued until two consecutive passes through the cards are made without any mistakes. The noun has been "learned." Or an octopus could be presented with an opportunity to open the door to some food. It is in the *conditioned* state (C) if it has learned to open

the door, otherwise it is in the *unconditioned* state (C'). The simplest model would consist of a series of trials where the state has a certain probability of changing from C' to C at each trial. Once in state C it would stay there. Express this as a series of coin tossings. What is the probability of being conditioned by the nth trial? What is the probability that it takes n trials to learn the task? What is the mean time to conditioning? This last is almost obvious if you keep in mind the remark in the text about rates and waiting times.

24. *More learning curves* A slightly more complicated model for an experiment in pair-associate learning is presented in ([6]; see [23]). In that experiment, 29 subjects were asked to match the numbers 1 and 2 to ten pairs of letters, "1" being correct for five pairs and "2" for the other five. In the model we assume that a stimulus element (SE) can be in one of two states in each trial: C (conditioned), meaning that the subject has learned the correct association to the SE and will give the correct response without fail, or C' (unconditioned). Thus, SE starts out in state C' and at each trial the subject samples SE: if C then the correct response is given, if C' then there is a probability of g for the correct response. At each trial, if SE is in state C it stays, but if it is in state C', it has a probability c of changing to C.

(a) Show that the probability of an error on the nth trial is

$$(1 - c)^{n-1}(1 - g).$$

(*Hint:* An error on the nth trial means that it must still be in state C'.) This gives the so-called *learning curve.* How would we find the parameters c and g in practice? (See [6] for data.)

(b) Let T be the total number of errors for a given stimulus. Show that the mean of T is $\frac{1-g}{c}$. Show that $P(T = 0) = \frac{gc}{1-g(1-c)}$. (*Hint:* $T = 0$ in infinitely many ways: the subject guesses the first k responses and then learns ($C' \to C$) at the kth step.)

(c) Although the data match the learning curve reasonably well for very simple tasks, these models, known as all-or-none models because the state is always completely conditioned or not at all, fall short of describing slightly more complicated experiments. For example, learning may take place through a culling: after many trials, the subject may not know the correct response, but is able to eliminate some as unsuitable, having sorted the possibilities. How might you allow a model to incorporate such a process?

25. *Retrieval curves and memory models* Recall from Problem 4 in chapter 4 that when subjects were asked to freely recall facts their responses followed an exponential relaxation retrieval curve; but when a constraint was placed on the recall process, the curve was linear.

   (a) Assume that there are a large number of facts to recall (i.e, infinitely many). Assume that the recollections form a random stream of events (this means, for example, that the time to the next event is exponentially distributed, with parameter λ, say). Argue that, if each recollection is new (this is the constraint), then we have a random Poisson distributed stream of events. Show that the mean number of responses is a linear function of time.

   (b) When the pool of items accessible to recall is finite, say M, then the number recalled cannot be Poisson distributed, because the Poisson distribution is defined on the nonnegative integers. Argue that the correct distribution in that case is the *truncated Poisson* distribution, where each integer less than M follows Poisson statistics, but M must be given all the weight left over, that is, $\sum_{k=M}^{\infty} (\lambda t)^k / k! e^{-\lambda t}$. Use Taylor's theorem to write this as an integral. Estimate (in terms of M) how much this changes the mean from what you calculated.

   (c) When the constraint is removed and subjects are allowed to recall any possible response even if it has already occurred, then we can still model by sampling a random stream, but not every response will be counted. It can be shown that the probability of correct responses being n by time is binomially distributed:

   $$P(N(N(t) = n) = \binom{M}{n} p(t)^n (1 - p(t))^{M-n},$$

   where $p(t) = 1 - e^{-\lambda t/M}$ (for details see [23]). Note that $E(N(t)) = Mp = M(1 - e^{-\lambda t/M})$.

Chapter 7

Random Processes

What is the origin of the exponential laws for relaxation? Why does such behavior appear in so many different contexts?

1   Processes: Poisson Points and Random Walks

Poisson Points

As seen in Chapter 6, rare, random events are distributed according to the Poisson distribution. Let us begin this chapter by casting that distribution in a slightly different light; this new characterization of the Poisson distribution will allow it to have even more versatility as a model.

In the discussion of Chapter 6, we derived the Poisson distribution via a limiting process of sampling from binomial trials. If the random experiment consists of sampling points from a finite interval of the real line, then the probability distribution is the binomial distribution. The experiment consisting of sampling from two disjoint subintervals gives the multinomial distribution, as in Exercise 18 of chapter 6, with the result that the random variables that represent counting in the separate subintervals are not independent. However, in the limit of an infinite interval, the resulting distribution is the product of two Poisson distributions for the separate counts in each subinterval. In other words, *these counting RVs are independent*. This is a fundamental property of the Poisson distribution.

Closely related to this notion of independence is the Markov, or *forgetting*, property. If the probability for a waiting time is independent of the starting time and so cannot depend on the previous history, and furthermore depends only on the length of the wait, then this probability is exponentially distributed. Conversely, exponential waiting times imply Poisson counts for the number of events.

A Poisson distribution can be characterized by two properties: independence on nonoverlapping intervals, as discussed above, and a well-defined density, or rate. A density implies the existence of the following limit:

$$\lim_{\Delta t \to 0} \frac{P\{\text{one point in } (t, t + \Delta t)\}}{\Delta t} = \lambda. \tag{7.1}$$

Thus,

$$P\{\text{one point in } (t, t + \Delta t)\} = \lambda \Delta t + o(\Delta t),$$

and

$$P\{\text{two points or more in } (t, t + \Delta t)\} = o(\Delta t).$$

EXERCISE 1 Explain why

$$P\{\text{no points in } (t, t + \Delta t)\} = 1 - \lambda \Delta t + o(\Delta t).$$

If $N(t)$ is a RV that counts the number of points in $(0, t)$, then, by letting

$$P_k(t) = P(N(t) = k),$$

we have

$$
\begin{aligned}
P_k(t + \Delta t) &= P(N(t + \Delta t) = k) \\
&= \sum_j P(N(t, t + \Delta t) = k - j) P(N(t) = j) \\
&= P\{\text{no points in } (t, t + \Delta t)\} P_k(t) \\
&\quad + P\{\text{one point in } (t, t + \Delta t)\} P_{k-1}(t) \\
&= (1 - \lambda \Delta t) P_k + \lambda \Delta t P_{k-1} + o(\lambda \Delta t).
\end{aligned}
$$

And so, dividing by Δt and letting $\Delta t \to 0$, we get

$$
\begin{aligned}
\dot{P}_k &= -\lambda P_k + \lambda P_{k-1}, \quad k = 1, 2, \ldots, \tag{7.2} \\
\dot{P}_0 &= -\lambda P_0. \tag{7.3}
\end{aligned}
$$

This is a system of ordinary differential equations for the $\{P_k\}$, the probability distribution. To determine the solution, we need to specify an initial condition.

EXERCISE 2 What is an appropriate initial condition?

This is an infinite system, but it can be easily solved, with a closed form representation.

EXERCISE 3 Solve the system (7.2, 7.3) by "scaling the ladder": Solve the initial value problem:

$$\dot{P}_0 = -\lambda P_0, \quad P_0(0) = 1.$$

The equation for P_1 is then just a nonhomogeneous scalar DE:

$$\dot{P}_1 + \lambda P_1 = \lambda P_0(t), \quad P_1(0) = 0.$$

Now continue up the ladder.

A particularly quick way to get the distribution $\{P_k(t)\}$ is to use the generating functions introduced in problem 13 of the previous chapter. The generating function for the distribution is defined to be

$$G(s,t) = \sum_{k=0}^{\infty} P_k(t)s^k.$$

Taking the time derivative, we get

$$\dot{G} = \sum_{k=0}^{\infty} \dot{P}_k s^k \tag{7.4}$$

$$= \sum_{k=0}^{\infty} \lambda P_{k-1}s^k - \lambda \sum_{k=0}^{\infty} P_k s^k \tag{7.5}$$

$$= \lambda(s-1) \sum_{k=0}^{\infty} P_k s^k \tag{7.6}$$

$$= \lambda(s-1)G. \tag{7.7}$$

Solving this with the initial condition

$$G(s,0) = \sum_{k=0}^{\infty} P_k(0)s^k = 1,$$

we get

$$G(s,t) = e^{(s-1)\lambda t}. \tag{7.8}$$

This is the generating function of a Poisson distribution. To see this, notice that, by Taylor's theorem,

$$P_k(t) = \frac{1}{k!}\frac{d^k}{ds^k}G(0,t) = \frac{1}{k!}(\lambda t)^k e^{-\lambda t}. \tag{7.9}$$

[Or you can arrive at this just by expanding (7.8) in a power series.]

The explicit assumptions that lead to the system (7.2,7.3) are commonly justifiable, and this opens up a whole range of related models. The following exercise presents an example of this type of modeling.

EXERCISE 4 *Epidemic* Model the spread of a contagious disease in a large population by assuming that the disease is contracted through random contacts with the infected part of the population. Assume that the infected population $N(t)$ is relatively small so that such contacts are rare. At each contact with the infected, a person has probability p of contracting the disease. Show that these assumptions lead to a Poisson process. What is the mean rate of infection,

$$m(t) = E(N(t))?$$

Is your answer reasonable? (You may want to consult Exercise 21 of chapter 4.)

EXERCISE 5 Suppose that the rate of the Poisson process, as defined in (7.1), is time dependent: $\lambda(t)$. How does that change anything?

Stochastic Processes

A **stochastic process** is any family or collection of random variables, but will typically be ordered by a parameter α. Thus, $X(\alpha)$ is a RV for each value of the parameter α. The parameter space will typically be a subset of either the nonnegative integers or the nonnegative real numbers. Since the process is usually thought of as unfolding in time, we call the former a **discrete-time process** and the latter a **continuous-time process**. The Poisson process is an example of a stochastic process. More correctly, it consists of several processes: there is the collection of random points themselves, $\{t_k\}, k = 1, 2, \ldots$; thus, t_k is a RV whose outcome is the kth point. Then there is the waiting time between two consecutive events, $W^{(k)} = t_k - t_{k-1}$, which is exponentially distributed, independently of k. (This shouldn't be confused with $W_N = t_N - t_0$, the wait until the Nth event, which has a gamma distribution.) And finally there is the counting process itself, $N(t)$, which gives the number of events that occur in $(0, t)$ in a sample. This latter is a continuous-time process, while the former two are discrete-time. A feature of the formulation of the Poisson process, via $N(t)$, that leads to the system (7.2, 7.3), is that it makes no explicit reference to the associated discrete-time processes.

The values attained by the RV $X(\alpha)$ are referred to as **states**. Thus, since $N(t)$ can take on the values $0, 1, 2, \ldots$, it is referred to as a

discrete-state process, while W_N, for example, which can have any value in $(0, \infty)$ is a **continuous-state process**.

Strictly speaking, by making these definitions regarding stochastic processes, we have not created anything new: a review of chapter 6 reveals that our modeling theme was binomial trials, which can certainly qualify as a stochastic process. But the present reformulation indicates a shift in our focus. In Chapter 6 with t fixed, $N(t)$ is a random variable: for each random sampling, it is specified by the number of events. But the system of equations (7.2, 7.3) describe the evolution of chances. Many of the examples and problems in Chapter 6 can be expressed in terms of stochastic processes. There was the RV M_N, the net magnetization in the system of dipoles, and the variable R_N, which represents net position in the random walk, where, in both cases, N was the total number of coin tosses. In each case, we asked for the sample distribution with fixed N. Now we ask how does the probability distribution R_N evolve via local dynamics, meaning how does it change from step to step: our concern is the process that is the sequence of RV, $\{R_1, R_2, \ldots\}$. Notice that

$$n \to R_n$$

specifies a discrete-time discrete-state process. We are interested in the distribution

$$P(R_m = k) = P_k(m),$$

so that $P_k(m)$ is the probability that the drunkard is k steps to the right after m steps (or, in other words, $k + m/2$ heads are seen in m coin tosses). By the Markov property, the probability distribution at step $m + 1$ follows from the distribution at step m:

$$P_k(m + 1) = P(R_{m+1} = k) = \sum_j P(R_{m+1} = k | R_m = j) P(R_m = j).$$

Recall the experiment: a fair coin is tossed and, with probability $\frac{1}{2}$ either way, steps of fixed length are made to either left or right. Thus, the only nonzero conditional probabilities are

$$P(R_{m+1} = k | R_m = k - 1) = \frac{1}{2} = P(R_{m+1} = k | R_m = k + 1).$$

And so

$$P(R_{m+1} = k) = \frac{1}{2} P(R_m = k - 1) + \frac{1}{2} P(R_m = k + 1),$$

or

$$P_k(m+1) = \frac{1}{2}P_{k-1}(m) + \frac{1}{2}P_{k+1}(m). \qquad (7.10)$$

This is a discrete-time dynamical system and should be compared to the continuous-time dynamical system (7.2, 7.3). These systems are first order, which follows from the fact that both have the Markov property: in the discrete case (7.10), it means that the distribution at the mth step is completely determined by the distribution at the $(m-1)$st step.

EXERCISE 6 Extend the random walk to two dimensions.
(i) *Movement along a square grid* R_m is now a vector (an ordered pair gives position on the grid. You have (in equilibrium) equal chances of moving in any of the four directions, NW, SW, NE, SE. Model this as a sequence of trials of double coin flips, say a quarter and a nickel. The state of the quarter determines N or S, while the state of the nickel determines E or W. Find the dynamics. (Note that there are four nonzero conditional probabilities.)
(ii) *Movement along a grid of equilateral triangles* Model this as a sequence of die rolls. Again find the dynamics. (See Figure 6.11.)

These are examples of **Markov processes**. It should be emphasized that we are dealing with deterministic dynamical systems, but that the state variables are probabilities; so even when the initial state is known [and we mean deterministically; i.e., $P_k(0) = \delta_{kl}, k = 0, 1, 2, \ldots$], the resulting evolution expresses the probabilities for the underlying system to be in its various states at consecutive steps.

2   Markov Chain Models

A discrete-time discrete-state process such as the random walk is called a **Markov chain**. The variable R_n is referred to as the state variable. The dynamics will follow once we specify the conditional probabilities

$$P_{kj} = P(R_{m+1} = k|R_m = j), \qquad (7.11)$$

known as *transition probabilities* or transitions.

CAVEAT: *Our notation does not match that found in the literature of Markov chains. In such books, the transition in (7.11) would be written as P_{jk}, with the subscripts transposed. It was decided to write it this way because a student who has had a course in vector analysis or matrix algebra is used to multiplying a column state vector, representing input to a system, to the right of a system matrix, to get an output* *vector. It should be pointed out that to match the notation of an input–output diagram we could write xA instead of the usual Ax (Figure 7.1). In any case, the translation is easily made by transposing any matrices that appear.*

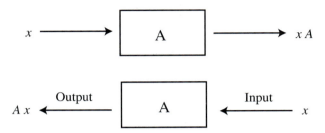

Figure 7.1 Our usual conventions of input-output systems and matrix algebra are not consistent.

In constructing Markov chain models it is convenient to refer to a pictorial diagram: states are represented by circles and an arrow from one state to another indicates a nonzero probability of transit from one to the other *in a single time step.* (Figure 7.2)

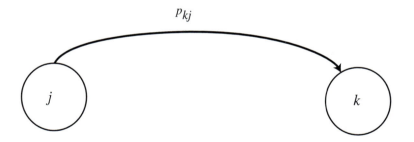

Figure 7.2 State transition diagram. P_{kj} is the conditional probability of entering state k from state j.

Corresponding to the transition probabilities, for a one-dimensional random walk,

$$P_{kj} = \begin{cases} \frac{1}{2} & j - k = 1 \text{ or } j + k = 1 \\ 0 & \text{otherwise} \end{cases},$$

is the diagram in Figure 7.3.

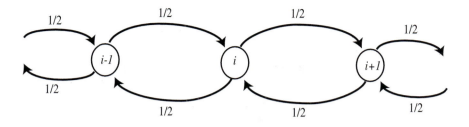

Figure 7.3 State transition diagram for the one-dimensional random walk.

The transition probabilities can be put in a matrix where j is the column index and k is the row index, so that the element at the k row of the j column is P_{kj}. For the random walk we have

$$P = \begin{vmatrix} \ddots & & \vdots & & \vdots & & \\ & 0 & 1/2 & 0 & 0 & \\ \cdots & 1/2 & 0 & 1/2 & 0 & \cdots \\ \cdots & 0 & 1/2 & 0 & 1/2 & \cdots \\ & 0 & 0 & 1/2 & 0 & \\ & & \vdots & & \vdots & & \ddots \end{vmatrix}.$$

The dynamics are

$$V_{m+1} = PV_m,$$

where V_m is matrix of unconditional probabilities at step m. In other words, the jth component is

$$V_{mj} = P(\text{state is in } j \text{ at step } m).$$

It is easy to see that the specification of the probability at any step depends only on knowing the distribution at $m = 0$. (Perfect knowledge of the initial condition, to be in state l for example, is written

$$V_{0j} = \delta_{jl}.)$$

EXERCISE 7 Show that

$$\sum_k P_{kj} = 1,$$

and explain that this is just what is needed so that probability distributions are transformed to probability distributions at each step. Show that 1 is always an eigenvalue of P, and that you can always find an eigenvector corresponding to eigenvalue 1 all of whose components are nonnegative (and so can be normalized to a probability distribution).

The matrix formulation is useful for state spaces that are finite. The random walk on a bounded domain with various boundary conditions is a clear prototype of Markov chain models. Modifications of the basic model appear in many of the models of this chapter as well as in Chapter 11, so we will break our teeth on what appears to be a toy model.

This simple model is a three-state random walk with reflection at each wall, whose diagram is shown in Figure 7.4. From this we can

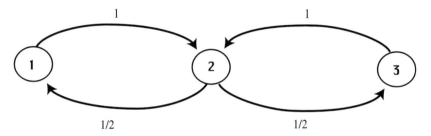

Figure 7.4 Three state random walk with reflection.

read off that the transition matrix is

$$P = \begin{pmatrix} 0 & 1/2 & 0 \\ 1 & 0 & 1 \\ 0 & 1/2 & 0 \end{pmatrix}. \tag{7.12}$$

Let us suppose that it is known that the drunkard starts at the left wall. (He has *zero* probability of being in the center or at the right wall and probability 1 of being at the left wall.) What is the probability of finding him at the left wall at any subsequent time?

The initial vector is

$$V_0 = \begin{pmatrix} 1 \\ 0 \\ 0 \end{pmatrix}.$$

We compute the subsequent vectors by multiplying with the state transition matrix:

$$v_1 = P \begin{pmatrix} 1 \\ 0 \\ 0 \end{pmatrix} = \begin{pmatrix} 0 \\ 1 \\ 0 \end{pmatrix},$$

$$v_2 = P \begin{pmatrix} 0 \\ 1 \\ 0 \end{pmatrix} = \begin{pmatrix} 1/2 \\ 0 \\ 1/2 \end{pmatrix},$$

and then we start repeating; indeed, we have the sequence

$$\begin{pmatrix} 1 \\ 0 \\ 0 \end{pmatrix}, \begin{pmatrix} 0 \\ 1 \\ 0 \end{pmatrix}, \begin{pmatrix} 1/2 \\ 0 \\ 1/2 \end{pmatrix}, \begin{pmatrix} 0 \\ 1 \\ 0 \end{pmatrix}, \begin{pmatrix} 1/2 \\ 0 \\ 1/2 \end{pmatrix}, \begin{pmatrix} 0 \\ 1 \\ 0 \end{pmatrix}, \dots$$

After the first step, the sequence locks into an alternation between two states. In fact, this type of behavior will occur for any choice of initial condition (except for one).

EXERCISE 8 Which is the one initial condition that will not give rise to an alternating sequence? What can we say about the the stability of that initial condition?

Since, for the matrix P,

$$P^3 = P, \quad P^4 = P^2, \quad \text{etc.},$$

the transition vectors will flip back and forth like a neon beer sign. Notice that we have a positive probability of reaching any state from any other state; this can be seen directly from the state transition diagram, since one can find a path joining any two states. (Such a process is called *ergodic*.)

EXERCISE 9 (i) Show that for any original distribution

$$\begin{pmatrix} p \\ q \\ r \end{pmatrix}$$

the alternating sequence is

$$\left\{ \begin{pmatrix} q/2 \\ p+r \\ q/2 \end{pmatrix}, \begin{pmatrix} p/2 + r/2 \\ q \\ p/2 + r/2 \end{pmatrix} \right\}.$$

(ii) Show that the eigenvalues are $\{1, -1, 0\}$. Why is there no initial distribution corresponding to the zero eigenvalue, that is, no valid state V such that

$$PV = 0?$$

(iii) By taking the average of the state over several cycles, we obtain the vector

$$\begin{pmatrix} 1/4 \\ 1/2 \\ 1/4 \end{pmatrix}.$$

What is this vector?

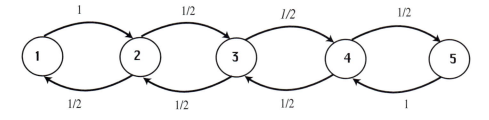

Figure 7.5

Extending to a five state system with the same reflecting boundary conditions, we obtain the state transition diagram in Figure 7.5, where the matrix is

$$P = \begin{pmatrix} 0 & 1/2 & 0 & 0 & 0 \\ 1 & 0 & 1/2 & 0 & 0 \\ 0 & 1/2 & 0 & 1/2 & 0 \\ 0 & 0 & 1/2 & 0 & 1 \\ 0 & 0 & 0 & 1/2 & 0 \end{pmatrix}.$$

The eigenvalues are $\{1, -1, 0, \frac{1}{\sqrt{2}}, \frac{-1}{\sqrt{2}}\}$. The -1 eigenvalue indicates that again we will have oscillation. And indeed, the transition matrix, after many steps, oscillates between

$$P_e = \begin{pmatrix} 1/4 & 0 & 1/4 & 0 & 1/4 \\ 0 & 1/2 & 0 & 1/2 & 0 \\ 1/2 & 0 & 1/2 & 0 & 1/2 \\ 0 & 1/2 & 0 & 1/2 & 0 \\ 1/4 & 0 & 1/4 & 0 & 1/4 \end{pmatrix}$$

and

$$P_o = \begin{pmatrix} 0 & 1/4 & 0 & 1/4 & 0 \\ 1/2 & 0 & 1/2 & 0 & 1/2 \\ 0 & 1/2 & 0 & 1/2 & 0 \\ 1/2 & 0 & 1/2 & 0 & 1/2 \\ 0 & 1/4 & 0 & 1/4 & 0 \end{pmatrix}.$$

If we average the state transition matrix over many cycles, we get

$$\frac{1}{2}(P_e + P_o) = \begin{pmatrix} 1/8 & 1/8 & 1/8 & 1/8 & 1/8 \\ 1/4 & 1/4 & 1/4 & 1/4 & 1/4 \\ 1/4 & 1/4 & 1/4 & 1/4 & 1/4 \\ 1/4 & 1/4 & 1/4 & 1/4 & 1/4 \\ 1/8 & 1/8 & 1/8 & 1/8 & 1/8 \end{pmatrix}.$$

EXERCISE 10 All the columns are

$$w = \begin{pmatrix} 1/8 \\ 1/4 \\ 1/4 \\ 1/4 \\ 1/8 \end{pmatrix}.$$

What is this vector?

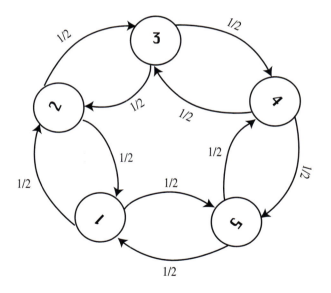

Figure 7.6 State transition matrix for periodic boundary conditions.

A contrasting symmetry is periodic boundary conditions. This state transition diagram for a five-state ring is shown in Figure 7.6, and the matrix is

$$P = \begin{pmatrix} 0 & 1/2 & 0 & 0 & 1/2 \\ 1/2 & 0 & 1/2 & 0 & 0 \\ 0 & 1/2 & 0 & 1/2 & 0 \\ 0 & 0 & 1/2 & 0 & 1/2 \\ 1/2 & 0 & 0 & 1/2 & 0 \end{pmatrix}.$$

As usual, 1 is an eigenvalue, but all the other eigenvalues satisfy

$$0 < |z| < 1.$$

What is the meaning of that?

Calculation reveals that the state transition matrix has a limit:

$$P^n \to \bar{P} = \begin{pmatrix} 1/5 & 1/5 & 1/5 & 1/5 & 1/5 \\ 1/5 & 1/5 & 1/5 & 1/5 & 1/5 \\ 1/5 & 1/5 & 1/5 & 1/5 & 1/5 \\ 1/5 & 1/5 & 1/5 & 1/5 & 1/5 \\ 1/5 & 1/5 & 1/5 & 1/5 & 1/5 \end{pmatrix}.$$

Notice that

$$\bar{P}w = w,$$

where

$$w = \begin{pmatrix} 1/5 \\ 1/5 \\ 1/5 \\ 1/5 \\ 1/5 \end{pmatrix}.$$

Any initial condition will tend to this state; if w_0 is any probability distribution, then

$$P^n w_0 \to w.$$

The vector w is the unique long-term *steady state*.

You can verify that

$$P^4 > 0$$

in the sense that all entries are positive. This means that every state has a chance to reach any other state in exactly four steps. It can be shown that when this occurs there will always be a stable steady state. Such a process is called *regular*. Regularity is a stronger property than ergodicity and expresses some kind of irreversibility. (For example, suppose that in the three-state system with reflection, matrix (7.12), you know that the mth step probability distribution is $(1/2, 0, 1/2)$. Is it possible to figure out what was the distribution at the $m - j$ step?)

The kth state is *absorbing* if

$$P_{kk} = 1$$

and *partially absorbing* if

$$P_{kk} > 0.$$

Let us consider now the case of a partially absorbing state in the three-state system (Figure 7.7).

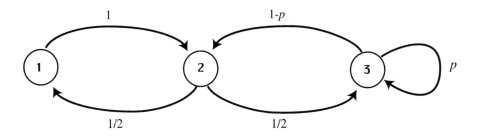

Figure 7.7 State transition diagram for system with partially absorbing state.

The system matrix is

$$P = \begin{pmatrix} 0 & 1/2 & 0 \\ 1 & 0 & 1-p \\ 0 & 1/2 & p \end{pmatrix}.$$

For $p = 0$, we have the previous case of reflection from both boundaries and the resulting oscillating probability distributions. If $0 < p < 1$, then P is regular, so there is a unique steady state w. However, if $p \sim 0$, then P is close to the matrix (7.12) and thus has an eigenvalue near -1; consequently, it will oscillate with extremely slow decay to the steady state w. (In the latter case, w will be close to a certain vector. Which one?)

EXERCISE 11 Show that the partially absorbing random walk whose state transition diagram is given by Figure 7.8 has identical properties to the ring.

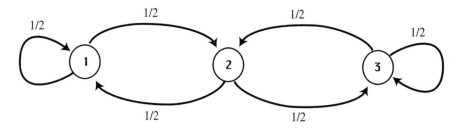

Figure 7.8

The Parking Lot Problem

A university is studying their parking situation. It has become con-
gested in the three lots on campus, and they are building an overflow
lot that is one-half mile from campus. The campus planners wish to
know how big to build the lot in order to accommodate the expected
overflow and whether the use of the lot will be so high that they should
offer a bus shuttle.

   One of the problems that they wish to eliminate is the unnecessary
traffic created by people who cruise around looking for a spot to park.
Let us concentrate on the flow problem for the time being. The university has conducted surveys and based on these has estimated transition
probabilities for movement.

EXERCISE 12 What could be the nature of such surveys? Notice that
although one could ask direct questions about the traffic, once the survey
actually goes out, it will probably have passed through several departments
and the set of questions will be both more comprehensive and less definite. For
example, there may be a question asking how likely drivers would be to use
the overflow lot when a shuttle is available, and so forth. What information
do you really need?

   These probabilities are averages over many different types of drivers:
some sit and wait while others quickly scan for open spaces and move
on.

   The states are denoted A, B, C for the present campus lots and K
for the overflow lot. There are two entrances to the university, near
the A and C lots, respectively. The transition probabilities are given
by the diagram in Figure 7.9.

EXERCISE 13 Comment on the assignment of probabilities. What could
we have learned from the surveys? Can you predict the long-term behavior?

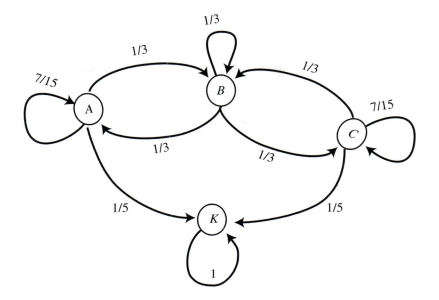

Figure 7.9 State transition diagram for the parking lot problem.

The transition matrix for this system is

$$P = \begin{pmatrix} 7/15 & 1/3 & 0 & 0 \\ 1/3 & 1/3 & 1/3 & 0 \\ 0 & 1/3 & 7/15 & 0 \\ 1/5 & 0 & 1/5 & 1 \end{pmatrix}.$$

Let us track the flow of an "average" driver. Because the survey shows that commuters are equally likely to enter campus through either entrance, the "average" driver has equal chances of being in lots A or C when first encountered and negligible chances of being in lot B. (Of course, there may be a significant number of drivers who would head directly to lot K, if it were built, in order not to have to bother searching for a space, even if they would have to walk.) So the "average" driver starts out with the distribution $[1/2, 0, 1/2, 0]$. After one time step, the distribution is

$$v_1 = Pv_0 = \begin{pmatrix} 7/30 \\ 1/3 \\ 7/30 \\ 1/5 \end{pmatrix}.$$

EXERCISE 14 What are the time scales? Is the size of a time step the time it takes to pass through one row of parked cars? To pass through every row in a lot?

After two time steps, the distribution is

$$v_2 = Pv_1 = \begin{pmatrix} 0.2200 \\ 0.2667 \\ 0.2200 \\ 0.2933 \end{pmatrix},$$

and after three time steps, we arrive at

$$v_3 = Pv_2 = \begin{pmatrix} 0.1916 \\ 0.2356 \\ 0.1916 \\ 0.3813 \end{pmatrix}.$$

Depending on what our time scale is, this may be as far as we want to take this crude model. Our "average" driver may have had a chance to visit every lot. From this we may wish to forecast lot usage. If we carry out further time steppings then we get

$$v_5 = \begin{pmatrix} 0.1471 \\ 0.1807 \\ 0.1471 \\ 0.5251 \end{pmatrix}, \quad v_{10} = \begin{pmatrix} 0.0759 \\ 0.0933 \\ 0.0759 \\ 0.7549 \end{pmatrix}, \text{ etc.,}$$

and eventually the fourth state absorbs all the weight.

The simple system diagram is not really suitable for long-term situations. One obvious criticism is that when people find open spots they don't get re-circulated but rather *stay* in a parked state. Spots open up at a certain rate, with people leaving, and so forth.

One remedy is to add extra states to represent parked states. These states, denoted AP, BP, CP are totally absorbing. In this way, the *macrostate* A is represented by the set of *microstates* $\{A, AP\}$. The new state vector is $[A, AP, B, BP, C, CP, K]$. And the transition probabilities can be put in the system diagram of Figure 7.10. There are now four absorbing states.

EXERCISE 15 How many eigenvectors correspond to the eigenvalue 1? (What is its multiplicity?)

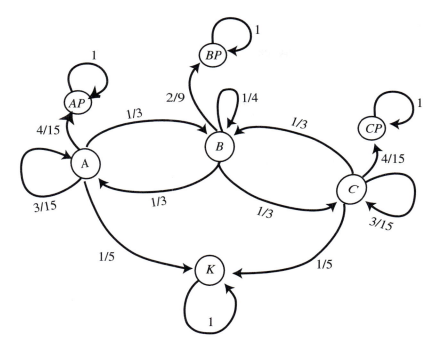

Figure 7.10

Suppose that a driver enters the parking lot system near lot A. Then his initial distribution is

$$w_0 = \begin{pmatrix} 1 \\ 0 \\ 0 \\ 0 \\ 0 \\ 0 \\ 0 \end{pmatrix},$$

reflecting the fact that, since we *know* he starts at lot A (and is not parked there), the probability of the first microstate is 1 (and all other probabilities are 0). What are the chances of finding him at any particular lot at ensuing times? We want to study the evolution of the distribution of probabilities that the driver is found in any particular lot. (We are eventually interested in the *frequency* of usage of the individual lots by many drivers; but you know how to interpret that frequency in terms of a probability of a single driver. Refer to the discussion in Chapter 6.) We compute the sequence

$$
v_1 = \begin{pmatrix} 0.2000 \\ 0.2667 \\ 0.3333 \\ 0 \\ 0 \\ 0 \\ 0.2000 \end{pmatrix}, \quad
v_2 = \begin{pmatrix} 0.1511 \\ 0.3200 \\ 0.1037 \\ 0.0741 \\ 0.1111 \\ 0 \\ 0.2900 \end{pmatrix}, \quad
v_3 = \begin{pmatrix} 0.0648 \\ 0.3603 \\ 0.0989 \\ 0.0971 \\ 0.0568 \\ 0.0292 \\ 0.2924 \end{pmatrix}, \ldots
$$

and after ten steps,

$$
v_{10} = \begin{pmatrix} 0.0027 \\ 0.4072 \\ 0.0034 \\ 0.1495 \\ 0.0027 \\ 0.0738 \\ 0.3608 \end{pmatrix},
$$

and we are within 3% of the steady-state distribution, which is

$$
\text{steady state} = \begin{pmatrix} 0 \\ 0.4091 \\ 0 \\ 0.1515 \\ 0 \\ 0.0758 \\ 0.3636 \end{pmatrix}.
$$

In consideration of the symmetry of the state transition diagram, we reach an analogous steady state if the initial vector is $w_0 = [0, 0, 0, 0, 1, 0, 0]$. What then is the steady state reached from $w_0 = [1/2, 0, 0, 0, 1/2, 0, 0]$? Since the system is linear, we simply take the average of the two previous cases, so in this case the steady state is

$$
\begin{pmatrix} 0 \\ 0.2424 \\ 0 \\ 0.1515 \\ 0 \\ 0.2424 \\ 0.3636 \end{pmatrix}.
$$

EXERCISE 16 (i) Notice that our system matrix satisfies $P^n \to \bar{P}$, but that the columns of $\bar{P}$ are not identical. Why not?
(ii) What recommendations do we make to the campus planners?

A finer modeling procedure would have to be more explicit about such things as times of passage. There are a number of directions to take. 1) One can add more microstates to represent different options open to the "average" driver. The model then begins to enter the realm of simulation, which we will discuss later in the chapter. 2) We can concern ourselves only with statistics: try to find explicit dynamics for means and higher moments. We could do this by explicitly keeping track of arrivals and departures. Continuous-time models of this sort will be considered in the next section of this chapter.

*Orbital Debris: Distribution

Since 1957, dozens of space agencies representing many more nations have placed satellites and other objects in earth orbit. A Vanguard spacecraft launched in March 1958 and still in orbit is one of approximately 25,000 objects that are large enough to be detected and catalogued, and there are many more than that too small to be seen. This small debris can result from spacecraft jettisons, solid rocket motor exhaust particles, fragmentation debris, peeling paint flakes, and other spacecraft surface deterioration. Some of these objects are racing around at speeds 10 to 100 times faster than a speeding bullet. A collision with a paint flake is thought to have caused a crater on a space shuttle window (see the photograph in [3]).

EXERCISE 17 Compare the kinetic energy of a 1-mm^3 paint flake moving at 10 km/s with that of a "speeding" bullet.

For the sake of future operations in earth orbit, it is important to know how the population evolves. This means both the spatial distribution of debris and its total number must be modeled. Models of the long-term evolution will have a different emphasis than such short-term models as are needed to predict the evolution of the bust-up clouds of fragments from spectacular collisions. It is interesting to note the modeling emphasis change in the late 1980s from long-term to short-term modeling, mainly as a result of the need to model the fragmentation clouds of Star Wars tests.

The broad picture then is that we have a continuum of possible orbits from ground level to the high geosynchronous orbits (GSO). We have a continuum of particle sizes from microscopic to large abandoned satellites. There is a wide range of velocities (up to 15 km/s). New missions create new debris, and more fragments are created (and eventually spread) through collisions, explosions, and deterioration. And particles in lower earth orbit (LEO) experience drag and eventually slow to lower than orbital speed.

EXERCISE 18 (i) How does orbital speed depend on orbital altitude for a circular orbit?
(ii) What happens if an object is placed at a certain altitude at a speed that is inferior to the orbital speed for that altitude?

One could envision a complicated deterministic model where each catalogued object was tracked with its mass and size noted, its orbit (apogee, perigee, and inclination) noted, and the whole population set in motion ruled by gravitation and suffering the effects of atmospheric drag. Every time there is a collision the resulting population must be recomputed and the evolution restarted. Even if one could compute such a model, there remains the problem of what to do with the uncatalogued small debris that can affect the orbits of larger objects as well as bring on destruction and explosion. This uncertainty means that any long-term evolution model will necessarily incorporate some probabilistic elements.

In order to get a handle on the long-term evolution of just the way things are distributed in LEO, let us consider a simple compartmental model. Suppose that our basic time step is one or more orbits so that the cloud from a collision resolves itself by placing particles into other compartments during a time step. Let us divide LEO into three regions Low (< 500 km), Medium ($500 - 1000$ km), and High ($1000 - 2000$ km). There will be two sizes of objects, Big (> 1 cm) and Small (< 1 cm), and, possibly, two momenta, fast and slow. [Why do we need different speeds when the altitude of an orbit will determine its speed? Due to collisions some objects will be sent into highly eccentric orbits. If the basic time step is long enough, then these objects will "behave" as though they are in lower (circular) orbit, mainly because that is where the drag is, and we can ignore the third category. However, remember that Kepler's law tells us that such objects spend most of their time in the higher orbit.]

The state variable is denoted X_{ijk} where the first index represents the orbital altitude, the second is size of the object, and the third gives the momentum bin, that is,

$$i \in \{L, M, H\}, \quad j \in \{B, S\}, \quad k \in \{f, s\}.$$

What does this variable represent? That depends on how it is used. We are going to use it to represent the *relative* amount of debris in the given compartment (at a given time). Now the 144 transition probabilities joining these 12 states must be estimated using mechanics, using known orbital data, by consulting short-term models of fragmentation clouds for resulting debris flux and atmospheric drag. In this context, note that the data on debris flux from collisions gives the probability of further collisions with the resulting debris and that a model for the atmospheric density must be supplied (recall from Chapter 3 that drag depends on the density of the air as well as the velocity relative to the air).

The issues of mission inputs and orbital decay must be addressed. These represent entry and exit from the debris population. There are several ways of dealing with these, none of which are completely satisfactory, thus betraying the inadequacies of our simple Markov chain model. One possibility is to create another state for objects in "limbo"; this state will absorb objects from orbital burn-out as well as inject objects into appropriate orbits. (In this way we could model *control* of the orbital environment.) Another possibility is to realize that since we are only concerned with the distribution of objects and do not intend to keep track of the total population, we can take the probability of death transition and spread it over the whole population. (This is done in an equitable fashion for the case of no mission inputs, but one can also reinject according to a nonequitable redistribution in order to simulate inputs.)

As an example, let us take the state represented by X_{LSf}. Because of its low altitude (where air density is greatest), small size (highest surface area per volume), and fast speed, an object in this state has a high probability of burn-up during a time step. If it doesn't burn up, it may end up in the same state, or drag could slow it down and it would enter state X_{LSs}. The partial flow diagram, which includes only flow *from* this state X_{LSf}, is given by Figure 7.11.

As there is a 1/3 probability of burning out in one step, this must be spread over all 12 states. In the case of no inputs, this means that we allow for an extra $\frac{1}{36}$ chance of entering any other state. We would spread this around disproportionately if there were planned orbital placements.

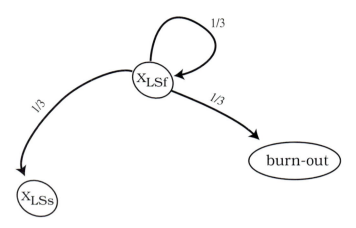

Figure 7.11 Partial flow diagram.

As a second example of the assignment of transition probabilities, consider flow from the state X_{HBs}. Collisions of large objects at high altitudes will tend to produce debris in lower altitudes and at lower speeds. This is reflected in the assignment of transition probabilities as indicated in the partial flow diagram of Figure 7.12.

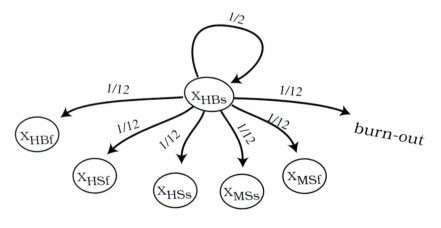

Figure 7.12

The $\frac{1}{12}$ probability of burn-out should be spread over the 12 states. The entire system matrix can be assembled from these partial (out-)flow diagrams. Instead of a complete 12×12 system matrix, we will complete the calculations using a 6×6 model, where we assume circular orbits, so that speed is determined by orbital altitude. The system matrix is shown in Figure 7.13 and the resulting evolution of distribution among the three altitude compartments is shown in Figure 7.14.

	LEO		MEO		HEO	
	Big x_1	Small x_2	Big x_3	Small x_4	Big x_5	Small x_6
x_1	0.3889	0.0833	0.2167	0.0278	0.0667	0.0278
x_2	0.3889	0.5833	0.2167	0.3611	0.0167	0.1944
x_3	0.0556	0.0833	0.3167	0.0278	0.0667	0.0278
x_4	0.0556	0.0833	0.2167	0.5278	0.0167	0.1944
x_5	0.0556	0.0833	0.0167	0.0278	0.6167	0.0278
x_6	0.0556	0.0833	0.0167	0.0278	0.2167	0.5278

Figure 7.13 The state transition matrix for the orbital debris distribution. The system is clearly regular and the powers of the matrix stabilize quickly, being within .01% of the steady state after fifteen iterations.

An important issue concerning orbital debris is the growth of the population. Will the present linear growth continue or will enough collisions swell the population to where it exceeds a critical size, where a "chain reaction" takes place with ensuing exponential growth? The model is this section cannot address such questions, so we will return to the orbital debris problem later.

3 Continuous-time Processes

Many populations, such as a culture of bacteria or any colony of cells, grow by random divisions of population members, who are also subject to random deaths. Let $X(t)$ be a RV giving the size of a population at time t. Growth by random divisions means, in this case, that each cell follows Poisson processes for division and death. That is, according to the Poisson hypotheses outlined in the first section of this chapter, we have the following transition probabilities:

$P\{$a given cell divides exactly once in $(t, t + \Delta t)\} = \alpha \Delta t + o(\Delta t)$,

$P\{$a given cell divides more than once in $(t, t + \Delta t)\} = o(\Delta t)$,

$$
w = \begin{bmatrix} 0.1061 \\ 0.3678 \\ 0.0786 \\ 0.1783 \\ 0.1223 \\ 0.1468 \end{bmatrix}
$$

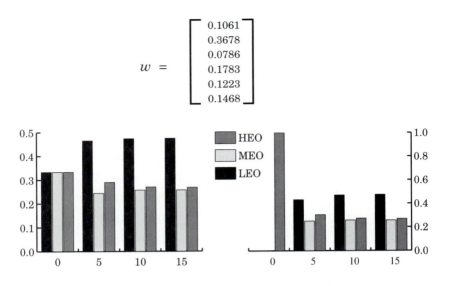

Figure 7.14 Orbital debris distribution simulation. The unique steady state vector is $\bar{w}$. The distributions in the three different orbital altitudes are displayed after $5, 10$, and 15 time steps for two different initial distributions. The evolution is non-monotone only because these are macrostates whose microstates approach steady state at different rates.

$$
P\{\text{a given cell dies in } (t, t + \Delta t)\} = \omega \Delta t + o(\Delta t),
$$

and so on.

EXERCISE 19 (i) What is the probability that a cell does not divide in $(t, t + \Delta t)$?
(ii) How would α and ω be found or estimated?

If we assume that the activities of separate individuals in the population are independent of each other, it is easy to derive the necessary parameters for the population as a whole. The population can increase by 1 if any particular cell divides; if n is the current population size, there are n ways for this to happen, so that

$$
\begin{aligned}
P\{\text{exactly 1 cell divides}\} \ &= P\{1^{st} \text{ cell divides}\} P\{\text{no other cell divides}\} \\
&\quad + P\{2^{nd} \text{ cell divides}\} P\{\text{no other cell divides}\} \\
&\quad + \cdots \\
&= n\alpha \Delta t + o(\Delta t).
\end{aligned}
$$

EXERCISE 20 A net increase of 1 will also be obtained if a cell divides twice (identifying the cell with one of its daughter cells) and a cell dies, or if two cells divide and a cell dies, and so on. You should be able to argue that, according to Poisson assumptions, the probabilities of all these events are $\circ(\Delta t)$.

Taking all this into account, then,

$$P\{\text{net increase of 1}\} = P\{\text{exactly 1 cell divides}\} + \circ(\Delta t).$$

Likewise, a net decrease is counted in the same way:

$P\{\text{net decrease of 1}\} =$

$\quad P\{\text{exactly 1 cell dies}\} \quad = P\{1^{st} \text{ cell dies}\}P\{\text{no other cell dies}\}$

$\qquad\qquad\qquad\qquad\qquad + P\{2^{nd} \text{ cell dies}\}P\{\text{no other cell dies}\}$

$\qquad\qquad\qquad\qquad\qquad + \cdots$

$\qquad\qquad\qquad\qquad = n\omega\Delta t + \circ(\Delta t).$

And the probability of no net change can be found analogously:

$P\{\text{zero net change}\} \quad = P\{\text{no cell divides}\}P\{\text{no cell dies}\}$

$\qquad\qquad\qquad\qquad + P\{\text{1 cell divides}\}P\{\text{1 cell dies}\}$

$\qquad\qquad\qquad\qquad + \cdots$

$\qquad\qquad\qquad = (1 - n\alpha\Delta t + \circ(\Delta t))(1 - n\omega\Delta t + \circ(\Delta t)) + \circ(\Delta t)$

$\qquad\qquad\qquad = 1 - n(\alpha + \omega)\Delta t + \circ(\Delta t).$

Putting all this together, if we set

$$p_n(t) = P(X(t) = n),$$

we have

$$\begin{aligned} p_n(t + \Delta t) \quad = \quad & (1 - n(\alpha + \omega)\Delta t)p_n(t) \\ & + (n - 1)\alpha\Delta t p_{n-1}(t) \\ & + (n + 1)\omega\Delta t p_{n+1}(t) \\ & + \circ(\Delta t), \end{aligned}$$

so in the limit $\Delta t \to 0$, we get the system of ordinary differential equations

$$\dot{p}_n = -n(\alpha + \omega)p_n + (n - 1)\alpha p_{n-1} + (n + 1)\omega p_{n+1}, \quad n = 1, 2, 3, \ldots. \tag{7.13}$$

If suitably interpreted, (7.13) also includes the dynamics for p_0:

$$\dot{p}_0 = \omega p_1. \tag{7.14}$$

Given suitable initial data, for instance given $p_n(0) = \delta_{nN_0}$ for a known initial population (or estimate therof), it is possible to solve the system (7.13, 7.14) explicitly. We shall not do that here (the computation of the probability generating function needs the theory of first-order partial differential equations and will be done in Chapter 10); rather, we will be content to compute certain statistics associated with the population. Particularly relevent (no surprise!) are the mean and variance, which can be written in terms of the population distribution as

$$\begin{aligned}
\bar{n} &= E(X(t)) = \sum np_n, & (7.15)\\
\sigma^2 &= E(X^2) - (\bar{n})^2 = \sum n^2 p_n - \left(\sum np_n\right)^2. & (7.16)
\end{aligned}$$

EXERCISE 21 Show that the mean satisfies the simple ODE

$$\frac{d\bar{n}}{dt} = (\alpha - \omega)\bar{n}. \tag{7.17}$$

Thus, by considering the population mean, we recover the deterministic population dynamics of equation (4.4) in chapter 4.

EXERCISE 22 (i) Derive a differential equation for

$$m_2 = E(X^2),$$

and use it to find σ^2. Show that when $\alpha = \omega$ we have $\sigma^2 = 2\alpha N_0 t$.
(ii) If $\alpha \simeq \omega$, then σ^2 is linear in t only for small times. Explain why the nondimensional number $\nu = \frac{\alpha}{\omega}$ is significant. Suppose that $\nu = 1 + \epsilon$. Estimate, in terms of ϵ, the time period over which the $\nu = 1$ statistics are valid.

EXERCISE 23 Suppose that growth of the population is by immigration, and not by division. How does that change the dynamics?

Service Facilities

The Air Force wants to establish a maintenance facility for a test wing in Alaska. The wing will consist of 100 aircraft. How many mechanics should be assigned to maintain and repair the planes? The nation's defense would presumably be compromised if aircraft are kept waiting long to be serviced, but, on the other hand, the Air Force wants to avoid paying for idle rep-airmen. On average, a plane will go λ^{-1} days without needing attention; if this time between pitstops is exponentially distributed, then the breakdown times form a sequence of Poisson points. Most repairs are minor, so we will start by assuming that repair time is exponentially distributed with density μ (but see below). Let

$$p_n(t) = P\{n \text{ aircraft are grounded at time } t\}.$$

Although the top brass has fixed the number of planes at 100 as being dictated by defense needs, let us denote it by a variable parameter M. The planes will be maintained by r support groups. A support group includes engineers, mechanics, technicians, and assistants. We want to know statistics concerning the number of grounded planes and the working schedules of support groups. For instance, how many will be without work, on average?

If $r = M$, there is the assurance of no wait, but that much personnel and equipment is expensive. On the other hand, if $r = 1$, then there will be large backups, based on repair data of modern aircraft.

Let $X(t)$ count the number of grounded planes at time t. If $X(t) = n < r$, then all grounded planes are being worked on and $r - n$ support groups are idle. If $X(t) = n > r$, then r planes are being worked on, $n - r$ planes are in line to the hangar, and $M - n$ planes are combat ready. The usual Poisson assumptions are in effect, so that the only relevant conditional probabilities,

$$P(X(t + \Delta t) = n | X(t) = k),$$

are $k = n - 1, n, n + 1$ (i.e., up to $o(\Delta t)$ there can only be changes of one unit in a time interval). The probability of increase of 1 is given by

$$
\begin{aligned}
P(X(t + \Delta t) = n \ \ & | X(t) = n - 1) \\
& = P\{1 \text{ net grounded in } (t, t + \Delta t)\} \\
& = P\{\text{exactly 1 malfunctions}\} P\{\text{none repaired}\} + o(\Delta t) \\
& = (M - n + 1)\lambda \Delta t + o(\Delta t),
\end{aligned}
$$

since the probability that any particular plane malfunctions in period $(t, t + \Delta t)$ is proportional to $\lambda \Delta t$, and there are $(M - (n - 1))$ planes flying and eligible for a mishap. In the same way,

$$\begin{aligned}
P(X(t + \Delta t) &= n | X(t) = n + 1) \\
&= P\{1 \text{ net repaired in } (t, t + \Delta t)\} \\
&= P\{\text{exactly 1 repaired}\} P\{\text{none break down}\} + o(\Delta t) \\
&= \begin{cases} (n + 1)\mu\Delta t, & n < r \\ r\mu\Delta t, & n \geq r \end{cases},
\end{aligned}$$

since at most r planes are being worked on at a time. And completely
analogously, we get

$$\begin{aligned}
P(X(t + \Delta t) &= n | X(t) = n) \\
&= P\{1 \text{ no net change in } (t, t + \Delta t)\} \\
&= P\{\text{no repairs}\} P\{\text{no break downs}\} + o(\Delta t) \\
&= \begin{cases} 1 - [(M - n)\lambda + n\mu]\Delta t + o(\Delta t), & n < r \\ 1 - [(M - n)\lambda + r\mu]\Delta t + o(\Delta t), & n \geq r \end{cases}
\end{aligned}$$

In the limit $\Delta t \to 0$, we get

$$\begin{aligned}
\dot{p}_n &= -[(M - n)\lambda + n\mu]p_n + (n + 1)\mu p_{n+1} + [M - n + 1]\lambda p_{n-1}, \\
&\qquad n = 1, \ldots, r - 1, \tag{7.18} \\
\dot{p}_n &= -[(M - n)\lambda + r\mu]p_n + r\mu p_{n+1} + [M - n + 1]\lambda p_{n-1}, \\
&\qquad n = r, \ldots, M - 1. \tag{7.19}
\end{aligned}$$

EXERCISE 24 (i) Show that, for $n = 0$,

$$\dot{p}_0 = -M\lambda p_0 + \mu p_1. \tag{7.20}$$

(ii) What is the equation for p_M?

The set of equations (7.18, 7.19, 7.20) constitutes a finite dimensional linear system of differential equations with constant coefficients,
so, given suitable initial data, a solution can be found in closed form.
However, this solution is too complicated to use to determine how certain statistics depend on the parameters r and M. For the problem at
hand, we are mainly interested in the steady-state solution. The matrix
defining the right side of (7.18, 7.19) has corank 1, so there is a one-
parameter null space and a unique nontrivial (normalized) probability
distribution solution.

Before we compute the steady state solution, let us define the model output variables that interest us. These are

$$G = \text{average number of planes}$$
$$\text{out of service}$$
$$= \sum_{n=0}^{M} n p_n, \tag{7.21}$$

and

$$D = \text{average number of planes}$$
$$\text{awaiting repairs}$$
$$= \sum_{n=r+1}^{M} (n - r) p_n, \tag{7.22}$$

and the average number of *idle* service units

$$J = \sum_{n=0}^{r-1} (r - n) p_n. \tag{7.23}$$

(*Note* : The average number of planes available for missions is $A = M - G$). We wish to relate these to our input variables λ, μ, M, r. Of course, λ and μ are computed from available data, and M is presumably fixed by defense needs, so r is the significant parameter of the model. The actual setting of r could be determined via optimization. For example, a defense-optimum solution could be to find minimum r that keeps D (or G) below some threshold. An economical solution could be to find the maximum r that keeps J below some tolerance. Actually, these functions are so closely related that it hardly matters how we phrase our actual criterion.

EXERCISE 25 Argue why the rates λ and μ are expected to appear in the form of the dimensionless parameter $\frac{\lambda}{\mu}$ in expressions for D and J.

Set $\dot{p}_0 = 0$ in (7.20):

$$0 = M \lambda p_0 + \mu p_1,$$

and solve for p_1:

$$p_1 = M \frac{\lambda}{\mu} p_0.$$

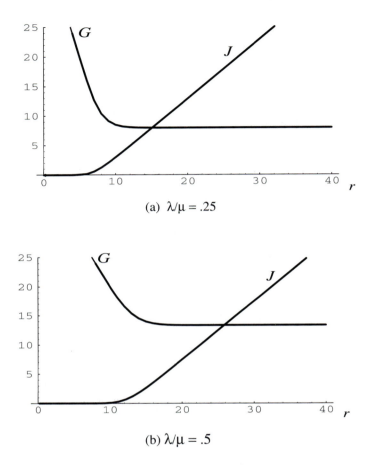

Figure 7.15 The functions G and J are plotted versus r.

In the equation resulting from $\dot{p}_1 = 0$, we can solve for p_2, and this procedure can be continued, so that

$$(n+1)\mu p_{n+1} = (M-n)\lambda p_n, \qquad (7.24)$$

for $n < r$, leading to

$$p_n = \binom{M}{n}(\frac{\lambda}{\mu})^n p_0, \quad n \le r. \qquad (7.25)$$

EXERCISE 26 Show that if $r = M$ then the explicit solution is found to be

$$p_n = \binom{M}{n}(\frac{\lambda}{\mu})^n (1 + \frac{\lambda}{\mu})^{-M}, \quad n = 0, \ldots, M.$$

If, in addition, $\lambda = \mu$, then

$$p_n = \binom{M}{n} 2^{-M}.$$

Interpret.

For $n \geq r$, the analogue of (7.24) is

$$r\mu p_{n+1} = (M - n)\lambda p_n,$$

leading to

$$p_n = \binom{M}{n} \frac{n(n-1)\cdots(r+1)}{r^{n-r}}(\frac{\lambda}{\mu})^n p_0, \quad n > r. \tag{7.26}$$

From the equation of normalization,

$$1 = \sum_{n=0}^{M} p_n,$$

we can solve for p_0 in terms of $r, M, \frac{\lambda}{\mu}$ and μ. Using this, we can find, via (7.25) and (7.26), $\{p_n\}$ and then compute values for G, D, and J. A simple computer program can take care of this. Figure 7.15 shows the results of this computation. It is clear that a good estimate of the parameter λ/μ is needed; then the correct number of repair units is practically dictated. It is interesting to see the sharp cutoff indicating the optimal choice.

In comparing these computations with the input data, we have to decide what to do with the fact that the repair times are *not really exponential*, but rather closer to a *delayed* exponential.

*Orbital Debris: Population Growth

For an understanding of the LEO environment, it is clearly inadequate to know just the distribution among the various altitude compartments. We also need to know the absolute number of objects in the population. Up to present times, the debris population has been growing at a linear rate. It is important to realize that, because of deterioration and collisions, the population can grow even with the cessation of all human activities in space, which, at any rate, is unlikely to happen because of the ever increasing need for new TV satellites. High-altitude orbits (> 2000 km) tend to be permanent, but because of the presence of atmospheric drag, objects in low earth orbit (LEO) are transitory in their orbits; they decelerate and eventually burn up. (There are a few objects that have reached the earth surface intact, but these are very rare, even when they are large). The growth of the population due to collisions is very much dependent on interaction in the population and thus depends in a nonlinear way on the population. Is there a "critical size" above which the chaining effect of collisions will lead to exponential growth?

We can model the orbital debris as a *vector* population, that is, using RVs $\{X_k\}$ where X_k counts the number of objects in the kth compartment. In this way,

$$X_k(t) = j$$

means that there are j things in the kth compartment at time t. Thus, a model with three orbital compartments and two size compartments would be represented by a 6-vector $\mathbf{X}(t) = (X_1(t), \ldots, X_6(t))$. Each RV can take on any integer value, so the state is a vector $\mathbf{x}$ that gives a point on the integer lattice in m-dimensional Euclidean space. For example, $(1, 2, 4, 10, 9, 8)$ means

1	in 1st compartment,
2	in 2nd compartment,
$\vdots$	
8	in 6th compartment.

We will denote the ith unit vector by

$$\hat{\mathbf{e}}_i = (\delta_i 1, \ldots, \delta_i m).$$

Instead of using multiple subscripts to denote probabilities, we will write them in functional form:

$$p(\mathbf{x}, t) = P(\mathbf{X}(t) = \mathbf{x}).$$

Besides inputs and decays from the population, for which the conditional probabilities are

$$P\{\mathbf{X}(t + \Delta t) = \mathbf{x} + \hat{\mathbf{e}}_i | \mathbf{X}(t) = \mathbf{x}\} = \lambda_i \Delta t + o(\Delta t),$$
$$P\{\mathbf{X}(t + \Delta t) = \mathbf{x} - \hat{\mathbf{e}}_i | \mathbf{X}(t) = \mathbf{x}\} = \mu_i \Delta t + o(\Delta t),$$

there are transfers from compartment to compartment. Under Poisson assumptions, only one object is transferred in a very small time interval. Thus, for $k \neq i$, we have

$$P\{\mathbf{X}(t + \Delta t) = \mathbf{x} + \hat{\mathbf{e}}_k - \hat{\mathbf{e}}_i | \mathbf{X}(t) = \mathbf{x}\} = \lambda_{ki}(\mathbf{x}, t)\Delta t + o(\Delta t). \quad (7.27)$$

Notice that these transfer rates are state and time dependent. The time dependence is due mainly to changes in the atmospheric density, the largest such effect is the 11-year sunspot cycle. The state dependence can be modeled in sophisticated ways using either kinetic theory or multiphase evolution models or more simply, using the ideas of the previous section of this chapter.

EXERCISE 27 Are Poisson assumptions valid at all here? Criticize them, keeping in mind the precise nature of 1) collisions, 2) explosions, 3) orbital decay.

As always in these Markov processes, we have

$$
\begin{aligned}
P\{\mathbf{X}(t + \Delta t) = \mathbf{x} \mid \mathbf{X}(t) = \mathbf{x}\} \\
&= P\{\text{no inputs}\}P\{\text{no decays}\}P\{\text{no transfers}\} \\
&= 1 - \left(\sum \lambda_k + \sum \mu_k + \sum\sum \lambda_{ki}\right)\Delta t + o(\Delta t).
\end{aligned}
$$

EXERCISE 28 Defining $X_0 = \sum_{k=1}^{m} X_k$, write down the dynamics for the probability distributions, where the vector $\mathbf{X} = (X_0, X_1, \ldots, X_m)$.

4 Beyond Markov

General Point Processes

By a point process, we mean the "throwing down" of points by some random process onto an underlying space X. It is easy to think of phenomena that fit this basic description.

1. Many of the stochastic processes that we've seen are "random streams" of points in time (the underlying space, the timeline, is R):

   (a) Poisson points

   (b) Cars passing a checkpoint on the highway

   (c) Chemical messengers arriving at a synapse

(d) Reversals of the earth's magnetic dipole moment

(e) Arrival of particles in a Geiger counter

(f) Arrival of customers at a pizza parlor counter

2. The underlying space can be R, but representing space instead of time:

   (a) The distribution of cars on the highway at a given time

   (b) The distribution of major movement along a fault line during a single earthquake

   (c) In pure mathematics: the distribution of the eigenvalues of random Hermitian matrices, or even the distribution of primes

3. The underlying space can be a plane:

   (a) "Blooms" in a Petri dish

   (b) The aerial bombing targets in a besieged city

   (c) The "hotspots" in an infrared scope,

   (d) The locations of a species of plant in a restricted ecosystem

4. The underlying space can be three-dimensional: the epicenters of earthquakes in the earth crust

5. A two-dimensional surface: the distribution of craters left by meteors on the surface of a moon.

Experimental data indicate that for the most part such processes do not have the statistics indicative of Poisson points with constant mean rate λ. Some may be Poisson with a nonconstant rate, but some cannot be modeled as Poisson points because the fundamental assumption of independence does not hold. Recall that a big concern with point processes is with the underlying *counting process*: given a subset A of the underlying space X, let

$$N(A) = \text{ number of points of the process in } A.$$

For example, we defined for Poisson points the counting process $N(t) = N([0, t])$, and that follows a Poisson distribution with parameter λt. A subsidiary concern is the relative spacing among elements of the process. For Poisson points the spacing between points follows an exponential distribution. This is special: usually, the distribution cannot be fully determined by analytic methods. One must usually be content to determine some statistics, that is, some of the lower moments of the distribution. We will concern ourselves solely with considerations of the *mean*. We will suppose a *mean rate* to exist, but will drop the independence property of Poisson points. Furthermore, we will specialize to one-dimensional processes. We say that the counting process $N(t)$ for a point process $\{t_k\}$ has **mean rate** (or *density* or *intensity*) $\lambda(t)$ if the mean of the counting process is always computed from the mean rate:

$$E(N(A)) = \int_A \lambda.$$

Notice that if $A = [0, t]$ then

$$\dot{m} = \frac{d}{dt} E(N(t)) = \lambda(t).$$

What are the ways of making new processes out of old, and what is the mean in each case?

1. For two independent processes on the same underlying space, we can consider the process obtained by combining them:

$$\{X_k\} \cup \{Y_k\}.$$

This process has the mean rate $\lambda_X + \lambda_Y$.

2. *Thinning or random deletion:* A point x_k is kept with probability p (and deleted with probability $1 - p$). In this case, $\lambda_p = p\lambda$.

3. *Markov shift:* A new process $\{X_k'\}$ is formed by shifting the points of a given process $\{X_k\}$ by samples from a RV T with density f:

$$x_k' = x_k + t_k.$$

The new process has the mean rate $\lambda' = \lambda * f$.

4. *Clustering:* Each point in a process gives rise to its own sequence of points. As an example, consider the aftershocks associated with a sequence of major earthquakes. Notice that thinning and shifting are special cases of clustering.

An example of shifting would be a Poisson stream $\{t_k\}$ with parameter λt, shifted by an exponential distribution with parameter μ. If the exponential RV is T, then each point in the new stream is

$$t_k' = t_k + \tau,$$

where τ is a sampling from T. A little thought will convince you that the independence assumption is conserved, and so this is a Poisson point process. However, the mean rate is not constant. Let's see if we can guess what this rate is. The "absolute number" of points doesn't change, but they are shifted forward by an average amount $\frac{1}{\mu}$ (recall that the mean of an exponential distribution is $\frac{1}{\mu}$). At time t, the count should represent the count from the original Poisson point process at time $t - \frac{1}{\mu}$. Thus, the intensity or mean rate should be

$$\lambda(t - \frac{1}{\mu}).$$

In fact, the true mean rate relaxes to this value.

EXERCISE 29 Show that the mean rate for this Markov-shifted Poisson process is

$$\lambda t + \frac{\lambda}{\mu}(e^{-\mu t} - 1).$$

This convergence to a Poisson process happens more generally (see [7], and [73]).

Queues

The idea of a queuing model is very natural. Think of customers entering an establishment where they can couple with a clerk if one is free; otherwise, they must wait in line. After being served, they exit. Queuing models are characterized by

1. a point process of *inputs*, or arrivals,

2. a distribution of residence times or service times, and

3. the number of servers

The notation $a/b/k$ is often used for queuing models. The first symbol a denotes the arrival process and is either M for a Poisson process (Markov) or G (general). The second symbol, b, is for the service times and is either M if they are exponentially distributed, or S if they are constant, or G. The third symbol gives the number of servers:

$$1 \leq k \leq \infty.$$

The service protocol must also be specified, but is usually *first-in, first-out*.

We cannot develop a theory of queues here (consult advanced texts such as [33]). We will simply point out that many of the models that we have seen can be reviewed as queuing models. Having made that connection, it is seen that this new way of looking at a phenomenon will suggest more comprehensive models, which, unfortunately, we will not be able to analyze.

Some examples of queueing models:

($M/M/1$) This is, in fact, identical with the birth–death process of Exercise 23. Let λ be the intensity of the arrival process and $1/\mu$ the mean service time. It is easy to see that the distribution

$$p_n(t) = P\{n \text{ customers in line at time } t\}$$

satisfies the following dynamics:

$$\dot{p}_k = -(\lambda + \mu)p_k + \lambda p_{k-1} + \mu p_{k+1}.$$

This is a system of linear differential equations with constant coefficients, and it is possible to find representations for the solutions of the initial-value problem. If the queue is functioning for a long time, we might expect it to reach a steady state.

EXERCISE 30 (i) Find the equilibrium distribution.
(ii) What is the probability of no wait?
(iii) Show that the conditional probability $P\{T \leq t | n$ already in line$\}$ is gamma distributed.
(iv) What is the distribution of waiting times? (*Hint* : condition on the number of customers already in line.) Does your answer surprise you?

($M/M/\infty$) This is the Poisson point process shifted by random exponentially distributed times. In the terminology of queuing models, we have already calculated the distribution of departure times.

($M/G/\infty$) This is a Markov-shifted Poisson point process.

EXERCISE 31 (i) Find the probability

$$p_k(t) = P\{ \text{ there are exactly } k \text{ in line at time } t\}.$$

(ii) Find the equilibrium distribution.

*Cell-cycle Modeling

Cell systems can be likened to a bee colony: there are queens and drones to whom falls the task of reproduction, but the vast majority of bees are the workers, whose daily round involves the upkeep of hive and production of honey. There is only one queen to a hive. There would be social chaos if there were more than one queen reigning or if there were too many drones competing for her attention. A similar division of labor holds in human cell systems: there are the *stem cells* concerned with reproduction and *differentiated cells* that have a specific job to do: skin cells, gut cells, nerve cells. A feedback mechanism keeps the stem cells active when a supply of cells is needed; otherwise, they are subject to inhibition (see Chapter 4). When the stem cells divide without regulation, the result is cancer. For example, bone marrow stem cells can either stay in a division cycle or differentiate to become other denizens of the blood, such as red blood cells, white blood cells, or platelet cells.

The ideal of chemotherapy is that by using drugs with effect specific enough to target those cells undergoing distrastrous unregulated division one may be able to kill "bad cells" without irremediably poisoning the others. But this is easier in principle than in practice. The dividing cells are not homogeneous in action, as would be the case if they were all either dividing or preparing to divide with the divisions occurring in synchrony. In fact, dividing cells go through a cycle of phases: after mitosis (M), there follows a preliminary growth phase (G_1), until DNA is replicated in a synthesis phase (S); then follows a secondary growth phase (G_2), during which enzymes untangle the replicated strands of DNA in preparation for the chromosome condensation and replication of the mitosis phase (M). It is possible to "peek in" on the cycle through various diagnostic means, using the characteristics of the different phases. For example, since thymidine is a component of DNA but not RNA, we can "watch" DNA synthesis via a thymidine tracer. In such a sampling procedure, a cell culture is exposed to radioactive thymidine and then one waits for it to appear; this is the sampled time of the G_2 phase ([31]). In another technique known as DNA fluorometry, one can distinguish cells with a single set of chromosomes (they are in the G_1 phase) from those with a double set (G_2, M) and those with a mixture (in the process of replication: S).

The drugs used in chemotherapy typically act in only one or two of the phases. One would like to know how to predict the entry to or exit from the phases in a cycle. For example, if one could know that at such and such a time there would be a large proportion of the "bad" cell population in a particular phase (X), then a drug specific to that phase could be used in such a way to reach its highest concentration at that time. The problem is made more difficult by growth inhibitions still working in the cycle: there is in some systems, such as bone marrow stem cells, another phase, (G_0), a resting phase, containing some pre-growth cells. These cells may be resistant to drugs while in this resting phase, and if there are many cells in this state, a low overall population may trigger recruitment of G_0 cells into the cycle. In this way the use of the drug will have precipitated an overall increase in the cancer population.

For a simple model we will follow ([34], [31], [35]). Suppose that cells are proliferating without differentiation (i.e., runaway stem cells: leukemia). We want to know the number of cells in each phase at a given time, $t > 0$, given an estimate of the distribution of the cell population at $t = 0$.

During mitosis, a cell divides and each daughter cell either (1) leaves the population, (2) enters a resting stage (G_0), or (3) begins the growth stage (G_1). A cell leaves the cycle either through death or differentiation. Thus, the cycle looks like Figure 7.16.

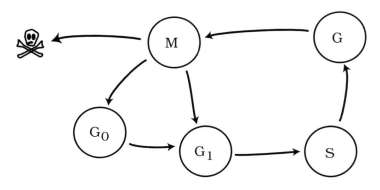

Figure 7.16 Flow diagram for the cell cycle.

Following ([34]), we will concentrate on a couple of time processes: (1) the instants that a cell leaves a particular phase, and (2) the instants that a cell enters a particular phase. The residence time of the ith phase, the time period between extrance to phase i and exit from phase i, is a RV R_i with density (pdf) given by $f_i(r)$. Thus, the relevant point processes are

1. $\{X_i^+\}$, the instants that cells enter phase i, and

2. $\{X_i^-\}$, the instants that cells leave phase i.

These processes have mean rates given by x_i^+ and x_i^-. So, if $N_i^+(t)$ counts the number of entrances to phase i by time t, then the expected number of such events is given by

$$E(N_i^+(t)) = \int_0^t x_i^+(s)\, ds.$$

Now, obviously, these two point processes are stochastically related to each other via the RVs $\{R_i\}$. There are two types of cells leaving phase i:

1. Cells that entered phase i at a previous time $t > 0$.

2. Cells that were already present in phase i at $t = 0$.

Cells of type 1 form a process that is random shifted by the residence time:

$$Y_i = X_i^+ + R_i.$$

This process has a mean rate given by

$$y_i = x_i^+ * f_i. \tag{7.28}$$

Cells of type 2 form a subprocess G_i of departure times with distribution density (pdf) $g_i(t)$. Since the process $\{X_i^-\}$ is just the disjoint union of the processes $\{Y_i\}$ and $\{G_i\}$, we have that its mean rate is

$$x_i^-(t) = y_i(t) + g_i(t). \tag{7.29}$$

Combining (7.28) and (7.29) then, we get

$$x_i^-(t) = (f_i * x_i^+)(t) + g_i(t).$$

We can close this to get equations for the $\{x_i^-\}$. From the flow diagram (Figure 7.16), we see that

$$\begin{aligned}
x_1^+ &= x_5^-, \\
x_2^+ &= x_1^-, \\
x_3^+ &= x_2^-.
\end{aligned}$$

Since there are actually two cells leaving M phase at each exit time, we should represent that as a process $2X_3^-$, with double the density, $2x_3^-$. A certain fraction (q_0) of these cells leave the cell population, another fraction (q_4) enter phase $4(G_0)$, so that

$$x_4^+ = q_4(2x_3^-),$$

while the remaining fraction (q_5) enters phase $5(G_1)$, so that the times of entering phase 5 are all times of exit from phase 4 plus a fraction (q_5) of the times of exit from phase 3. This could be denoted suggestively as

$$\{X_5^+\} = \{X_4^-\} \cup q_5\{2X_3^-\},$$

or, in terms of the mean rates, as

$$x_5^+ = x_4^- + q_5 2x_3^-. \tag{7.30}$$

Now we have a system of integral equations for the variables, $x_i = x_i^-$:

$$
\begin{aligned}
x_1 &= f_1 * x_5 + g_1, \\
x_2 &= f_2 * x_1 + g_2, \\
x_3 &= f_3 * x_2 + g_3, \\
x_4 &= f_4 * 2q_4 x_3 + g_4, \\
x_5 &= f_5 * x_4 + f_5 * 2q_5 x_3 + g_4.
\end{aligned}
$$

The statistics for the number of cells in each phase can be computed from the solutions to this system. If $N_i(t)$ is a RV that counts the number of cells in the ith phase, then the mean number of cells in phase i is

$$m_i(t) = \overline{N_i(t)} = m_i(0) + \int_0^t x_i^+(s) - x_i^-(s)\,ds. \tag{7.31}$$

Or, differentiating, we get that m_i satifies the equation

$$\dot{m}_i = x_i^+ - x_i^-. \tag{7.32}$$

Combining this equation with the system, we get

$$
\begin{aligned}
\dot{m}_1 &= x_5 - x_1, & (7.33)\\
\dot{m}_2 &= x_1 - x_2, & (7.34)\\
\dot{m}_3 &= x_2 - x_3, & (7.35)\\
\dot{m}_4 &= 2q_4 x_3 - x_4, & (7.36)\\
\dot{m}_5 &= x_4 + 2q_5 x_3 - x_5. & (7.37)
\end{aligned}
$$

To get a feel for what this system of equations might turn out to be as well as to understand the range of models that this outline covers, let us consider the special case where the times of passage (both f_k and g_k) are exponentially distributed with a parameter μ_k that depends only on the phase. (Then all departures and arrivals are Poisson streams.) In that case, it is easy to see (by differentiating) that the system for the mean rates uncouples, and we get

$$\dot{x}_k = \mu_k(x_k - \mu_k), \quad k = 1, 2, 3, 4, 5,$$

so each mean rate undergoes exponential relaxation to its equilibrium value. Thus, the long time process is a homogeneous Poisson process, which shouldn't surprise you after the discussion at the beginning of this section. (A formulation of the cell cycle from the viewpoint of a Poisson process is left for the problem section.) The discussion at the beginning of this section shows that, in this case, we have

$$x_k^- = \mu_k m_k,$$

and so the system (7.33) becomes

$$\dot{m}_1 = \mu_5 m_5 - \mu_1 m_1, \tag{7.38}$$
$$\dot{m}_2 = \mu_1 m_1 - \mu_2 m_2, \tag{7.39}$$
$$\dot{m}_3 = \mu_2 m_2 - \mu_3 m_3, \tag{7.40}$$
$$\dot{m}_4 = 2q_4\mu_3 m_3 - \mu_4 m_4, \tag{7.41}$$
$$\dot{m}_5 = \mu_4 m_4 + 2q_5\mu_3 m_3 - \mu_5 m_5. \tag{7.42}$$

EXERCISE 32 Show that similar equations can be derived under the assumption that entrances and exits are Poisson distributed but with a nonconstant rate.

There are ways of estimating these times of passage in real systems. Unfortunately, "real" in this context usually means data from cell cultures grown in the lab, and this data is known to conflict strongly with data from (rarely available) cells grown in an animal (but, *in vivo veritas*). The distribution of times of passage in many cases look like the distributions for repairs times in the service facilities problem, but in other cases are more symmetric.

5   *Simulation and the Monte Carlo Method

It has been indicated at several places in this chapter that it is sometimes necessary to use a computer to complete the analysis of a model.

The term *simulation* is used in two senses. One refers to the computer solution of equations and the general plan of making inference from a qualitative, and often visual, study of these solutions. The other sense is the estimation of certain parameters by sampling appropriate random variables. This latter technique is used sometimes even when the parameter does not arise in a probabilistic model. (An example of this is the Monte Carlo method of integration mentioned at the end of Chapter 6.)

In practice, these two uses of simulation are typically combined. There are even special high-level computing languages that combine differential equation solvers and statistical samplers with visual interfaces.

We can illustrate a possible simulation procedure by a *direct* simulation of the service facility. Recall that airplanes break down at a rate that is Poisson distributed and are then attended to by one of r different service groups. A sequence of breakdown times

$$b_1, b_2, b_3, \ldots,$$

is easily generated, since the interval $b_k - b_{k-1}$ is exponentially distributed, independently of k. These times can be generated in advance and stored or else generated as needed. At each breakdown time, one must check for free repair groups to which the airplane may be assigned; a repair time is then calculated. If there aren't any free groups, then it must be put into the queue. If the repair times are distributed according to a graph such as in Figure 7.17, they are perhaps best simulated using a rejection-type method (see Chapter 6).

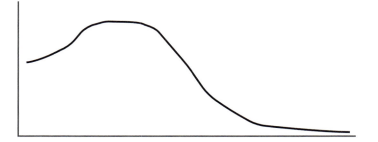

Figure 7.17 Distribution of residence times in repair queue.

It is possible in simulation to consider complications that are way beyond analytical treatment. For example, in this service facility we can distinguish different types of servers, and their associated repair times can be distributed differently. One can even distinguish among types of arrivals, that is, among different types of malfunctions, and have different servers specializing in each of these problems. In short, we can have in our computer a simulacrum of reality to whatever degree our resources can handle. A comprehensive simulation of this sort typically will rely on visual diagnostics rather than statistical diagnostics, showing the contents of different populations and service queues for different time steps, although statistics can be kept. A simplified simulation to estimate, for example, the mean queue length can be arrived at by

Doing N_1 times:
>   Stepping through N_2 arrivals:
>   >   check if any departures since last arrival
>   >   >   if so, decrease Q by 1.
>   >   If all service groups are busy $(Q > 0)$,
>   >   >   place in Q (increase by 1);
>   >   >   otherwise, generate a departure time.

Let's suppose that we are interested in the average queue size. We can store this value at the end of each sample trial: $Q = Q(N_2)$; so that we a sequence

$$Q_1(N_2), Q_2(N_2), \ldots, Q_{N_1}(N_2),$$

and we can use the estimate

$$\frac{Q_1 + Q_2 + \cdots + Q_{N_1}}{N_1}$$

for the average queue length. Methods of statistical inference can be used to determine confidence intervals and the like (see the problem section of Chapter 6).

When we get to systems of the order and complexity of real matter (10^6 individual atoms, say), then direct simulation is out of the question. A direct simulation would involve the laws of molecular mechanics and all interactions (although we could prepare the calculation by making a million atoms look like just a few molecules).

For example, equilibrium thermodynamics says that a system at temperature T has energy E with a probability proportional to $e^{-E/kT}$. Thus, if ϕ is some parameter defined on phase space (position $\times$ momentum), its mean is calculated from

$$\bar{\phi} = \frac{\int \phi(p, q) e^{-E/kT} dp dq}{\int e^{-E/kT} dp dq}.$$

A Monte Carlo method could be used to calculate this through sampling.

As an example of a Monte Carlo simulation in statistical mechanics, let's look at the Ising model. Imagine an array of magnetic dipoles as in Chapter 6, but in two or three dimensions. The equilibrium state of a particular dipole [either Up (+1) or Down (-1)] is not just randomly distributed according to the binomial distribution, but is determined in relation to its nearest neighbors on the lattice, taking into account the probability $e^{-E/kT}$. The Ising energy is

$$E = -J \sum \sum S_k S_i - B \sum S_k,$$

where $S_k = \pm 1$ are the dipole states, and the double sum is taken over pairs of nearest neighbors only. For example, on a plane grid, there are only four terms involving a particular index n: (N, E, S, W) (Figure 7.18).

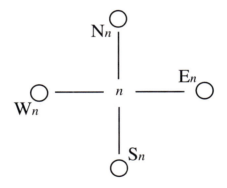

Figure 7.18

Each term is either $-J$ if the two neighbors have the same sign (point in the same direction) or $+J$ if they are oppositely oriented. Thus, changing the relative orientation between two next-neighbors changes the energy by the amount $2J$.

EXERCISE 33 Show that the minimum-energy configuration for the lattice at $T = 0$ (minimizing, that is, E given above) is a uniform lattice with all spins pointing the same way.

For various reasons, the equilibrium configuration at $T > 0$ will depart from the uniform minimum-energy configuration. One is that for $T > 0$ there is a finite probability that a spin will change its orientation

even if it is not, according to E, energetically favorable to do so. The
other reason is that, in the approach to equilibrium, local centers of
collusion will abut one another: this is the growth of *domains*. A Monte
Carlo simulation can deal directly with the fluctuations and furnishes,
as well, a way to observe the formation of domains.

The computational method is quite simple: one cycles through all
the sites and checks the energy change ΔE that would result from
changing the orientation of the spin at that site. If $\Delta E < 0$, then go
ahead and flip; otherwise, compare $p = e^{-\Delta E/kT}$ to a random number
$0 < \gamma < 1$, and flip if and only if $p > \gamma$. (This latter is a *fluctuation*, a
random push up-energy-hill.) Keep cycling until little further change
is noticed. (This is an example of a relaxation method.) Recall that
the magnetization is $M = \sum S_k$. A parameter that can be calculated
from the Monte Carlo simulation is the magnetization density

$$m = M/N,$$

where N is the number of sites. During the simulation, we can start
with a random orientation and compute $M(T)$ for many runs and calculate an estimate.

An interesting feature of the Ising system (which qualifies it for
a simplified model of true magnetism) is the existence of a transition
temperature T_C. Above this temperature, the magnetism is zero; but
for $T < T_C, M = M(T) > 0$ (Figure 7.19). This T_C is difficult to

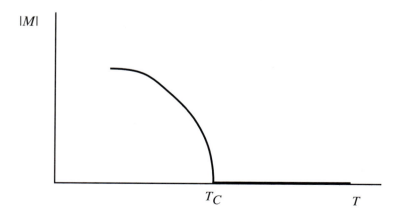

Figure 7.19 The ferromagnetic phase transition.

find solely by simulation and other methods are needed ([86]). For real magnetic materials, T_C is called the Weiss–Curie temperature; the value of T_C for iron is 1050 K (see [11]).

The Ising-type models are related to the so-called cellular automata (CA). These are discrete dynamical systems defined on a grid. The state at each site of the grid can take on one of finitely many values, and this state, at time $t+1$, is determined by the states of its neighbors at time t. This determination can be functional, in which case this is a deterministic cellular automaton, or else involve some probabilitic decisions, which makes it a stochastic cellular automaton. Also closely related to these types of simulation are diffusion and percolation, discussion of which we leave to the problems and Chapter 11.

6   Problems and Recommended Reading

1. Derive the mean rates for the point processes in section 4.

2. *Learning models* This problem is another look at the models of Problems 21 and m 22 of Chapter 6. Those original learning models were a sequence of steps among the two states: C, the conditioned state, and C', the unconditioned state. Notice that C is an absorbing state. Associated to the state vector

$$\begin{pmatrix} C \\ C' \end{pmatrix}$$

is the error vector $(0, 1 - g)$, where g is the guessing probability. What operation on the state gives the probability for error? What is the matrix for transition between states? What is the limit of the powers of the transition matrix? Supposing that there is a small probability of forgetting, of transition from C to C', what is the limit of the powers of the resulting transition matrix?

3. *More learning models*

   (a) A generalization of the learning model of the previous problem can be found by allowing an intermediate state, I, that one must pass through in order to reach C. What is the appropriate state transition matrix? What is the probability of error? How does this model compare with the one in the previous problem?

(b) Another generalization is arrived at by supposing that there are two (or more) stimulus elements (SE) involved in the learning of a task (pattern). Each can be in either the conditioned or unconditioned state. Associated to each is its own guessing probability. This is now a four-state model. What is the state transition matrix? Compare qualitatively the learning curves produced by this model with the simpler model.

4. *Chemical reaction* Suppose that a molecule of a species A reacts with a molecule of B to form $A + B$. Let $X(t)$ be the amount of A present at time t. Model this with a death-type Markov process, where the transition probability is state dependent: we assume the Poisson hypotheses, where the probability of any pair of molecules reacting in some small time interval is proportional to the product of the amounts of reactants available. What are the resulting dynamics? What equation determines the mean evolution?

5. Consider telephone calls on a network with infinitely many lines. Suppose that the starts of calls are Poisson points with mean rate λ, and the terminations of calls are Poisson points with mean rate μ, and they are independent processes. If $P_n(t)$ is the probability that exactly n lines are in use at time t, derive the following dynamics

$$\dot{P}_n = -(\lambda + n\mu)P_n + \lambda P_{n-1} + (n+1)\mu P_{n+1}.$$

Show that the mean number of calls, m, satisfies

$$\dot{m} = \lambda - \mu m,$$

and thus show exponential relaxation to the value $\frac{\lambda}{\mu}$. What is the steady-state distribution, $\bar{P}_n$? Notice that this is a birth–death process; it is also a queuing process. What if only a finite number of lines are available? Note that here, if no lines are available, there is typically no "queue", but that a request will be rejected and must call back later. Further discussion of such problems can be found in the book [18].

6. *A loss for words* Can we decide when two languages diverged by looking at the loss of relative cognates? Two words from separate languages are *cognates* if they have a similar structure and

meaning. Examples are *käse* and *queso* in German and Spanish, and *bratr* and *brother* in Czech and English. Model this as a death process, the population is the collection of cognates shared by the two languages. A word "dies" when it is replaced by a noncognate; the death rate is Poisson distributed with parameter μt. What are the appropriate dynamics? How could we estimate the parameter μt? Is the implied assumption of independence of the two languages really valid? Notice that we have here an inverse problem: we know the population of cognates at the present time, and we want to know how long it is since the two languages diverged. Given an observation of the present state of the shared-cognate population, choose t that makes the probability of this observation the greatest. For details and references, consult [23]. See also [20].

7. Express the single-server queue (M/M/1) as a birth–death process with birthrate λ and deathrate μ. If $P_n(t)$ is the number in the queue at time t, show that the equilibrium (steady-state) distribution is

$$\bar{P}_n = P_n(\infty) = \{\frac{\lambda_0 \lambda_1 \ldots \lambda_{n-1}}{\mu_1 \mu_2 \ldots \mu_n}\} = (1 - \frac{\lambda}{\mu})(\frac{\lambda}{\mu})^n.$$

8. Carry out the simulation of the orbital debris population time evolution.

9. The parks commission is planning to open a beach to the public. The beach is on an island. Roads will have to be built, as well as a parking lot and a connecting bridge to the mainland. One of the issues in deciding which kind of bridge to build is the frequency of traffic backups on the bridge, due to the fill-up of the parking lot. Model this as a queuing process. Do the arrivals form a stationary process (where the properties are independent of time)? What are the parameters of interest in this problem?

10. *Pattern Recognition* A machine has to learn to classify patterns as belonging to one of two classes. It doesn't know the rules for membership in the classes, but during a sequence of trials it is told whether its guess at each trial is correct or not. The machine is "learning" if the frequency of errors decreases with increasing time. Suppose that there is a real function, f, that splits the two classes in the sense that the smallest value assigned

to a member of the first class is strictly greater than the largest value assigned to a member of the second class. If the machine had perfect knowledge of this function, it could make perfect guesses. One way to "learn" is to turn the guessing trials into a sequence of estimates for this function. What is a simple situation where something like that can be programmed to wor? A general convergence result is presented in [2].

11. A simple Monte Carlo procedure to estimate the time that comets stay in elliptical orbit about the sun is presented in [26]. Each pass through the solar system allows perturbations by Jupiter and Saturn to pump it up to a hyperbolic orbit, from which it never returns. The energies of elliptical orbits are taken to be negative, while hyperbolic orbits have positive energy. The energy of an elliptical orbit is inversely proportional to its semimajor axis, and the cube of the semimajor axis is proportional to the square of the period of the orbit. If we assume that the perturbation at each pass, η, is a normally distributed RV, then the sequence of energies is a sequence of RVs:

$$-\varepsilon_0, \quad -\varepsilon_1 = \varepsilon_0 + \eta_1, \ldots$$

The total lifetime of the elliptic orbit is proportional to

$$G = \sum_{1=0}^{T-1} \varepsilon_i^{-3/2},$$

where T is when energy changes sign. What is the distribution of G, given ε_0?

Chapter 8

Complex Systems

Mrs. Sova is sitting in a cafe in the warehouse district. A ventilation grill across the street has grabbed her attention. The slats on the grill are dancing together in a rhythmical way. Sometimes they move apart, sometimes together, but there is a definite period to it, although it is complicated. Surely a big machine is breathing in and out in a complicated manner. When she crosses the street to investigate, she is surprised to find that the exhaust is coming out in a steady stream.

1 Coupled Oscillators

This chapter will be concerned with the modeling of phenomena composed of two subsystems. When these subsystems are not interacting with each other, their behavior is simply and easily modeled. This is in contrast with the complicated and somewhat surprising behavior that the coupled system can display. Coupling just two simple systems in this way affords models for a wide range of physical and biological phenomena. We will investigate nonlinear systems and review bifurcation, hysteresis, and harmonics. We will look at the circumstances of lock-in or synchronization. And we will see a system that exhibits nonperiodic, or chaotic, action.

But first let's tend to the mathematics of two simple oscillators with a linear coupling.

Figure 8.1 shows two pendula that swing about a common rod. They are coupled tightly to the rod, so that the rotation of one pendulum will, through the elastic torsion in the rod, exert a torque on the other pendulum. The force balance for each pendulum is

$$l\ddot{\theta} = -g\sin\theta + \Upsilon/m, \tag{8.1}$$

where l is length of the pendulum, and m is its mass. The torque Υ is exerted on the pendulum by the elastic restoring force. We have the system of equations

$$l_1\ddot{\theta}_1 + g\sin\theta_1 = \Upsilon_{21}/m_1,$$
$$l_2\ddot{\theta}_2 + g\sin\theta_2 = \Upsilon_{12}/m_2,$$

where Υ_{21} is the torque on oscillator 1 via the connection. Symmetry dictates that we have

$$\Upsilon_{21} = -\Upsilon_{12}.$$

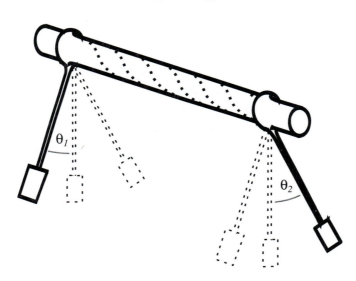

Figure 8.1 Torsion pendulum.

Let us suppose that the elastic restoring force is linear so that it satisfies Hooke's law:

$$\Upsilon_{21} = k(\theta_2 - \theta_1).$$

(Of course, the restoring force could be a nonlinear function of the difference of the two angles. Also, it could depend on the relative velocities, effectively acting as a damper.)

Thus, we have the coupled system of oscillators:

$$l_1\ddot{\theta}_1 + g\sin\theta_1 = \Omega_1(\theta_2 - \theta_1), \qquad (8.2)$$
$$l_2\ddot{\theta}_2 + g\sin\theta_2 = \Omega_2(\theta_1 - \theta_2), \qquad (8.3)$$

where

$$\Omega_j = k/m_j.$$

The phase space of this system is four dimensional. The time trajectory of any motion traces out a curve in this phase space:

$$t \rightarrow (\theta_1(t), \theta_2(t), \dot{\theta}_1(t), \dot{\theta}_2(t)).$$

This is a conservative system. The potential energy is the sum of the individual potentials plus the interaction potential:

$$
\begin{aligned}
U &= U_1 + U_2 + \frac{k}{2}(\theta_2 - \theta_1)^2 \\
&= m_1 g(1 - \cos\theta_1) + m_2 g(1 - \cos\theta_2) + \frac{k}{2}(\theta_2 - \theta_1)^2
\end{aligned}
$$

The kinetic energy is the sum of the individual kinetic energies:

$$
\begin{aligned}
T &= T_1 + T_2 \\
&= \frac{1}{2}m_1 l_1 \dot{\theta}_1^2 + \frac{1}{2}m_2 l_2 \dot{\theta}_2^2.
\end{aligned}
$$

The total energy is

$$E = T + U.$$

A simple computation resulting in

$$\frac{dE}{dt} = 0$$

shows that energy is conserved. Can we conclude, as we did for a conservative system with one degree of freedom, that the system trajectories are closed curves? No. However, when the energy is conserved,

$$E = E_c,$$

then it is easy to show that the trajectories must stay in a region that satisfies

$$U \leq E_c. \tag{8.4}$$

This says that, whatever the actual range of the trajectory its extent must be such that its projection onto its first two coordinates satisfies (8.4).

EXERCISE 1 Does (8.4) necessarily specify a bounded region?

We can get a hint of the phase space behavior by looking at the *linear case*, which we will do later in this section.

Studying the energy is particularly useful when there is damping. Let's suppose that the force balance for each oscillator, (8.1), includes on the right side a viscous damping force that is proportional to the velocity:

$$F_{\text{damping}} = -\lambda\dot\theta.$$

Then, when we differentiate the energy, we get

$$\frac{dE}{dt} = -\lambda_1\dot\theta_1^2 - \lambda_2\dot\theta_2^2,$$

which means that

$$\frac{dE}{dt} \le 0$$

along all the trajectories. In fact, the energy must necessarily decrease, except at those times when

$$\lambda_1\dot\theta_1^2 + \lambda_2\dot\theta_2^2 = 0.$$

During oscillation, this will happen only at isolated instances, so over any finite time interval, the energy will have decreased. What can we conclude concerning the trajectories? Since the energy is always decreasing the motion is always eventually headed for a minimum of the energy. These minima in phase space are all equivalent and satisfy

$$\theta_1 = \theta_2 = \dot\theta_1 = \dot\theta_2 = 0.$$

In other words, the motion dies out.

EXERCISE 2   Why can we have

$$\frac{dE}{dt} = 0$$

only at isolated instances?

If the only damping comes from the rod that joins each pendulum, then the viscous damping force is proportional to the relative velocities of the two pendula. In this case the mathematical analysis is only a little more subtle. To be specific, suppose that the damping coefficient is λ, so that the term

$$\lambda(\dot\theta_2 - \dot\theta_1)$$

is added to the right side of (8.1) for oscillator 1, and the term

$$\lambda(\dot\theta_1 - \dot\theta_2)$$

is added to the right side of (8.1) for oscillator 2. Then, when we differentiate the energy we get

$$\frac{dE}{dt} = \lambda(\dot\theta_2 - \dot\theta_1)\dot\theta_1 + \lambda(\dot\theta_1 - \dot\theta_2)\dot\theta_2$$
$$= -\lambda(\dot\theta_1^2 + \dot\theta_2^2) + 2\lambda\dot\theta_2\dot\theta_1$$
$$= -\lambda(\dot\theta_1 - \dot\theta_2)^2.$$

So the energy is strictly decreasing, except when $\dot\theta_1 = \dot\theta_2$. Now this means that the energy will decrease until the oscillators are moving at the same speed everywhere. This says slightly more than that they are moving at the same frequency, which is an example of the general phenomenon of synchronization. In this case they satisfy the stronger condition that one oscillator is simply displaced from the other. Moreover, the angular separation between the oscillators is not arbitrary. To see this, note that if

$$\dot\theta_1 = \dot\theta_2$$

over some time interval, then, necesarily,

$$\ddot\theta_1 = \ddot\theta_2$$

over that same interval. But then we can solve for the angular displacements in the equations of motion, leading to a fixed relation between them.

EXERCISE 3 Find the equations of motion of the pendulum–spring system of Figure 8.2(a) and the double mass–spring system of Figure 8.2(b).

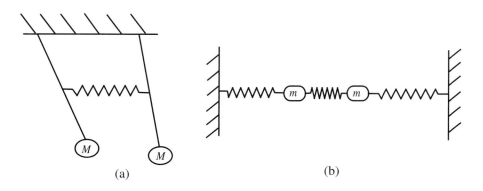

(a) (b)

Figure 8.2 Coupled oscillators.

It turns out to be very difficult to analyze two coupled nonlinear oscillators so let's specialize to the case of small oscillations, so that: the equations are linear, they have linear coupling, and they and are undamped.

In other words, we consider two harmonic oscillators coupled in a symmetric way:

$$\ddot{x}_1 + \omega_1^2 x_1 = -K(x_1 - x_2) \qquad (8.5)$$
$$\ddot{x}_2 + \omega_2^2 x_2 = -K(x_2 - x_1) \qquad (8.6)$$

This could be a linearization of the system (8.2,8.3). Of course, the system is conservative. The "energy" is

$$
\begin{aligned}
E &= T_1 + T_2 + U_1 + U_2 + \frac{K}{2}(x_1 - x_2)^2 \\
&= \frac{1}{2}\dot{x}_1^2 + \frac{1}{2}\dot{x}_2^2 + \frac{\omega_1^2}{2}x_1^2 + \frac{\omega_2^2}{2}x_2^2 + \frac{K}{2}(x_1 - x_2)^2.
\end{aligned}
$$

Again,
$$\frac{dE}{dt} = 0,$$

but as was pointed out earlier, we can't infer from this that the trajectories are closed. In fact, this need not be the case, even when the oscillators are uncoupled, when $K = 0$. Of course, when $K = 0$, then each oscillator is a harmonic oscillator and we know that it follows a closed trajectory in its two-dimensional phase plane. This means that the trajectory of the combined motion, which is in the four-dimensional phase space, must project onto ellipses in the coordinate planes

$$\{(x_1, x_2, \dot{x}_1, \dot{x}_2) : x_2 = 0, \dot{x}_2 = 0\},$$
$$\{(x_1, x_2, \dot{x}_1, \dot{x}_2) : x_1 = 0, \dot{x}_1 = 0\}.$$

However, the projections onto the (x_1, x_2)-plane are not necessarily ellipses.

If the oscillators are identical,

$$\omega_1 = \omega_2 = 1,$$

then the equation
$$E = E_c$$

is the equation for a 3-sphere. So the phase trajectories are great circles on that sphere (and are thus closed). A projection of a circle is necessarily an ellipse, and so the projections onto any plane are ellipses. So the movement in the (x_1, x_2)-plane is an ellipse. This ellipse must lie inside the circle, $U = E_c$. Furthermore, since neither potential energy by itself can exceed the total energy, this ellipse must be in the rectangle (see Figure 8.3):

$$U_1 \le E_c, \quad U_2 \le E_c.$$

In this plane, the two coordinates are

$$x_1(t) = A_1 \cos(t - \phi_1), \quad x_2(t) = A_2 \cos(t - \phi_2).$$

Everything depends on $\phi_1 - \phi_2$. Suppose that $\phi_1 = \phi_2$. Then

$$x_1/x_2 = A_1/A_2 = \text{ constant},$$

so the projection lies on a line. In fact, it is the diagonal of the box in Figure 8.3.

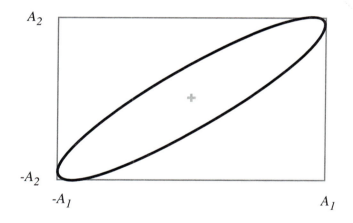

Figure 8.3 Lissajous figure.

Now suppose that $\phi_1 - \phi_2$ is not zero, but is a small positive number. Starting from some point near the center, we find that the coordinate x_1, with the smaller phase delay, will reach the edge of the box first and start back while x_2 is still going forward; after a while x_2 will reach the edge. In this way a skinny ellipse from corner to corner is traced out (Figure 8.3). When $\phi_1 - \phi_2 = \pi/2$, we get the full ellipse that touches the midpoint of each edge of the box (Figure 8.4).

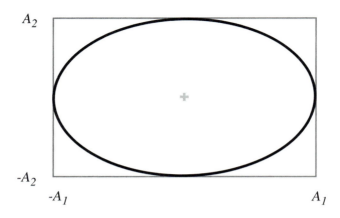

Figure 8.4

Now suppose that the frequencies are slightly different,

$$\omega_1 = 1, \quad \omega_2 = 1 + \epsilon,$$

so that the coordinates are now

$$x_1(t) = A_1 \cos(t - \phi_1), \quad x_2(t) = A_2 \cos(\omega_2 t - \phi_2).$$

Then, over one complete period of oscillator 1,

$$0 \le t \le 2\pi,$$

the projection traced out is close to an ellipse, but it doesn't quite close up, because

$$\phi_1(2\pi) - \phi_2(2\pi) = \phi_1(0) - \phi_2(0) + 2\pi\epsilon,$$

or

$$\Delta\phi|_{t=2\pi} = \Delta\phi|_{t=0} + 2\pi\epsilon.$$

The next time by it will miss by the same factor; so eventually the curve drifts, like the precession of an elliptical orbit (Figure 8.5).

These types of parametrized curves are called *Lissajous figures*. They are easily drawn on an oscilloscope by applying different sinusoidal voltages to the two perpendicular deflection plates.

When $\omega_2 = \omega_1$, the Lissajous figure will be a figure eight or, for special initial phase differences, a parabola. The line and the parabola are

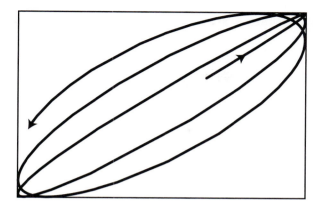

Figure 8.5 Open Lissajous figure.

the first of a sequence of Chebyshev polynomials that can be generated when

$$\omega_2/\omega_1 = \text{integer}.$$

Actually, whenever

$$\omega_2/\omega_1 = \text{rational},$$

the curve will be closed. But if the rational number can only be represented as the ratio of very big integers, then the figure will look like a wanderering curve. If

$$\omega_2/\omega_1 = \text{irrational},$$

then the Lissajous figure completely fills the rectangle as $t \to \infty$.

Now, back to the symmetrically coupled system (8.5–8.6) where $K \neq 0$.

It turns out that such a system can be decoupled very easily by a change of coordinates. To motivate this coordinate change, we will introduce the idea of normal modes of oscillation. Two coupled oscillators will be in a **normal mode** if they are

1. frequency locked, that is, oscillating at the same frequency, and

2. either exactly in phase or exactly anti-phase.

This means that the ratio of the instantaneous amplitudes is constant. The normal modes of any linear system can be found sometimes by intuitive reasoning or, more generally, by finding the transformation that turns the energy into a homogeneous quadratic form. In the present case of the system (8.5–8.6), it is sufficient to point out that we can call upon symmetry to decouple the system by a rotation. By introducing the coordinates

$$q_1 = \frac{x_1 + x_2}{\sqrt{2}}, \quad q_2 = \frac{x_2 - x_1}{\sqrt{2}},$$

the resulting system is decoupled:

$$\ddot{q}_1 + \Omega_1^2 q_1 = 0, \tag{8.7}$$
$$\ddot{q}_2 + \Omega_1^2 q_2 = 0. \tag{8.8}$$

The kinetic and potential energies in these coordinates are

$$T = \frac{1}{2}\dot{q}_1^2 + \frac{1}{2}\dot{q}_2^2$$

and

$$U = \frac{1}{2}\Omega_1^2 q_1{}^2 + \frac{1}{2}\Omega_2^2 q_2^2.$$

The decomposition into normal modes is equivalent to the linear algebra concept of simultaneous diagonalization of two quadratic forms. The general result for any number, n, of coupled oscillators, can be phrased in terms of matrices. Let B be a square $n \times n$ matrix, and A a positive definite matrix of the same order. Suppose that a system's kinetic energy is

$$T = \frac{1}{2}A\mathbf{x} \cdot \mathbf{x},$$

and potential energy is

$$U = \frac{1}{2}B\mathbf{x} \cdot \mathbf{x}.$$

Then a matrix C can be found so that the linear change of coordinates,

$$\mathbf{q} = C\mathbf{x},$$

diagonalizes the quadratic forms determined by A and B. In the new coordinates, the kinetic and potential energies are

$$T = \frac{1}{2}\sum_{i=1}^{n} \dot{q}_i^2, \quad U = \frac{1}{2}\sum_{i=1}^{n} \lambda_i q_i^2.$$

The values (characteristic frequencies) $\{\lambda_i\}$ are the eigenvalues of B with respect to A: this means that they are the solutions to the equation

$$\det(B - \lambda A) = 0. \tag{8.9}$$

(An example given below will illustrate all this.) The roots of the polynomial (8.9) are real (since A and B are symmetric and A is positive) and so the characteristic frequencies always exist. However, without further restrictions they could be negative; such values are not properly "frequencies" since the corresponding decoupled equation is not a harmonic oscillator.

Notice that for the symmetric system (8.5–8.6), one normal mode is given by the first system (8.7), that is, when $q_2 \equiv 0$. This corresponds to the two oscillators being in phase at all times (perfect synchronization). In the other normal mode, which corresponds to $q_1 \equiv 0$, the oscillators are in anti-phase.

EXERCISE 4 (i) Solve the quadratic equation (8.9) for the symmetric system (8.5–8.6). Show that the characteristic frequencies have the following properties:
(a) For very slight damping, $K \approx 0$, the frequency of the in-phase normal-mode oscillation is close to the smaller of the ω_1, ω_2, while the frequency of the anti-phase normal-mode oscillation is close to the greater of the ω_1, ω_2.
(b) As $K \to \infty$, the frequency of the in-phase motion reaches a limiting value, while the frequency of the anti-phase motion increases without bound. Give a physical interpretation of this.
(ii) *Beating modes* Suppose that $\omega_2 \approx \omega_1$ and $K \approx 0$. Show that if the two oscillators start out in phase then

$$x_1 \approx A(\sin \omega_1 t + \sin \omega_2 t), \quad x_2 \approx A(\sin \omega_1 t - \sin \omega_2 t).$$

Notice that for a stretch of time the two sine terms will positively add, in which case x_1 will be large and x_2 will be small; but eventually the relative phases will drift apart and the sine terms will cancel each other. Then x_2 will be large and x_1 will be small. Draw the graph of x_1 versu t. This is *beats*.

An illustrative example

Consider the double mass–spring system of Figure 8.6. What is the equation of motion? The masses are equal and the springs are equivalent, but is the system symmetric? It is natural to introduce the coordinates y_1 and y_2, which are the displacements of the top mass and the bottom mass, respectively. Then the kinetic energy of the system is just the sum of the kinetic energies of each mass:

$$T = \frac{1}{2}\dot{y}_1^2 + \frac{1}{2}\dot{y}_2^2.$$

Potential energy is stored in each spring. The stretching of the top spring is given by the displacement of the top mass, while the stretch-

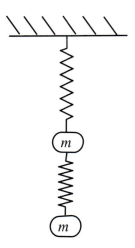

Figure 8.6

ing of the bottom spring is given by the relative displacements of the masses, so

$$U = \frac{k}{2}y_1^2 + \frac{k}{2}(y_1 - y_2)^2.$$

The conservation of energy,

$$\frac{dE}{dt} = 0,$$

implies the equations of motion:

$$m\ddot{y}_1 + ky_1 = -k(y_1 - y_2), \qquad (8.10)$$
$$m\ddot{y}_2 = -k(y_2 - y_1). \qquad (8.11)$$

EXERCISE 5 Interpret the system (8.10–8.11) as an expression of Newton's law.

Define

$$\omega_0^2 = k/m,$$

and look for a solution of the form

$$y_1 = C_1 e^{i\omega t}, \qquad (8.12)$$
$$y_2 = C_2 e^{i\omega t}. \qquad (8.13)$$

This will be a normal mode and ω will be a characteristic frequency. If we substitute these values into the system (8.10–8.11), we get the system of equations

$$(-m\omega^2 + 2k)A - kB = 0, \tag{8.14}$$
$$-kA + (-m\omega^2 + 2k)B = 0. \tag{8.15}$$

This is a homogeneous linear system, so the determinant of the coefficient matrix must be zero. This is the characteristic equation

$$\text{determinant} = (\omega^2)^2 + 3\omega_0\omega^2 + \omega_0^2,$$

which is just the equation (8.9) appropriate to this example. The solutions are

$$\Omega_1^2 = \frac{3 - \sqrt{5}}{2}\omega_0^2, \tag{8.16}$$
$$\Omega_2^2 = \frac{3 + \sqrt{5}}{2}\omega_0^2. \tag{8.17}$$

EXERCISE 6 In a normal mode (8.12), the ratio of the instantaneous amplitudes is a constant,
$$C_2/C_1.$$
Show that in this example this ratio of amplitudes is ϕ, the aspect ratio of the golden rectangle. Notice that the frequencies are ratios of the ratios. Is this system symmetric?

2   *Biological Rhythms

A sense of rhythm is as essential to life and as fundamental as other functions, such as energy transduction, reproduction, and metabolism. Some biological rhythms are an important part of physical well-being, while others are seen as indicating malfunction of the organism. Good rhythms are represented by the healthy heartbeat or the puffs and pants of unmindful breathing. Examples of bad rhythms are palsy, stuttering, hiccups, or even the rare case of uncontrolled sneezing. Some bacteria carry out reproduction by division according to a rigorous timing: the *Escherichia coli* in your gut divide every 40 minutes.

Many of the rhythms of living organisms are related to the Earth's rotation, either having been established during millions of years of evolution to match a 24 hour day or by being directly reset by a cue of the passage of the day, such as the rising sun. These daily rhythms with a period of about 24 hours are termed *circadian rhythms* from "circa-" meaning "about" and "dia" for "day." Examples can be found in all body systems. In the immune system, white blood cells are known to work more effectively at night. Asthma sufferers typically fare worse around midnight. The enzyme alcohol dehydrogenase, which breaks down alcohol in the blood, also has a circadian period, but the phase relationship for women is different than for men. It seems that for men the enzyme is most active several hours after sunrise, while for women it is hardest at work several hours before sunrise. This may be one explanation of why one is less likely to see women stumbling around at bar time.

A natural question to ask concerning these biorhythms is whether they are all inherent to the organism, having evolved that way, or whether they are all resettable, and whether they are independent rhythms or are coupled together, responding in some way to a "pace-maker." In the course of experiments involving a bean plant, *Phaseolus*, which turns its leaves according to a circadian rhythm, evidence was obtained that these rhythms can be entrained by stimuli. The question the experimenters wanted to answer is whether the bean leaf is only responding to a daily stimulus or whether the rhythm is inherent. First, by placing the bean plants in a darkened room and noticing that the rhythmic pattern of opening and closing continued, it was seen that there really is an inherent clock.[1] These observations led to the search for a biological explanation of rhythmicity. On the other hand, it was found that the rhythm could also respond to a stimulus. By subjecting the bean plants to a stimulus with a different frequency, for example, by subjecting them to 25 hours of light followed by 25 hours of darkness, it was found that their bio-rhythm could be *entrained* to the new frequency. However, they couldn't be entrained to periods far different from their own. They tended to ignore stimuli that had a frequency much different than their own 24-hour rhythm, and they would go back to following their built-in rhythm.

Similar experiments have been done on human subjects by having them live in laboratories or in caves and depriving them of the usual diurnal time clues. The human body has many circadian rhythms. In one type of experiment, two of these rhythms are monitored, the cycle of sleeping and waking and the rise and fall of body temperature. The

---

[1] Of course, this doesn't rule out the possibility that the plant senses the earth's rotation by other means, gravitational or magnetic.

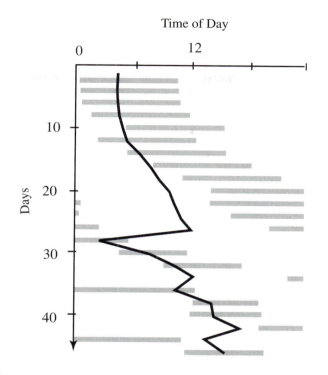

Figure 8.7 Results of an experiment of a human subject deprived of diurnal time cues. The shaded bars are sleep and the solid line is the minimum daily body temperature. Adapted from [38]

data gathered from these experiments, displayed in simplified form in Figure 8.7, showed that there was a loss of synchronization between these rhythms over time, indicating that their fundamental periods were different. It takes somewhere between 1 month and 3 months for a subject to display desynchronization of these two rhythms. In fact, there was some evidence that the sleep–wake cycle could lock-in on a period that was two-thirds the period of variation of the body temperature. It is likely that they were governed by different pacemakers. On the other hand, some of the evidence obtained in these isolation studies showed that, when certain phase relationships existed between the cycles, a beatlike phenomenon occurred, as though the two rhythms were trying to get in phase. This indicated that, although they might be individual oscillators, the two rhythms were coupled somehow.

The term **pacemaker** is usually used to refer to a group of cells that generates oscillations in the presence of a constant input. Pacemaker cells are found in the heart, in smooth muscle systems (like the diaphragm), in the central nervous system, and in hormonal systems. These pacemakers convert a steady input, such as metabolic energy, into a periodic output, much as gravity is used by a pendulum. Such a system, being a rudimentary clock, is specified most simply by its phase angle, θ. There are two physical interpretations of such a *simple clock*: (1) It is the swinging hand on the clock face, or a rotating pendulum, and the phase is the angle that the hand makes with some reference position; (2) it is the phase angle of an oscillator, any reference to the amplitude of the oscillator having been suppressed. For example, the phase angle of a harmonic oscillator satisfies

$$\frac{d\theta}{dt} = \dot{\theta} = \omega_0,$$

where ω_0 is the natural frequency of oscillation. The physical picture we get is of a clock hand rotating at a fixed speed like a wall clock. A simple clock can rotate at varying speeds or can "get stuck" at a certain angle. It can be subjected to regular stimulations that periodically slow it down or speed it up. The simplest such model is

$$\dot{\theta} = \omega - \beta \cos \theta + F,$$

where F is an irregular "stimulus" or "cue," and the other term represents a regular (feedback) force. This is like a pendulum, treating the amplitude as irrelevant. The relative sizes of the parameters ω and β determine the behavior. Suppose, for the moment, that $F = 0$. If $\omega = \beta$, the clock acts roughly like a pendulum that is released at its upper equilibrium and then turns, moving quickly at the bottom of the circle, slowing down as it comes back to the equilibrium (Figure 8.8a). This is the critical case. If $\omega > \beta$, then the phase continuously advances, only slowing down near the "shadow equilibrium" according to the size of $\omega - \beta$ (Figure 8.8b). This is the running mode. If $\omega < \beta$, then the phase angle heads directly (the shortest way—forward or backward) to a fixed angle of rest (Figure 8.8c). This is the stopped mode. It is easy to add the effect of the force F: the result is either a speeding up, a slowing down, or stopping of the clock. There is no real entrainment between the phase and the regular force.

Notice that the response of an impulsive stimulus is simply to instantaneously reset the phase angle, but, once reset, the motion is

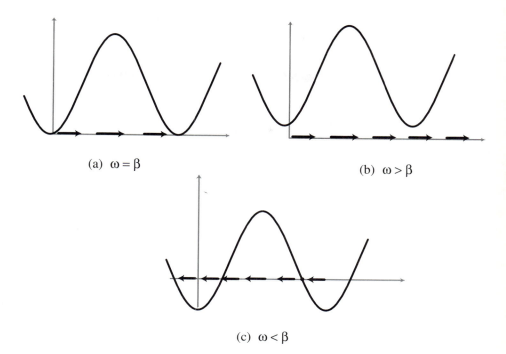

(a) $\omega = \beta$

(b) $\omega > \beta$

(c) $\omega < \beta$

Figure 8.8 Rotating simple clocks.

much the same as if there had been no stimulus. There is some evidence that the frequency as well as phase of some biological rhythms can be changed by a short impulsive stimulus.

For these reasons, it has been proposed to handle the pacemakers of circadian rhythms by self-excited oscillators of the Van der Pol type. Let Y represent the activity of the pacemaker that governs the sleep–wake cycle and X represent the pacemaker for the core body temperature cycle. In [38] their evolution was modeled by coupled Van der Pol oscillators. Recall that the Van der Pol equation is an oscillator equation

$$\ddot{z} + r\dot{z} + \omega^2 z = 0,$$

where the damping coefficient depends on the state:

$$r = -\lambda_0 + \lambda_1 z^2.$$

This nonlinear equation exhibits self-excited oscillations. As this is the characteristic of the pacemaker, we will model each pacemaker by such an equation. The frequency of each oscillator can be estimated by experiment. The equations used are

$$\ddot{X} + \mu_x(X^2 - 1)\dot{X} + \omega_x^2 X = C_{Y \to X}\dot{Y}, \tag{8.18}$$
$$\ddot{Y} + \mu_y(Y^2 - 1)\dot{Y} + \omega_y^2 Y = C_{X \to Y}\dot{X}, \tag{8.19}$$

where ω_x and ω_y are the natural frequencies and μ_x and μ_y are relaxation constants. These four parameters were felt sufficient because of the difficulty of estimation of parameters such as λ_0 and λ_1. The relative size of these parameters determines the size of the limit cycle. The inclusion of the full set of parameters would imply that we had some way of directly comparing the amplitude of the limit cycle of oscillation of X with that of Y. It is not clear what quantitative measures the variables X and Y represent. X can fairly well represent a rectal temperature, but the sleep–wake cycle is more of an all-or-nothing variable, so the sinusoidal nature implied by the equations is rather artificial.

EXERCISE 7 Based on the last remark, would you expect the relaxation coefficients, μ_1, μ_2 to be small or large?

The $C_{Y \to X}$, $C_{X \to Y}$ are coupling coefficients, the effect of Y on X, and vice versa. The particular form of coupling was chosen because 1) being linear, it is the simplest, and 2) it gives the correct phase relation between the oscillators. The parameters are subject to estimation using experimental data ([38]). Briefly, the natural frequencies were taken to be the frequencies during desynchronization. It is clearly a difficult problem to estimate the true frequencies from observations, since the coupling is always present, even during desynchronization. The relative size of the coupling coefficients was estimated by comparing the ratio of the differences between the estimated natural frequencies and the frequency at which the two osillators are synchronized:

$$\frac{C_{X \to Y}}{C_{Y \to X}} = \frac{|\omega_y - \omega_s|}{|\omega_x - \omega_s|}$$

This ratio was set to be 4.

The absolute size of the coupling was determined by the difference of frequencies at which the oscillators became desynchronized. (This was about 15% of the unit, which was chosen in terms of a period of 1 day.) In principle, the coefficients μ_1 and μ_2, can be found by determining relaxation times for each oscillator: how this is to be done in practice is not clear. In [38] these constants are set to be quite low, $\mu \approx 0.1$. Since X represents the stronger oscillation, ω_x is easy to determine; it is close to 1 (in units of 2π radians per day). Simulations of this model were carried out in [38], and some of the results are summarized in Figure 8.9. Figure 8.9a shows the results when the the periodic stimulus is

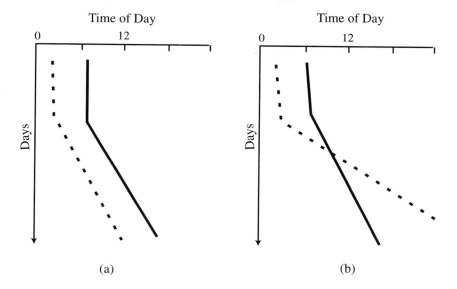

(a)   (b)

Figure 8.9 Results of the simulation of the coupled oscillators. The lines represent curves of constant phase for X (dashed) and Y (solid). The dashed line is taken to represent the time of minimum body temperature and the solid curve is the mid-point of the sleep cycle. In (a) the forcing is applied to X for 5 days, then removed. In (b) the forcing is applied to Y for 5 days then removed. Adapted from [38]

effected by applying a sinusoidal forcing to the X equation only for 5 days and then removing it. There is only a slight phase advance. Figure 8.9b shows the results of the same simulation when the forcing is applied to Y instead. In that case the Y oscillator undergoes a strong phase advance directly after the stimulus is removed. This is taken to imply that the Y pacemaker, that responsible directly for the sleep–wake cycle, is the one that receives the environmental cues.

EXERCISE 8 The authors were able to reproduce much of the experimental data by slowly decreasing (over the course of the simulation) the parameter ω_y. Interpret.

3   *Swaying Smokestacks

Everybody has noticed the steady oscillations of a stop sign in a steady wind. Tall industrial smokestacks and skyscrapers can experience the same sort of slow oscillations of not inconsiderable amplitude during a strong wind. The sway in offices and apartments of the John Hancock building in Chicago is sometimes on the order of feet. Flow-induced oscillations can be especially serious when the relative weight of fluid to structure is high. This can be the case with overhead cables or just about any structure submerged in water. Thus, flow-induced oscillations are a regular source of trouble for bridge pilings, drill casings on floating oil rigs, and periscopes. One possible scenario for the Tacoma Narrows bridge disaster is that the water current induced oscillations of the bridge pilings, with the effect of stimulating the span vibrations to the extent that the wind-induced vibrations were in the (nonlinear) resonant range. The oscillations of these long or tall structures originate in the alternate shedding of vortices from opposite sides of these structures. The period of the vortex shedding in most cases depends only on the free-stream flow velocity and the size of the structure. Shed vortices are even observed in the cloud patterns leeward of ocean islands. The vortex structure, known as a street, is very important, and the most successful vibration models take this into account. (The wake flow, having its own frequency, can cause further trouble to structures downstream, where it can interact with the shedding of vortices. Such a situation is thought to be responsible for the periodic de-tailing of certain jet fighters.)

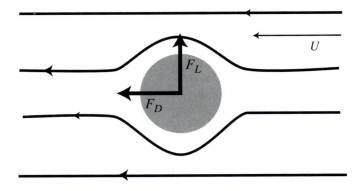

Figure 8.10 Flow about cylinder cross-section. Vertical displacements are transverse, horizontal displacements are longitudinal.

There is a very large variety of flow-induced oscillations (see [10]). We are concerned here with modeling the specific case of the transverse oscillations of a single cylinder in a cross flow. Any body will move in a direction transverse to the mean flow whenever it experiences a force in that direction. Such a transverse force is called *lift*; any force in the flow direction is *drag* (Figure 8.10).

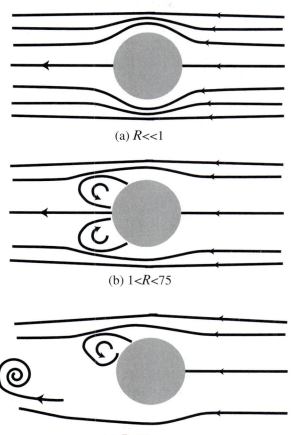

(a) $R \ll 1$

(b) $1 < R < 75$

(c) $R > 75$

Figure 8.11 Flow about cylinder cross section for various values of the Reynolds number.

The actual flow pattern past a cylinder depends on the mean velocity of the free stream and is usually expressed as a function of the

Reynold's number:

$$R = \frac{\rho U d}{\mu},$$

where ρ is the fluid density, μ is the fluid viscosity, U is the mean flow velocity, and d is some characteristic length, taken here to be the diameter of the cylinder cross section. For very low R, the flow does not separate from the surface of the cylinder (Figure 8.11a). This creeping flow is symmetric and so there is no force on the cylinder cross section. As the flow velocity increases, there is a point (typically $R \sim 1$) where the flow separates from the surface, forming a recirculatory bubble behind the cylinder (Figure 8.11b). However, since this flow is still symmetric about a horizontal axis, there is no lift force. At a higher value of the Reynolds number (typically about $R = 50$), this symmetric flow becomes unstable, and vortices are shed from the top and bottom surfaces in alternate sequence (Figure 8.11c). This flow remains stable up to very high values of the Reynolds number (10^5) unless the boundary layer on the cylinder becomes turbulent. This will happen if, for instance, the surface of the cylinder is rough. (Many interesting photographs of vortex shedding can be found in [17].)

It should be pointed out at this point that the flow is really three dimensional and the shedding of vortices may not be correlated from one cross section of the cylinder to the next. Of course, it is precisely such correlation that gives rise to the greatest effect. For this reason, vanes are added to structures such as overhead power lines and smokestacks to reduce the correlation of vortex shedding from one cross-section to another.

While investigating the aeolian tones (or "singing") of wires in 1878, Strouhal noticed that the frequency of vortex shedding depended only on the mean flow velocity, and that over a wide range of flow velocities this relation is linear. With the relevant parameters d, U and frequency of vortex shedding, f_s, we can form the nondimensional number

$$S = f_s \frac{d}{U},$$

called the Strouhal number. A linear relation between vortex-shedding frequency and flow velocity is reflected by a constant Strouhal number. Typical data obtained from experiments on stationary or near-stationary cylinders can be expressed as a plot of S versus R (Figure 8.12). For a wide range of Reynolds numbers, the Strouhal number is constant. The actual value of S depends on the turbulence quality

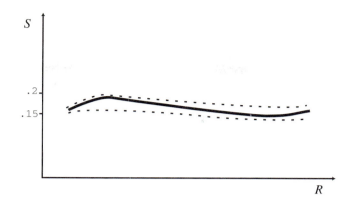

Figure 8.12 Strouhal number versus Reynolds number for flow about a stationary right circular cylinder. Adapted from Chen.

of the flow, but the variation in any case is not great. The value for circular cylinders is in the range $.15 - .2$.

The most striking thing about the free flow–oscillations of a cylinder is the phenomenon known as *lock-in*. In the regime where the Strouhal number is constant for the stationary cylinder, the plot of vortex-shedding frequency versus Reynolds number should be displayed as a straight line. Instead, experimental data for freely oscillating cylinders typically show the plot of Figure 8.13a. The data points follow a straight line (corresponding here to $S = 0.18$) only up to a certain point; then the plot flattens out, and the frequency is constant over a wide range, at the end of which the frequency jumps back up to follow the original line. What happens is that the mechanical structure has a natural frequency of oscillation, f_0, and the frequency of the plateau is f_0. In other words, the cylinder natural frequency "captures" the vortex-shedding frequency and the latter "locks in" to the former. Furthermore, as the frequency increase follows the velocity increase, the amplitude of oscillation of the cylinder is also increasing, but when lock-in occurs, the rate of increase of the amplitude is suddenly drastic (Figure 8.13b). The amplitude reaches a maximum in the interior of the lock-in region, and near the end of lock-in the amplitude drops catastrophically. In many cases, the amplitude for these high flow velocities is negligible. The phenomenon is thus seen to possess many of the characteristics of resonance. As Figure 8.13b indicates, there is hysteresis: if the flow velocity is decreased from higher values, the amplitude obtained is not the same high amplitude as obtains in the opposite direction.

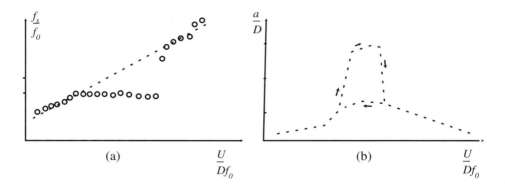

Figure 8.13 Experimental data for flow-induced oscillations of cylinders exhibiting "lock-in". In (a) Strouhal frequency is plotted versus flow velocity; a constant Strouhal number corresponds to the straight line. Hysteresis in amplitude of oscillation is indicated by the plot in (b). Adapted from [10] and [41].

The lock-in behavior is quite universal, seen both in the field and in the laboratory. Many different kinds of setups are possible, ranging from rigid cylinders whose transverse motions are restricted to wires with full three-dimensional motion. Of course, in engineering practice there is a continuum of structural setups ranging between these extremes. The behavior illustrated by Figure 8.13 is, to some extent, common to all these structures. There are always parameter values for which there are multiple amplitudes, a threshold needed to obtain the higher amplitude solution. In general, there is a double hysteresis. In Figure 8.13 the lock-in region corresponds to the instability of the small-amplitude solution; however, there are experimental situations where both solutions are locally stable in the lock-in region (Figure 8.14). The range of the lock-in is related to the maximum amplitude possible. These high-amplitude motions are thought to increase the correlation of vortex shedding spanwise along the cylinder. There are mitigating factors, the extent of which is hard to assess. These include ambient turbulence in the mean-stream flow and the excitation of higher modes.

In setting out to fashion a mathematical model for this phenomenon,

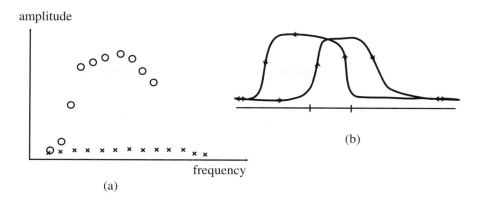

(a)

(b)

Figure 8.14 These figures, adapted from [41], illustrate some of the experimental results obtained with the flow-induced vibrating cylinder. In (a) is
illustrated the possibility of two solutions, a small amplitude solution and a
large amplitude (resonant) solution. Double hysteresis is exhibited in (b).

it will be useful to stick to explaining specific experimental results, such
as the existence of multiple stable periodic solutions. Certainly, we have
a vibrating elastic structure coupled to a time-periodic fluid flow. But
what models should we use for each of the subsystems?

Dealing first with the elastic system, it is clear that the internal
damping of the structure will prevent any sustained oscillation in the
absence of fluid forcing, so it seems reasonable to suppose that when
uncoupled from the fluid this subsystem is a damped linear system.
Furthermore, the damping of structures is such that higher modes are
usually more heavily damped than the first mode, so for a first model
we can suppose that all the motion is in the first mode of vibration
(Figure 8.15). By way of these considerations, we have reduced the
model to that of a linear oscillator with one degree of freedom:

$$m\ddot{y} + 2\lambda\dot{y} + ky = F_L, \tag{8.20}$$

where F_L is the fluid dynamic lift force that is acting on the cylinder.
The constants, k and λ, represent the elastic restoring force and the
internal damping (that not due to the fluid) of the cylinder.

A simple laboratory model of this system is shown in Figure 8.16.
What is the force, F_L? If we know the variables that represent the flow,
that is, the velocity and density, everywhere about the cylinder, then
the total force can be found by an integration. There are partial differential equations that are believed to be adequate (continuum) models

Figure 8.15 First mode of vibration.

for fluids. If we had in hand an appropriate solution to one of these equations, then that would do here. The problem is that appropriate solutions to the vortex-shedding flow field equations are to be had only by the numerical computation of the equations starting from some initial flow. Besides the fact that these numerical solutions are extremely crude, even when performed on today's most sophisticated computers, the solving of initial-value problems is not the best way to evaluate models (see below).

It is best to keep in mind the type of solution that we are interested in. The experiments show a periodic oscillation of the cylinder in a periodic flow. Our experience with linear equations leads us to expect that the right side of (8.20) will be a periodic function. From our adventures in dimensional analysis, we expect that we can write this in the form

$$
\begin{aligned}
F_L &= \text{lift force per unit length} \\
&= \frac{1}{2}\rho U^2 C_L \sin(\omega_s t),
\end{aligned}
\tag{8.21}
$$

where C_L is the nondimensional lift coefficient. Here we have endowed the forcing function with a known frequency, ω_s, the angular frequency of the vortex shedding,

$$\omega_s = 2\pi f_s;$$

but we expect that the vortex-shedding system has its own dynamics. The simplest model to incorporate this would be to suppose a variable, ζ (a time-varying lift coefficient), that satisfies a one-degree-of-freedom oscillator equation. Of course, this equation will be nonlinear because it will oscillate even when it is isolated from other systems (as the experiments with a stationary cylinder have shown). Since it is linear,

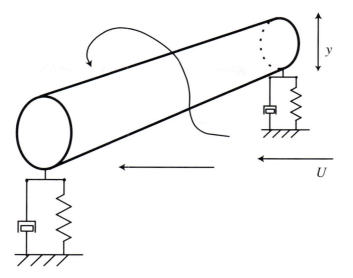

Figure 8.16 Laboratory model of vibrating cylinder.

the expression for this variable that gives the force F_L,

$$C_L \sin(\omega_s t),$$

will not be correct except in special circumstances where we can justify a linearization.

We know examples where such equations model self-excited systems: the Van der Pol equation is one. An approach open to us is that we use such an equation as our model and then match solutions with experiment in order to identify parameters in the model. Of course, with many free parameters in the model we can fit any data!

It will be better to come upon a nonlinear oscillator by using more of the physics of the system.

First, the proper equations will not be found by simply grafting the variable ζ to the right side of (8.20) as the expression for F_L, (8.21), would seem to indicate. The reason is that the fluid has elastic and damping effects on the cylinder motion, as well as the periodic forcing. Moreover, the mass that appears on the left side of (8.20) will be augmented by a virtual, or added mass, of the fluid. To see where this factor comes from, imagine dragging the cylinder through the water. Ignoring the drag force, the cylinder will still require a larger force to accelerate it through the water than to move it through air or vacuum. This is because one must push some of the fluid along with it; this fluid has mass, and this mass must be added to give the cylinder a greater inertia.

Any body sitting immersed in water has its usual mass. This mass gives the initial inertial resistance to acceleration, but, as motion begins, it becomes harder and harder to accelerate the body, that is, its *effective mass* increases. To find the form of this added mass, let us consider the impulse required to change the (translational) momentum of the cylinder by a given amount. The impulse will be due to a force, F, acting during a short time, Δt, resulting in an increment in the momentum, $\Delta P = P(t + \Delta t) - P(t)$. This momentum change can only come about by a change in the velocity, so we have the balance

$$F\Delta t \propto \Delta U.$$

In a two-dimensional flow with density ρ (such as that about a cross section of a cylinder), the impulse will be give in terms of a force per unit length, F. Meanwhile,

$$\rho U,$$

gives the momentum per unit volume, so, by dimensional considerations, the balance should be

$$F\Delta t = \rho A \, \Delta U,$$

where A is an appropriate cross-sectional area. For the circular cylinder, this is

$$A = \pi a^2,$$

where a is the radius of the cylinder.

There have been many versions of fluid-oscillator models. Such models can be as simple as incorporating the added mass and added damping as parameters and presuming a Van der Pol oscillator, all these parameters to be determined by matching with data. The process here is slightly circular, since we would like a model to actually explain the experimental data. A model that incorporates the fluid physics by force balance in a control volume, without the added difficulty of directly using fluid variables is presented in [30], which has the virtue of giving, through the coupling with the elastic equation (8.20), the added mass and the added damping. Before we go through that derivation, let us review what we should expect the form of the lift coefficient equation to be.

For stationary cylinders and for small lift, the natural frequency of the the fluid oscillator is simple harmonic at the Strouhal wake frequency:

$$\ddot{\zeta} + \omega_s^2 \zeta = 0.$$

At low frequencies of cylinder vibration, the lift coefficient should be proportional to the velocity of vibration and to the fluid velocity. At high frequencies the lift coefficient should express the added mass effect.

Figure 8.17

Following [30], we will develop a fluid-oscillator equation by considering the balance of fluid dynamic forces in the "control box" shown in Figure 8.17. Since we assume that the vortices are correlated along the spanwise length of the cylinder, which implies a two-dimensional flow, variables represent quantities that are to be taken as *per unit length*.

The equation for the force balance in the vertical (y) direction is

$$P_y - F_y = \frac{dJ_y}{dt} + S_y, \qquad (8.22)$$

where P_y is the pressure force on the top and bottom sides of the control box, F_y is the force that the fluid exerts on the cylinder (the negative of the reactionary force that the cylinder exerts on the fluid), S_y is the momentum flow out the back of the control surface, and J_y is the vertical momentum within the control box.

EXERCISE 9 Show that the dimension of S_y, momentum flow (per unit length), is

$$MLT^{-2}.$$

By taking the control box big enough, it can easily be established that

$$P_y = 0. \tag{8.23}$$

The total (vertical) momentum (per length) in the box is found by integrating the momentum density over the box:

$$J_y = \int \int \rho v \, dx dy,$$

where v is the vertical component of the fluid velocity. This vertical velocity is to be averaged over the cross section in order to arrive at the "hidden" variable, $\dot{z}$, of our fluid oscillator:

$$J_y = a_0 \rho \dot{z} d^2, \tag{8.24}$$

where d is the cylinder cross section and a_0 is a proportionality constant.

If we assume that the flow upstream of the cylinder has no vertical oscillation component, then the momentum flow is to be evaluated across a surface that splits two vortices, with an even number of vortices left in the control box, as in Figure 8.17. The momentum is delivered across the surface by the translational motion of the vortices. The momentum (per unit length) in a vortex is

$$\int \int_{\text{vortex}} \rho \mathbf{u} \, dx dy.$$

If we assume that the flow is (mainly) tangential in a vortex, then we can write this momentum as

$$\rho d \oint_{\substack{\text{about a} \\ \text{vortex}}} \mathbf{u} \cdot d\mathbf{r} = \rho \, d\Gamma,$$

where Γ is known as the *circulation*. Thus, on averaging over a cycle, the momentum flow has the magnitude of

$$|S_y|_{\text{avg}} = \rho \Gamma w,$$

where w is the translational velocity of the vortex street. It is reasonable to expect that

$$\Gamma = K|\dot{z}|d,$$

where K is a proportionality constant. But, from Figure 8.17, we see that S_y will lag $\dot{z}$ by one-quarter cycle. Thus

$$S_y = \rho d K(\omega_s z) w - a_1 \rho U d\dot{z} + a_2 \rho \dot{z}^3 d/U, \qquad (8.25)$$

where a_1 and a_2 are proportionality constants for the fluctuating terms that are zero over a cycle. The net fluid force on the cylinder can be written as

$$F_y = a_3 \rho d^2 U \dot{\alpha} + a_4 \rho d U^2 \alpha,$$

where a_3 and a_4 are dimensionless constants of proportionality, and α is the angle between the free stream and the normal component of the fluid impinging on the cylinder. This angle is, for small displacements, approximately $(\dot{z} - \dot{y})/U$. Thus,

$$F_y = a_3 \rho d^2 (\ddot{z} - \ddot{y}) + a_4 \rho d U (\dot{z} - \dot{y}). \qquad (8.26)$$

Putting together all equations (8.23) through (8.26) into the force balance equation (8.22), we get the equation for the fluid mechanical oscillator:

$$\ddot{z} + K' \frac{w}{d} \omega_s z = (z_1' - z_4') \frac{U}{d} \dot{z} - a_2' \frac{\dot{z}^3}{Ud} + a_3' \ddot{y} + a_4' \frac{U}{d} \dot{y}, \qquad (8.27)$$

where

$$K' = K/(a_0 + a_3),$$

and

$$a_i' = a_i/(a_0 + a_3), \quad i = 1, 2, 3, 4.$$

EXERCISE 10 (i) Use the expression for the fluid–cylinder coupling force to write out the complete equation for the elastic oscillator and obtain the complete system of equations.
(ii) Verify that the equations have the correct form in the limits discussed above.

Notice that the small-amplitude oscillations over a stationary cylinder will oscillate at the correct shedding frequency when

$$\omega_s^2 = K' \frac{w}{d} \omega_s.$$

Comparing this to the expression for Strouhal frequency,

$$\omega_s = 2\pi S \frac{U}{d},$$

where S is the Strouhal frequency, we see that experimental values for S and $\frac{U}{d}$ will determine the parameter K'. It is shown in [30] that all of the model parameters but one can be determined from the data for the special cases of stationary and harmonically forced cylinders. With the specification of one free parameter this model can be used for numerical studies that can be compared to experiments on complicated motion. (For other studies, consult [41] and [54].)

EXERCISE 11 Show that one way of investigating the possibility of the lock-in phenomenon is to reduce the system of equations to a force Van der Pol oscillator.

4   *Dynamo Theory

What is the origin of the earth's magnetic field? Or those large-scale magnetic fields found throughout the universe, in planets, stars, and galazies? It was thought for a long time that the earth's magnetic field was due to ferromagnetism. But it is now known that the temperatures in the iron core of the earth are too hot to support magnetism. It is believed that the magnetic field is induced by currents that are generated by the fluid motion of the molten core. This induction can be explained in broad form if not in detail by some simple models where the conducting fluid in the rotating body is replaced by a rotating solid conductor. These models, known as disk dynamos, will be examined here. One of the more sophisticated of these, the Rikitake two-disk dynamo, indicates a possible dynamical reason for an interesting archaeological fact: the earth's magnetic field, which is predominately of a dipole nature, has reversed polarity many (more than 200) times in the the past two hundred million years. These reversals have occurred without any apparent periodicity, in a seemingly "chaotic" way. The average time period between reversals is about 1 million years, but the pole stayed north between about 119 and 82 million years ago.

To see how magnetic fields can be generated by moving conductors, let us recall that a current moving through a wire will generate a magnetic field that wraps itself around the wire; conversely, if a wire is moved transversely through an already existing magnetic field that has a nontrivial projection perpendicular to the axis of the wire, then a current is induced in the wire. Moreover, since the current will be energetic, it will take work to push the wire through the field; this work must be done to oppose the Lorentz force, which is directed perpendicular to both the field and the velocity of the conduction. The force takes the form

$$\mathbf{F} = q(\mathbf{v} \times \mathbf{B})$$

or

$$dF = I\, d\mathbf{r} \times \mathbf{B}$$

where $d\mathbf{r}$ is directed along the wire, I is current, and $\mathbf{B}$ the magnetic field. Suppose that now we take a length, r, of wire and spin it at one of its ends about an axis perpendicular to a magnetic field (Figure 8.18). What torque must be applied to keep the wire spinning? The

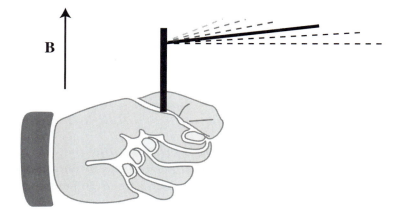

Figure 8.18 A steady field induces a current in a moving conductor.

torque on each infinitesimal segment of the wire is $s\, dF$, where s is the distance from the segment to the pivot and dF is as given above. Thus,

$$G = \int_0^r sIB\, ds = \frac{I}{2}r^2 B = \frac{1}{2\pi}\Phi I, \qquad (8.28)$$

where Φ is the magnetic flux,

$$\Phi = \int_S \mathbf{B} \cdot d\hat{\mathbf{n}}$$

and S is the surface formed by the rotating wire.

We can suppose that if the spinning wire is replaced by a spinning disk then the situation should be the same. Faraday did this, setting up a current between the axle and the edge of the disk (Figure 8.19); the wire closes the loop so that current will flow.

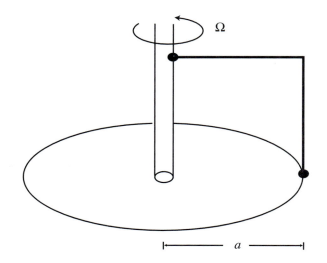

Figure 8.19 Faraday disk dynamo. The wire closes the loop so that current will flow.

Let us suppose that the disk rotates at a constant angular velocity Ω, and let's calculate the size of the current set up and the torque necessary to keep it all going. The Lorentz force per unit charge is $F/q = \mathbf{v} \times \mathbf{B}$, and

$$\mathbf{v} = r\Omega\hat{\phi},$$

which means that

$$\mathbf{v} \times \mathbf{B} = r\Omega B\hat{r},$$

so that a potential difference (voltage) is set up:

$$V = \int_0^a \frac{\mathbf{F}}{q} \cdot d\mathbf{r} = \frac{a^2}{2}\Omega B, \qquad (8.29)$$

where the outer edge is positive with respect to the spindle. The disk has a resistance R, and so in the steady-state situation the current is

$$I = \frac{V}{R} = \frac{a^2}{2R}\Omega B = \frac{\Phi\Omega}{2\pi R}.$$

To get the disk spinning, we must apply a torque G to overcome resistance by the Lorentz forces; recall from (8.28) that the torque is proportional to the product of the current and the magnetic flux:

$$G = \frac{1}{2\pi}\Phi I = \left(\frac{R}{\Omega}\right)I^2.$$

EXERCISE 12 Show that the quantity $M = R/\Omega$ has dimensions of inductance.

We need both the mechanical torque and the magnetic field together to keep up the supply of current. But remember that a current always gives rise to a magnetic field. Thus the current flowing through the loop will produce a field.

EXERCISE 13 We neglected the effect that this may have. Can you make estimates of the effect of the field induced by the loop in Figure 8.19?

Note that the field induced by the loop in Figure 8.19 is not a significant reinforcement of the applied field. Can we change the configuration of the wire loop so that the induced field will reinforce the applied field? Or so that it can *replace* the applied field, so that only the *torque* (mechanical force) need be supplied to run the whole thing? *Indeed*: such a self-inducting dynamo is shown in Figure 8.20.

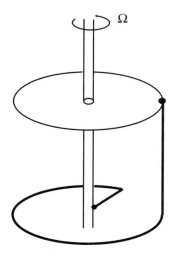

Figure 8.20 A self-induction dynamo.

Let us develop the dynamical equations for this disk dynamo and show that a nontrivial steady-state solution exists. If C is the moment of inertia of the disk, then

$$C\dot{\Omega} = \text{ total torque.}$$

The total torque is the applied torque minus the retarding torque due to the Lorentz forces, which we found to be $\frac{1}{2\pi}\Phi I = MI^2$; thus,

$$C\dot{\Omega} = G - MI^2$$

is the mechanical equation.

If L is the self-inductance of the wire loop, then, from what we have derived [see 8.29], the circuit equation for the current is

$$L\dot{I} + RI = \frac{a^2}{2}\Omega B = MI\Omega,$$

so the coupled system of equations is

$$
\begin{aligned}
C\dot{\Omega} &= G = MI^2, \\
L\dot{I} &= -RI + MI\Omega.
\end{aligned}
$$

The steady-state solution satisfies

$$G = MI^2, \tag{8.30}$$
$$R = M\Omega, \tag{8.31}$$

which means that

$$\Omega = \frac{R}{M}, \qquad I = \sqrt{G/M},$$

give a "characteristic" frequency, and a "characteristic" current for this dynamo.

EXERCISE 14 What is the power supplied by the rotatory torque? What is the power dissipated by the moving charge? Equate these two expressions and show that it is consistent with (8.30, 8.31). (*Hint:* use the "characteristic" frequency and current).

EXERCISE 15 Investigate the local and global stability of this (unique) equilibrium. (*Hint:* construct a Liapunov function.)

It should be apparent after doing the last exercise that the single-disk dynamo is not capable of replicating the field reversals that have been found in the earth's archaeological record. Rikitake proposed a two-disk dynamo. It turns out that the currents in this model can change sign: there are oscillatory solutions. Our discussion of Rikitake's model follows that of Cook and Roberts ([12]). Rikitake's model dynamo is pictured in Figure 8.21.

The equations are set up as we did before for the single-disk dynamo, except that now there are the two frequencies, Ω_1 and Ω_2, and

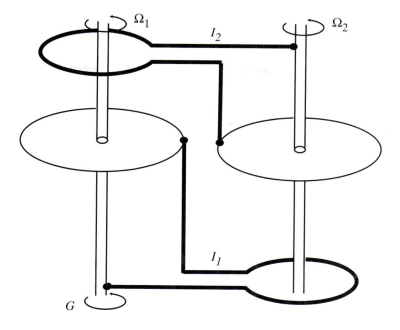

Figure 8.21 The Rikitake two-disk dynamo.

the two currents I_1 and I_2; this makes a four dimensional system. From the picture we see that the flux of the field induced by the current I_1 will cut the second disk, and vice versa; thus, the retarding torques will be MI_1I_2. Also, the voltage across each disk will be set up by the magnetic field generated by the opposite coil. The voltage across disk 1 will be $M\Omega_1 I_2$ and the voltage across disk 2 will be $M\Omega_2 I_1$. Thus, the complete system of four equations is

$$\begin{aligned} L\dot{I}_1 + RI_1 &= M\Omega_1 I_2, & (8.32)\\ L\dot{I}_2 + RI_2 &= M\Omega_2 I_1, & (8.33) \end{aligned}$$

$$\begin{aligned} C\dot{\Omega}_1 &= G - MI_1 I_2, & (8.34)\\ C\dot{\Omega}_2 &= G - MI_1 I_2. & (8.35) \end{aligned}$$

Since the equations for Ω_1 and Ω_2 are identical,

$$\dot{\Omega}_1 = \dot{\Omega}_2,$$

which when integrated gives

$$\Omega_1 - \Omega_2 = \sqrt{\frac{GL}{CM}} A, \tag{8.36}$$

where A is a constant to be determined (by the initial conditions, for example).

EXERCISE 16 Show that A is dimensionless.

The relation (8.36) defines a hyperplane upon which the solution of the equations (8.32) to (8.34) is constrained to lie. As fundamental variables to describe the dynamics on this hyperplane, we can use I_1 and I_2 and either of the angular velocities or even some combination such as $\Omega_1 + \Omega_2$, as long as it is not constant on the plane. On the hyperplane there are two equilibrium solutions (rest points), because the equations are quadratic. To find these equilibria, we solve simultaneously the equations

$$\begin{aligned} RI_1 &= M\Omega_1 I_2, \\ RI_2 &= M\Omega_2 I_1, \\ G &= MI_1 I_2. \end{aligned}$$

Then $I_1 I_2 = G/M$, and so

$$\begin{aligned} I_1 &= \pm K\sqrt{\frac{G}{M}}, \\ I_2 &= \pm K^{-1}\sqrt{\frac{G}{M}}, \end{aligned}$$

where K is to be determined. (A nondimensionalization carried out below will disclose the physical significance of K.)

Now

$$\frac{R}{M}I_1^2 = \Omega_1 I_1 I_2 = \Omega_1 G/M$$

and

$$\frac{R}{M}I_2^2 = \Omega_2 G/M,$$

which means that

$$\Omega_1 = K^2(R/M), \qquad \Omega_2 = K^{-2}\left(\frac{R}{M}\right).$$

Subtracting one equation from the other, we get

$$\frac{R}{M}\left(I_1^2 - I_2^2\right) = \frac{G}{M}\left(\Omega_1 - \Omega_2\right) = \sqrt{\frac{G}{M}\frac{G}{M}}\sqrt{\frac{L}{C}}A,$$

which shows that K can be determined from the constant of integration, A. Note that the choice of the particular equilibrium point corresponds to a particular orientation of the currents, and so the two rest points can be regarded as opposite polarities of the magnetic field.

In order to put the equations in a dimensionless form, we need to identify the "characteristic scales". From the simple dynamo, we know that $\sqrt{GM}$ is a characteristic current and that R/M is a typical (steady state) angular velocity. In the absence of opposing induction it would take a time τ_m to accelerate the disk to this speed, where τ_m is determined by

$$\frac{R}{M} = \Omega = \frac{G}{C}\tau_m,$$

and so

$$\tau_m = \frac{C}{G}\frac{R}{M}.$$

In the absence of a rotation that supplies a voltage, the current through the loop would decay as

$$e^{-\frac{R}{L}t},$$

and so a characteristic time of relaxation is

$$\tau_e = L/R.$$

An important nondimensional parameter is

$$\mu^2 = \tau_m/\tau_e = \frac{CR^2}{GLM},$$

the ratio of the mechanical time scale to the electromagnetic relaxation time.

EXERCISE 17 Show that μ^2 can be interpreted as the ratio of stored mechanical energy to the stored electromagnetic energy.

We will choose a time τ measured in units of $\sqrt{\tau_m \tau_e}$, the geometric mean of the two characteristic times. We will measure current in units of $\sqrt{G/M}$, that is,

$$I = \sqrt{\frac{G}{M}}X,$$

where X is the nondimensionalized current. From the relation (8.36) above, we can write the frequency

$$\Omega = \sqrt{\frac{GL}{CM}}Y,$$

where Y is the nondimensionalized angular velocity. Thus, to nondimensionalize, we get

$$I_i = \sqrt{\frac{G}{M}}X_i, \qquad \Omega_i = \sqrt{\frac{GL}{CM}}Y_i, \qquad Y_1 - Y_2 = A.$$

If $Y = Y_1$, then our equations are

$$\begin{aligned}
\dot{X}_1 &= \mu X_1 + Y X_2, \\
\dot{X}_2 &= -\mu X_2 + (Y - A) X_1, \\
\dot{Y} &= 1 - X_1 X_2.
\end{aligned}$$

EXERCISE 18 Show that the equilibrium solutions are

$$X_1 = \pm K, \quad X_2 = \pm\frac{1}{K}, \quad Y = Y_1 = \mu K^2, \quad Y_2 = \mu K^{-2}.$$

EXERCISE 19 Linearize about one of the equilibrium points and show that the linearized equation has one root that is real negative and a pair of complex conjugate roots. Argue that near each rest point there is a two-dimensional center manifold, and that it is stable in the sense that trajectories nearby are attracted to this manifold.

Cook and Roberts [12] perform a local Liapunov stability analysis to determine that the rest points are unstable. A contraction of phase space indicates the existence of an attractor. Numerical simulations indicate that, globally, the trajectories follow a strange attractor (Figure 8.22).

5   Problems and Recommended Reading

1. Explain why the study of a weakly nonlinear system with a limit cycle can be reduced to an equation of the form

$$\frac{d\theta}{dt} = \omega + \text{ small term}.$$

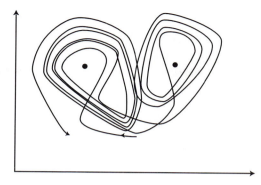

Figure 8.22 Numerical solution of the coupled-disk equations indicates that the system evolves near a strange attractor.

2. Suppose that

$$\dot{\mathbf{x}} = \mathbf{F}(\mathbf{x})$$

is a three-dimensional dynamical system, and let

$$\mathbf{x}(t, \mathbf{y})$$

be the solution starting at the vector $\mathbf{y}$:

$$\mathbf{x}(0, \mathbf{y}) = \mathbf{y}.$$

The ω-limit set of a point $\mathbf{y}$ is

$$\omega(\mathbf{y}) = \{\mathbf{z} : |\mathbf{x}(t_n, \mathbf{y}) - \mathbf{z}| \to 0, t_n \to \infty\}.$$

An *invariant set* S satisfies the property that the solution starting at any point in S stays in S for positive and negative times:

$$\mathbf{x}(t, \mathbf{y}) \in S, \quad \text{for } \mathbf{y} \in S \text{ for all } t \in R.$$

An *attracting region* A is a closed invariant set A together with a neighborhood, $A \subset U$, that is invariant for positive times and that satisfies

$$\mathbf{x}(t, \mathbf{y}) \to A, \quad \text{as } t \to \infty, \text{ for all } \mathbf{y} \in U.$$

A *trapping region* is a region that is invariant for positive times. Show that if

$$\operatorname{div} \mathbf{F} < 0$$

then a trapping region contains an attracting region that cannot have an interior.

3. A *Liapunov function* for a dynamical system on a set $U \subset R^n$ is a positive function, $V : U \to R, V > 0$, which is decreasing along trajectories of the dynamical system:

$$\frac{dV}{dt} < 0.$$

Give a geometrical condition for a vector field to have a Liapunov function. Show that if a function is a Liapunov function on $R^n \setminus \{0\}$ then the origin is asymptotically attracting for the dynamical system. Can you generalize the notion of Liapunov function?

4. *Malarial rhythm* A striking characteristic of malaria is a fever with a pronounced periodicity, having a period of a day and a half. Body temperature, presumably governed by the X pacemaker, has a decidedly diurnal natural frequency, but other body rhythms such as urine calcium excretion and plasma growth hormone tend to occur in cycles of a period close to 34 hours ([49]). These rhythms are presumably governed by the Y pacemaker. How must the model presented in the text be modified so that body temperature (X) is observed with a period close to that of Y? An interesting popular account of biological rhythms is [59].

5. *The female cycle* A typical women of childbearing age who is not pregnant or breast-feeding releases an egg every 28 days. This rhythm is the result of two interacting systems, the ovary and the pituitary gland. In a cyclical fashion, the pituitary secretes a pair of hormones, a luteinizing hormone and a follicle-stimulating hormone, while the ovary secretes estrogen and progesterone. It is believed that the maintenance of this cycle is ensured by the mutual inhibition of these two systems. In the first part of the cycle, when stimulated, the ovarian follicle produces estrogen; in the second phase, the follicle ovulates and makes progesterone. These sex hormones inhibit the the secretions of the pituitary gland. The concentration of these hormones decays with a relaxation time under two weeks and, freed from inhibition, the pituitary starts making follicle-stimulating hormone, and so on. Model this cycle as a relaxaton oscillator. (See the references on this subject in [85].)

6. *Running and puffing* Locomotion in limbed organisms occurs naturally without conscious feedback according to a *pattern gen-*

erator located somewhere in the nervous system. From the point of view of rhythmicity, many-limbed organisms can exhibit many qualitatively different types of gaits. However, in bipeds, the rhythm is straightforward: left–right, left–right, ... Meanwhile, there is the rhythm of breathing: in–out, in–out, ... It would seem that efficiency dictates the latter rhythm to entrain to the former. How does this happen? In [79] this was modeled by representing the diaphragm as a simple linear oscillator that is coupled to three simple clocks, I, X, and M, representing inspiration, expiration, and mechanoreception, respectively. Explain why the diaphragm should be treated as a mechancial oscillator:

$$m\ddot{\delta} + \dot{\delta} + \delta = F.$$

The forcing function is the effect of the gait. What form should it have? It is supposed that these variables interact by simple inhibition and excitation according to Figure 8.23. Write down equations for this system. Can you estimate parameters? In a simulation in [27] they found that the breathing cycle locked-in to three times the running frequency.

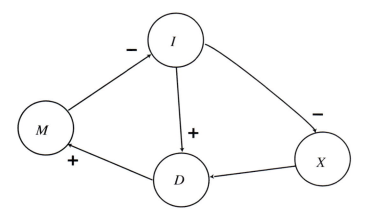

Figure 8.23 The (-) are inhibiting relations. The (+) are excitatory relations.

Chapter 9

Snakes and Chains

The behavior of large systems seems not to be related to the behavior of its constituent parts. Can every complex system be explained by accounting for the interaction of subsystems? Or is there a critical size, above which a system is "greater than the sum of its parts."

1   Snakes and Chains

Consider automobile traffic on the highway. We can think of this as a chain whose links are the individual cars. Each car has independent dynamics, characterized by its velocity, which is controlled by an autonomous driver, yet the dynamics of the other cars influence the driver. For example, the driver may wish to keep a constant separation from the other cars, so this will be the major factor in determining the velocity. The interactions between cars in traffic of moderate spacing are strong enough that, when seen at night, a line of cars sometimes seems to be a glowing snake. This leads to a different way of describing traffic: as a continuous "smear" of cars. Is there something to be gained by this approach? A simplification? Are the two types of models equivalent? As with all questions in modeling, it depends on what you want the model for. Let the model grow out of the questions you are asking. Let us consider some questions that may occur to someone who is interested in traffic control.

If a car at the front of a line of traffic must slow down quickly for a short time, as in traversing a work area, will this create a disturbance in the line of cars that may be amplified as it is passed down the line, eventually causing an accident? Suppose that a stop sign is being considered for a particular intersection. How is traffic density affected? How is the density affected at different times of day? These questions, which on the face are only slightly different, will demand quite different modeling approaches. Let's consider some other questions that call for similar models.

A crystal is made up of individual atoms that interact with each other via atomic forces. As the distance between atoms is so small (10^{-7} mm), crystals can be thought of as solid bodies characterized by mass densities and forces spread out as body and surface stresses.

When we look at a galaxy through a telescope, we see a swirling fluid entity, like water. But the distances between the individual stars making up the swirl are of the order of 10^{12}.

To get back to traffic flow: Suppose that the lead car slows down. Is this slowing down transmitted to the other cars so that each one will in turn slow down? The answer will depend on a mix of characteristic times, distances, and velocities. To a mild disturbance (slowing down for an exiting car or a bear running across the road), the second car in line can make a simple adjustment and an even slighter adjustment than that is needed by the third car. Cars farther down the line may not have even noticed the disturbance. On the other hand, a disturbance of a more sustained nature (slowing down for a yellow flag) will propagate quickly down the line, with possibly disastrous consequences. In this latter case of a propagating disturbance, we say that a *wave* has been formed. Wave nature is prevalent and not always 'disturbing': the sound generated by a lute or the reflected light of the moon involve wave propagation. In order to allay confusion, let's introduce some notation that is used in describing waves.

We'll call a **wave** any change of state propagating through space. (For the time being, we will only notice one-dimensional waves going along a snake or chain.) A wave is characterized by a distance, the wavelength, and a velocity, the wave speed. The **wavelength** is the minimum distance separating two elements of the snake or chain that are in the same state. In the case of traffic flow, the state of a car is its velocity. If all the cars in line have the same velocity, then there is no wave (if you prefer, you can say that this is a wave whose wavelength is infinite.)

What determines the speed of a wave? Let's look at a simple example of information transmission. Xanthippus carried the news of the victory at the battle of Marathon 26 miles to Athens. He must have run at quite a clip because it killed him to do it, and yet it probably took him at least two or three hours. Suppose that, instead of individual runners, the Athenians had set up a sequence of stations every quarter-mile, each manned by a soldier with a signal flag in line of sight with the next. After receiving the message, each soldier in line would pass it on to the next. How fast can the message get to Athens? Suppose that it takes 2 seconds for a mesage to be signaled and immediately passed on. The time separating samestates (flag up to flag up) is

$$T = 2\,\text{s},$$

and the wavelength is

$$\lambda = \frac{1}{4}\,\text{mi},$$

so the speed is

$$c = \frac{\lambda}{T} = \frac{1}{8}\,\text{mi/s}. \tag{9.1}$$

The message could be transmitted from Marathon to Athens in 208 seconds, about 3 1/2 minutes, without one soldier changing position.

Suppose that the message was more complicated so that T was longer. Then, by (9.1), the wave speed would be reduced, unless the wavelength were made longer. This can be done in the obvious way by placing the soldiers farther apart. But it can also be achieved by having each soldier not wait until he finished receiving a message to begin transmission, but to transmit simultaneously with the reception. In this way, the wavelength of the transmittal can be many times the interparticle spacing.

A related example: In the 1980s, human "waves" became popular in the bowl-shaped sports stadiums. Here the change of state is from sitting to standing. Let d be the interfan distance, t_u be the time it takes to stand up, and t_r be the time between when the fan notices her neighbor standing and when she begins to stand. Thus, the wave speed is

$$c = \frac{d}{t_u + t_r}.$$

If $d = 2$ ft, $t_u = 1$ s, and $t_r = 0.2$ s, then

$$\begin{aligned} c &= \frac{2}{1.2}\,\text{ft/s} \\ &\approx 1.2\,\text{mph}. \end{aligned}$$

The wave can be speeded up by decreasing t_r; or the fan can respond not to her nearest neighbor but the next nearest. (Which parameter is changed in that case?)

Can we find a dynamical model for these waves? That is, can we describe the propagation with equations?

Let $y_n(t)$ be the height at time t of the nth fan. The nth fan moves in response to the $(n-1)$st fan's position, and the move is such that the fans tend to align themselves. Let's try

$$\dot{y}_n = -\lambda(y_n - y_{n-1}),$$

where λ is some adjustable factor. This seems right:

$$\text{if } y_{n-1} > y_n, \quad \text{then } \dot{y}_n > 0;$$
$$\text{if } y_n > y_{n-1}, \quad \text{then } \dot{y}_n < 0.$$

EXERCISE 1 Solve the system of equations with all y_n initially zero and then bloop the first one. Show that, although something seems to be propagating, the quantity

$$\max_{t>0} |y_n(t)|$$

is a decreasing function of n. How far could such a fan wave go before effectively dying out? What's missing in this model?

The problem, as you probably realize if you think of this model as a negative feedback problem, is that the correct equations must incorporate a loop–delay in the form of the reaction time. Although our dimensional analysis allowed us to predict that a decreased reaction time could speed up the propagation, it couldn't explain what the dynamics discloses: that the disturbance amplitude decays in space. The system of ODEs implies that, even though only y_1 is nonzero at $t = 0$, every other link in the chain is *instantaneously* affected, and this is obviously wrong. We'll have to investigate the effect of a delay, though, intuitively, it is apparent that it will keep everything from "happening at once"; otherwise, every state carrier has instantaneous knowledge of what others down the line are doing.

A Line of Cars

A driver will accelerate or decelerate in order to maintain a constant distance between herself and the car ahead. Let x_n denote the position of the nth car. The desideratum that

$$x_n - x_{n-1} = \text{ constant},$$

is the same as

$$\dot{x}_{n-1} - \dot{x}_n = 0.$$

So if the driver is to change acceleration level according to some stimulus, it seems reasonable to have that stimulus formed from the quantity

$$\dot{x}_{n-1} - \dot{x}_n,$$

since any nonzero value of that variable will represent a deviation from the desired state of affairs. The equation of motion for the nth car should take the form

$$\ddot{x}_n = f(\dot{x}_{n-1} - \dot{x}_n).$$

The function f is the response function, and, as we now understand, it should incorporate a time delay. The simplest model, arrived at by choosing a linear response function,

$$f(z) = \lambda z,$$

is

$$\dot{u}_n(t) = \lambda(u_{n-1}(t - \tau) - u_n(t - \tau)), \tag{9.2}$$

where $u_n = \dot{x}_n$. The differences in velocities are input and result in accelerations after a time delay.

How is the sensitivity λ determined? (An often-quoted value for λ is 0.37 s^{-1}. Of course, τ is a highly variable parameter, depending on the condition of the road, as well as the age and emotional state of the driver.)

If the lead car must slow down suddenly, will this disturbance propagate? Is there danger of an accident down the line? The answers should depend on the values of λ and τ. Intuitively, we should expect a safe distance to be associated with a maximum τ and a minimum λ.

Let us suppose that up to time $t = 0$ the cars are equally spaced,

$$x_n(0) - x_{n-1}(0) = \bar{d}$$

(and thus have a common velocity, $\bar{U}$). Then the spacing as a function of time is

$$
\begin{aligned}
d_n(t) = x_n(t) - x_{n-1}(t) &= \bar{d} + \int_0^t u_n(s) - u_{n-1}(s)\, ds \\
&= \bar{d} - \frac{1}{\lambda} \int_0^t \dot{u}_n(s + \tau)\, ds \\
&= \bar{d} - \frac{1}{\lambda} u_n(s + \tau)|_0^t \\
&= \bar{d} - \frac{1}{\lambda} u_n(t + \tau) + \frac{1}{\lambda} u_n(\tau).
\end{aligned}
$$

The condition for an accident to occur is that $d_n(t) = 0$ for some n at some time t.

Suppose that the lead car comes to a stop at time t_0 so that $u_1(t) = 0$ for $t > t_0$. The second car will start to slow at $t_0 + \tau$, so $u_2(t') = 0$ for some $t' > t_0 + \tau$. At that time

$$d_2(t') = \bar{d} - 0 + \bar{U}/\lambda.$$

This means that there will be a collision if

$$\bar{d} < \frac{\bar{U}}{\lambda}. \tag{9.3}$$

This agrees with the rule of thumb that separation distance on the highway should be proportional to the velocity of traffic. A practical rule, such as *one car spacing for every* 20 *miles per hour*, cannot be calculated from this directly: if you plug in reasonable values for λ and $\bar{U}$ you will find that this criterion puts you too close for safety.

EXERCISE 2 Why doesn't the criterion, (9.3), depend on the reaction time, τ_r? By taking into account possibilities for τ_r, show that there is still the possibility of collisions even when (9.3) is satisfied.

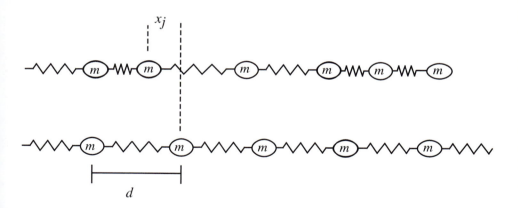

Figure 9.1 A chain of masses connected by elastic springs presents a simple model for a one-dimensional monatomic crystal.

Crystal Vibrations

A simple model for a monatomic crystal consists of point masses connected each to each by springs that represent the chemical bonds. Now,

to describe this explicitly we would need to model long-range interactions, which means that each point mass would interact with not
just nearest neighbors but next nearest, and so on. In addition, the
interaction force is nonlinear. And, moreover, we would need a three-
dimensional array of masses. The model we will analyze is truly simplified (Figure 9.1): it is a one-dimensional chain of masses, only nearest
neighbors are directly interacting, and the interaction force is linear.
Let's see what information we can get from this simple model. Under
our assumptions, it is easy to write down the equations of motion for
any of the point masses. Let x_j be the displacement of the jt mass
from its equilibrium position in the chain; then

$$m\ddot{x}_j = -k(x_j - x_{j-1}) + k(x_{j+1} - x_j)$$

or

$$m\ddot{x}_j + 2kx_j = k(x_{j+1} + x_{j-1}), \quad j = 1, \dots, N. \tag{9.4}$$

(The potential energy for a linear nearest-neighbor interaction is

$$U = \sum_{j=2}^{N} \frac{k}{2}(x_j - x_{j-1})^2 .$$

The equilibrium distance between the masses will be denoted d.)

This is a bunch of coupled oscillators, much like the coupled oscillators we saw in Chapter 8, but now there are a lot of them. Since
the equations are linear, we can "solve" them by finding the normal
modes. If you recall, in the *normal modes* each oscillator is vibrating
at the same frequency.

The normal modes constitute a basis for the solutions of the system
of ODEs. The IVP can be solved by finding the right coefficients to be
used in a superposition of normal modes.

The actual normal modes will depend on the boundary conditions.
(What were the boundary conditions for the two coupled oscillators of
Chapter 8?) Suppose you say *I don't care about the boundary, I only
want to see what vibrations are intrinsic to the crystal deep down inside
it.* Then one can just as well suppose that we have

$$x_{N+1} = x_1.$$

This doesn't mean that we are modeling a ring, but that we ignore any
boundary effects; we are only interested in behavior that cannot be due
to boundary conditions. (The real reason is that the math is easy.)

Let's look for wavelike solutions; disturbances propagating blip-blip-blip down the line (or around the ring). Since the oscillators are linear, imagine each one doing simple harmonic motion, all motions identical, except there is a delay in time between when an oscillator is doing something and when the next one is doing that same thing; this can be represented by a delay in phase:

$$x_p = Ae^{i(\omega t - \phi_{p-1})}, \tag{9.5}$$

where $\phi_0 = 0$ for the first oscillator.

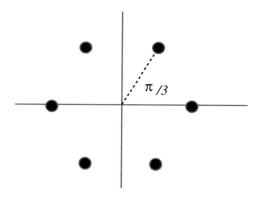

Figure 9.2 The 6 sixth roots of unity in complex plane.

Our ring condition is $x_{N+1} = x_1$, which amounts to

$$e^{i\phi_N} = e^{i\phi_0} = 1.$$

Let's allow a constant phase delay (or perfect symmetry) in our oscillator:

$$\phi_p = p\phi,$$

so that

$$e^{iN\phi} = 1.$$

This means that ϕ is the phase of a (complex) Nth root of unity; there are N distinct solutions,

$$\phi \in \{\frac{n}{N}2\pi\}_{n=1}^{N}.$$

Figure 9.2 shows the six 6th roots of unity in the complex plane. One solution is $z = 1$ itself, for which the phase is $\phi = 0$ (there are infinitely many equivalent solutions, $\phi = 2k\pi, k = 0, 1, 2, \ldots$). In the mode corresponding to this solution, all the oscillators are in phase. This motion would be simply a rotation of the ring of oscillators if it were a physical ring, but this doesn't represent any true oscillation mode of the crystal lattice, so we'll call it the *trivial solution*.

We catalog the possible (nontrivial) solutions for the case $N = 6$:

If $\phi = \pi/3$, then

$$\phi_p = p\phi = 2\pi,$$

only if $p = 6$; thus *all* oscillators are out of phase.

If $\phi = 2\pi/3$, then $\phi_p = 2\pi$ if $p = 3$ or 6, so every third oscillator is in phase; that is, x_1 and x_4 are in phase, x_2 and x_5 are in phase, and x_3 and x_6 are in phase (Figure 9.3a).

If $\phi = \pi$, then $\phi_k = 2\pi$, for $k = 2, 4, 6$, so every other oscillator is in phase: x_1 is in phase with x_3 and x_5, x_2 is in phase with x_4 and x_6 (Figure 9.3b).

If $\phi = 4\pi/3$, then $\phi_p = 4\pi$ for $p = 3, 6$, so that in this case x_1 and x_4 are in phase, x_2 and x_5 are in phase, and x_3 and x_6 are in phase. (You might think that $\phi = 2\pi/3$ and $\phi = 4\pi/3$ represent the same solution, but they are distinct modes of vibration, as Figure 9.3c indicates.)

If $\phi = 5\pi/3$, then all oscillators are out of phase.

The *wavelength* is, we recall, the minimum distance separating two carriers in phase. We have identified three distinct wavelengths:

$\lambda_1 = 6d$ corresponding to $\phi = \pi/3, 5\pi/3$,
$\lambda_2 = 3d$ corresponding to $\phi = 2\pi/3, 4\pi/3$,
$\lambda_3 = 2d$ corresponding to $\phi = \pi$.

If you want, you can think of the trivial motion, $\phi = 0$, as having an infinite wavelength.

EXERCISE 3 Try the example with $N = 5$. Show that although there are four distinct modes of vibration, they all have the same (nontrivial) wavelength, $\lambda = 5d$.

The coupled oscillator of chapter 8 had a normal mode with the same natural frequency as each individual oscillator, $\omega_0 = \sqrt{k/m}$, as well as one higher frequency. Do all the modes of the chain have different frequencies? To answer this, we have to look at the equations of motion. [That's right, thus far we have used absolutely no information from the dynamics. What we have derived so far is pure geometry, from the assumption of a solution of the form (9.5). Now we have to show that there really are solutions of that form.]

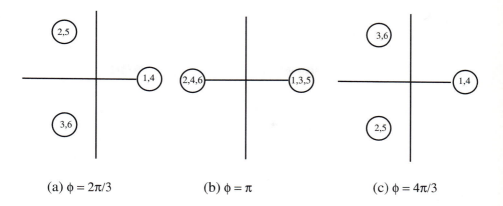

(a) $\phi = 2\pi/3$ (b) $\phi = \pi$ (c) $\phi = 4\pi/3$

Figure 9.3 Phases for $N = 6$ ring of oscillators.

If we plug the (supposed) solution,

$$x_p = e^{i(\omega t - p\phi)},$$

into the equations of motion, we get

$$-m\omega^2 + 2k - k(e^{i\phi} + e^{-i\phi}) = 0,$$

or

$$\begin{aligned} m\omega^2 &= 2k(1 - \cos\phi) \\ &= 4k\sin^2\left(\frac{\phi}{2}\right). \end{aligned}$$

Thus, the frequencies are

$$\omega_n = 2\omega_0 \left| \sin\frac{n\pi}{N} \right|, \quad n = 1, \ldots, N.$$

EXERCISE 4 Show that there are $N/2$ distinct frequencies. [As long as N is even, that is. There are $(N - 1)/2$ if N is odd, and we are excluding the possibility $\omega = 0$.] Why are there two distinct frequencies for $N = 5$? (*Hint:* Look at Exercise 3.)

Let's think of the different modes of vibration in terms of "waves": the wave speed will depend on the wavelength and the frequency. Recall that the wavelength, λ_n, of the nth vibrational mode is the distance between same-phase oscillators, and the frequency determines the time interval separating the same state at the same oscillator (i.e., the time of the disturbance):

$$T_n = \frac{2\pi}{\omega_n}.$$

Then the wavespeed is calculated as

$$c_n = \frac{\lambda_n}{T_n} = \frac{(N/n)d}{(2\pi/2\omega_0)|\sin\frac{n\pi}{N}|}, \quad n = 1, \ldots, \frac{N}{2}$$

$$= \frac{d\omega_0}{\pi}\frac{N}{n}|\sin\frac{n\pi}{N}|, \quad n = 1, \ldots, \frac{N}{2}.$$

Notice that these velocities are all different.

EXERCISE 5 How many different wave speeds are there on the ring with six oscillators? With five?

Fixed-end Boundary Condition

Suppose that we impose the boundary condition

$$x_0 = x_{N+1} = 0.$$

Since the motions are continuous, we expect that for many modes the end masses will have less amplitude. With the ring condition, we had the solution

$$x_p = Ae^{ip\phi}e^{i\omega t},$$

where all the oscillators had the same amplitude of oscillation. This is impossible for the new boundary conditions, since the amplitude should go smoothly to zero at $p = 0$ and $p = N + 1$. So let's try a periodic function with zeros:

$$x_p = A\sin(p\phi)e^{i\omega t}.$$

Then the use of the boundary condition

$$x_{N+1} = 0$$

gives the equation

$$\sin[(N + 1)\phi] = 0,$$

so

$$\phi = \frac{n\pi}{N + 1}, \quad n = 1, 2, 3, \ldots.$$

Note that in this case the minimum phase delay is approximately half that of the minimum phase delay in the case of a ring.

What are the wavelengths? The wavelength is the minimum distance between two masses that would be in phase. So let's look for the oscillators that are in phase:

$$\phi_p \equiv \phi_q \Longrightarrow p - q = \frac{2(N+1)}{n}, \quad n = 1, \ldots, N.$$

This generates the N distinct wavelengths

$$\{\lambda_n\} = \{\frac{2(N+1)}{n} d\}_{n=1}^{N}.$$

Notice that not all the values $p - q$ determined in this way are necessarily integers, so there may not actually be two masses separated by a distance λ_j; but this is still the length scale characterizing the jth mode.

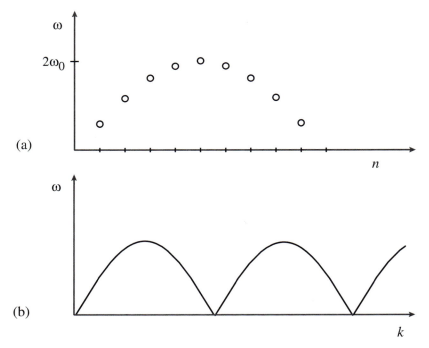

Figure 9.4 (a) Frequencies as given by (6). (b) A dispersion relation.

Again, the dynamics determines the frequencies through the equation (9.4). Now we get the characteristic equation

$$\omega = 2\omega_0 |\sin \frac{\phi}{2}|,$$

which gives the frequencies

$$\omega_n = 2\omega_0 \left| \sin \frac{n\pi}{2(N+1)} \right|. \tag{9.6}$$

Now there are N distinct frequencies, and all the frequencies are lower than twice the "natural" oscillator frequency,

$$\omega_n < 2\omega_0,$$

or, in other words, strictly less than the highest frequency in a *ring* of oscillators. Figure 9.4a shows all the frequencies plotted according to (9.6). Since the inverse wavelengths are proportional to the index n, we can plot frequency versus inverse wavelength for all values of the latter parameter (Figure 9.4b).

Forced Vibrations

Now suppose that we take the right end of the chain and push it back and forth rhythmically (Figure 9.5). That is, the new boundary condi-

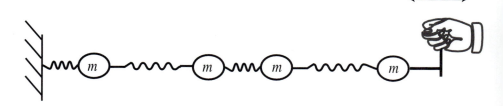

Figure 9.5 Forced chain.

tions are

$$x_0 = 0, \quad x_{N+1} = C \cos \omega t.$$

From our previous exposure to forced vibrations, we should expect that a steady-state solution should exist with the frequency ω, unless it is one of the natural modes of vibration, in which case there should be a resonance. Let's then look for a solution of the form

$$x_p = A_p \cos(\omega t) = A \sin(p\theta) \cos(\omega t).$$

Plugging this into the equation of motion (9.4), gives the condition

$$\cos \theta = 1 - \frac{1}{2} \left(\frac{\omega}{\omega_0} \right)^2.$$

For any $\omega \leq 2\omega_0$, we can solve this for θ; this gives us a well-defined "steady-state mode" with an amplitude of

$$A = \frac{C}{\sin(N+1)\theta}.$$

EXERCISE 6 (i) For which values of ω do we have something like resonance? What happens for $\omega = \omega_0$? Notice that the answer depends on N.
(ii) Show that if $\omega \ll \omega_0$ then the wavelength corresponding to that mode is effectively infinite. (*Hint:* Find an equation for A_p and show that there is no curvature, no local extrema allowed.)
(iii) Show that if $\omega = \sqrt{2}\omega_0$ then $\lambda = 4d$
(iv) Show that if $\omega = 2\omega_0$ then $\lambda = 2d$

What happens if $\omega > 2\omega_0$? Is there a steady-state solution? In that case, the equation

$$\cos\theta = 1 - \frac{1}{2}(\frac{\omega}{\omega_0})^2$$

has no real solution. However, it does have a complex solution. Write

$$\cos\theta = -(1+\beta), \quad \beta > 0.$$

Then

$$\theta = \pi + i\gamma, \quad \gamma > 0.$$

Thus,

$$x_p = A_p e^{i\omega t},$$

with

$$A_p = A\sin(p\pi + ip\gamma) = A(-1)^p i \sinh(p\gamma),$$

so that we have a solution of the form

$$x_p = C\frac{\sinh(p\gamma)}{\sinh[(N+1)\gamma]}(-1)^{p-(N+1)}e^{i\omega t}.$$

Make note of two things:

1. The wavelength, λ, is always $2d$, so adjacent masses are always out-of-phase. In other words, phase is independent of position along the chain.

2. The amplitude, $A\sinh(p\gamma)$, is strictly decreasing to zero as we go along the chain from $p = N+1$ to $p = 0$. In fact, since

$$\omega \gg \omega_0 \implies \beta \gg 1$$

and using
$$\sinh x = \frac{e^x - e^{-x}}{2},$$

we have
$$\frac{\sinh(p\gamma)}{\sinh[(N+1)\gamma]} \approx e^{[p-(N+1)]\gamma}.$$

In other words, $A_N \approx e^{\gamma}C$ or, generally,

$$A_p/A_{p+1} \approx e^{-\gamma}.$$

For high frequencies, most of the line is completely unresponsive. Only the first few masses have any significant amplitude (Figure 9.6). The

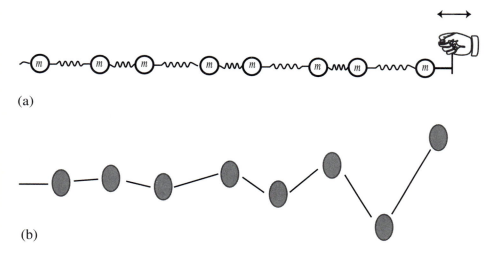

(a)

(b)

Figure 9.6 (a) High-frequency forcing of a long chain. (b) Same forcing, but visualized as a transverse wave.

critical frequency,
$$\omega_c = 2\omega_0,$$

is called the *cut-off frequency*. This line of coupled masses acts as a low-pass filter.

Transient phenomena

For short times after the shaking of the chain begins, the quiescent chain responds gradually (depending on ω). For low frequencies, the initial disturbance is propagated at a wave speed of

$$C = C_{min} \approx \omega_0 \frac{\pi d}{N + 1}.$$

When the disturbance reaches the other end of the chain, it is reflected and turns back to interact with incoming to set up a standing wave. It is hard to study this phenomenon in the chain, but we will show in what follows that for low frequencies the chain can be replaced by a continuum snake. In Chapter 10, the snake will be analyzed. When forced at high frequencies, the solution relaxes to the final "waveform."

Crowded chains

What happens when the number of oscillators in a chain of a given length is made to increase so that the chain becomes finer? Notice that if we double the number of oscillators so that

$$d \rightarrow d/2,$$

then, since the total mass should stay the same, we must divide the mass among the oscillators so that

$$m \rightarrow m/2.$$

Explicitly, if we denote

$$L = \text{ total length}$$

and

$$M = \text{ total mass,}$$

then

$$d = \frac{L}{N + 1}, \quad m = M/N, \text{ as } N \rightarrow \infty.$$

For large N, we have that

$$m/d = \frac{M}{L} \frac{N + 1}{N} \approx \frac{M}{L} = \mu,$$

where μ is the mass per unit length, or the *mass density*.

Besides m and d, the other parameter important to the chain dynamics is the spring stiffnes, k. How does that scale with N? An investigation of the frequencies allows us to see what happens.

In high-frequency motions, the influence of the boundaries extends only for a small region near the ends. We should expect this to mean that for high frequencies we can replace the "fixed-end model" by a "ring model". In that case,

$$\omega_N \to 2\omega_0, \quad N \to \infty,$$

as $2\omega_0$ is the highest frequency for a ring.

For low frequencies, we have the approximate relation

$$\begin{aligned}\omega_n &\approx 2\omega_0 \frac{n\pi}{2(N+1)} \\ &= 2(\frac{k}{m})^{1/2} \frac{n\pi}{2(N+1)}.\end{aligned}$$

Because of the scalings,

$$m/d \to \mu, \quad (N+1)d \to L,$$

we must have that

$$kd \to \kappa,$$

where κ is a paramater characterizing the stiffnes of the infinite-oscillator chain. Notice that κ has dimensions of force. In this way, we have for very large N,

$$\begin{aligned}\omega_n &\approx (\frac{kd}{m/d})^{1/2} \frac{n\pi}{d(N+1)} \\ &\approx \frac{\pi}{L}(\frac{\kappa}{\mu})^{1/2} n \\ &= n\omega_1.\end{aligned}$$

Thus, the low frequencies are all multiples of a *fundamental* lowest frequency,

$$\omega_1 = \frac{\pi}{L}(\frac{\kappa}{\mu})^{1/2}.$$

What is the significance of this fundamental?

If our one-dimensional chain is to be a model for a solid crystal, which is three dimensional, it can only be that we are interested in vibrational disturbances that are restricted to motion along the direction represented by the chain. Let A be the cross-sectional area of the crystal in any plane perpendicular to the chain. (We are thinking of our solid as being what is usually called a *rod*.) Then

$$\mu/A = \rho$$

is the mass density in the crystal. Also,

$$\kappa/A = E$$

is the modulus of elasticity (Young's modulus). Then

$$\sqrt{\frac{\kappa}{\mu}} = \sqrt{\frac{\kappa/A}{\mu/A}} = \sqrt{\frac{E}{\rho}}.$$

But we know from dimensional analysis that $c = \sqrt{\frac{E}{\rho}}$ is the characteristic wave speed for longitudinal motions in a solid. This means that

$$T_L = L/c = L/\sqrt{\kappa/\mu}$$

is the time for a disturbance to propagate the length of the solid. Comparing with the expression for ω_1, we see that T_L is one-half the period of the slowest oscillation.

EXERCISE 7 Can you think of an intuitive explanation for that?

A Snake

It is interesting that for the discrete chain there are all these different wave speeds, but that there is only one characteristic speed for the limiting case of infinitely many oscillators. This characteristic parameter appears naturally if we look at the limit of the dynamics, (9.4), as the number of oscillators goes to infinity. In order to minimize confusion in introducing continuous variables, let us use x to denote distance along the rod, which extends from $x = 0$ to $x = L$. Let $y(x)$ denote the displacement from equilibrium of the oscillator, which at equilibrium is at the point x. In other words, for the discrete oscillator, y_j is displacement of the j oscillator, which is in equilibrium ($y_j = 0$) at the point $x_j = jd$. Or, in other words, the connection between the discrete and continuum variables is

$$y_j = y(x_j).$$

Then we can write the dynamics of the chain (9.4) as

$$\ddot{y}_j = \frac{k}{m}(y_{j+1} - 2y_j + y_{j-1}).$$

By introducing the appropriate scalings of the variables m and k, we get

$$\ddot{y}_j = \frac{kd}{m/d} \frac{y_{j+1} - 2y_j + y_{j-1}}{d^2}. \tag{9.7}$$

In the limit,

$$d \to 0,$$

as our chain becomes a snake, we get the partial differential equation

$$\frac{\partial^2 y}{\partial t^2} = \frac{\kappa}{\mu} \frac{\partial^2 y}{\partial x^2}. \tag{9.8}$$

(Alternative notations for partial derivatives, such as $f_u = \frac{\partial f}{\partial u}$, will regularly appear in the remainder of this book. In this way, the equation (9.8) can also be written

$$y_{tt} = \frac{\kappa}{\mu} y_{xx}.) \tag{9.9}$$

Since

$$\sqrt{\frac{\kappa}{\mu}} = \sqrt{\frac{E}{\rho}} = c,$$

we can write this as

$$y_{tt} = c^2 y_{xx}. \tag{9.10}$$

This is the equation known as the one-dimensional wave equation. Continuum models based on the use of partial differential equations will be the subject of chapters 10 and 11.

EXERCISE 8 Using Taylor expansions, show that indeed

$$\frac{y_{j+1} - 2y_j + y_{j-1}}{d^2} \to y_{xx},$$

in the limit as $d \to 0$.

2   *Filters and Ladders

It is common in some circles to refer to linear systems as *filters*. The idea is that the input has desired components as well as undesired components and that after passing through the system (filter) the output should reproduce the desired part of the input while suppressing the undesired part. Because certain familiarity with the language of filters is useful to the young modeler, we will look at a restricted class of filters, whose characteristics are that they 'pass' certain frequency components in the input, while suppressing others. If one pass through a filter does a pretty good job, then two passes should do better, and this leads to the idea of chaining filters to improve the desired effect.

An example of this is seen in the line of vibrating masses. A single linear oscillator, characterized by a frequency, ω_c, will, according to its frequency-response curve, show little response to very high frequencies. For a long chain of masses, there is a sharp cutoff so that frequencies above that cutoff do not oscillate the chain at all.

Suppose that we take a certain signal, say a voltage, and we want to remove the high frequencies from it. We could pass this signal through a simple circuit, for example the RC circuit of Figure 9.7, and measure the resulting voltage across the capacitor. Previously, when studying

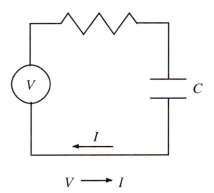

Figure 9.7

this circuit, we had thought of the input–output relationship as

$$V \to I,$$

but now we think of $W = V_C$ as the output; so we could write it like

$$(V, I) \to W.$$

We could draw the circuit with two loops as in Figure 9.8, but we are only interested in the case where there is no current through the second loop. And so filter circuits are usually drawn as in Figure 9.9.

Since no current is drawn on the output,

$$I_C = C\dot{W}.$$

But, by tracing the voltage drops around the first loop,

$$
\begin{aligned}
V &= RI + \frac{1}{C}\int I_C \\
&= RI + W = RC\dot{W} + W.
\end{aligned}
$$

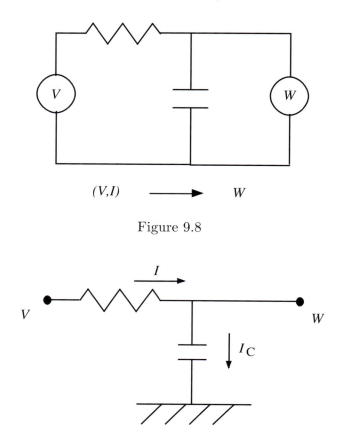

Figure 9.8

Figure 9.9 What current is drawn from right-hand terminal?

We now take the Fourier transform of this equation or, equivalently, look for the response of the form

$$\hat{W}e^{i\omega t}$$

to an input of the form

$$\hat{V}e^{i\omega t}.$$

Then the amplitudes are related by

$$\hat{V} = (i\omega RC + 1)\hat{W}.$$

The ratio of output to input,

$$\frac{\hat{W}}{\hat{V}} = T(\omega) = \frac{1}{1 + i\omega RC}, \tag{9.11}$$

characterizes the filter. The function T is called a *transfer function*. The transfer function in (9.11) is an example of a **low-pass filter**. In Figure 9.10 we plot the squared amplitude, $|T|^2$, of the transfer function, which represents the transfer of power. Signals of very low frequency are transferred with practically no change in amplitude, while high-frequency signals are highly attenuated.

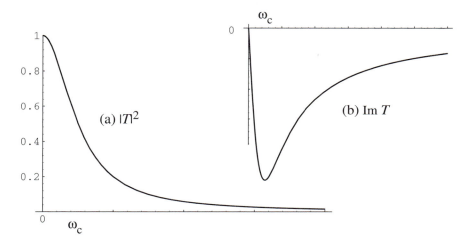

Figure 9.10 Low-pass filter.

EXERCISE 9 (i) Find the real and imaginary parts of the function T in (9.11), and show that the phase is

$$\tan \phi = -\omega RC.$$

(ii) Show that the frequency where the power, $|T|^2$, drops by half is

$$\omega_c = \frac{1}{RC}.$$

Notice that applying the function T (in 9.11) in the frequency domain is equivalent to the operation of solving the differential equation

$$V = RC\dot{W} + W.$$

For this reason the function T (in 9.11) is sometimes called an *integrator*.

The frequency where the power drops by half is the conventional marking place for the *bandwidth*. For the filter in (9.11) this is

$$\omega_c = \frac{1}{RC}.$$

The cut-off frequency, ω_c, is sometimes called the 3 dB point, since 3 dB$\approx 10\log_{10} 2$.[1]

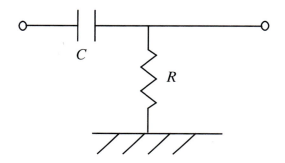

Figure 9.11

The circuit of Figure 9.11 functions as a **high-pass filter** for voltage. If we assume that the output port is an open circuit, then the output voltage, W, is related to the input current by

$$W = RI.$$

Then

$$\dot{V} = R\dot{I} + \frac{1}{C}I = \dot{W} + \frac{1}{RC}W,$$

so that, transforming as we did above, we get

$$i\omega\hat{V} = \hat{W}(i\omega) + \frac{1}{RC}W,$$

resulting in the transfer function

$$\frac{\hat{W}}{\hat{V}} = T(\omega) = \frac{i\omega}{\frac{1}{RC} + i\omega}.$$

The power, $|T|^2$, is plotted in Figure 9.12. The cut-off frequency is again $\omega_c = \frac{1}{RC}$.

---

[1]Recall that a signal is k dB (decibels) with respect to a reference value if it is a factor $10^{k/10}$ of it. And $10^{3/10} \approx 2$.

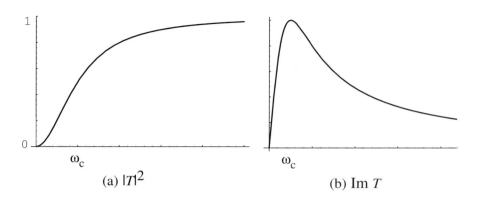

(a) $|T|^2$

(b) Im T

Figure 9.12 High-pass filter.

If we pass a signal through the low-pass filter of Figure 9.9, then all frequencies above ω_c will be reduced in amplitude by at least half. So, if we pass the signal once more through the same filter, we expect that all the frequencies above ω_c will be attenuated by a factor of one fourth. Since the filter acts as a multiplication, the double-pass filter transfer function is

$$T^2(\omega) = \frac{1}{(1 + i\omega RC)^2}.$$

By repassing n times, we get a power response like

$$|T^n|^2 = \frac{1}{(1 + \omega^2 R^2 C^2)^n}.$$

EXERCISE 10 Show that the cut-off frequency for the n-pass filter is

$$\omega_c^2 = \frac{2^{1/n} - 1}{R^2 C^2},$$

so that $w_c \to \infty$ as $n \to 0$.

It may seem that an implementation of the double-pass filter would be the circuit of Figure 9.13. But this is not quite right. The reason is that to get the first part of the circuit to act as a low-pass filter we specified that there would be no current output to the next section, but obviously we must supply current to the next section to change the voltage.

EXERCISE 11 Solve the circuit in Figure 9.13 and obtain the overall tranfer function. Show that it is a low-pass filter, with cut-off frequency lower than is obtained with only one section.

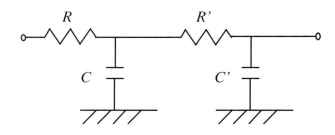

Figure 9.13 Double filter.

If we consider a long ladder of such sections, then the transfer function of the ladder is closer to the function T^n. Actually, if we go to an infinite ladder, the calculation becomes easier, as we will see below.

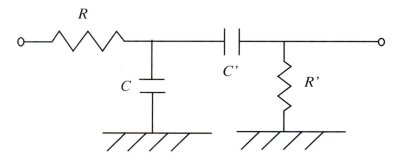

Figure 9.14 Band-pass filter.

EXERCISE 12 The circuit of Figure 9.14, is made up of a low-pass filter followed by a high-pass filter. Solve the circuit to show that very low frequencies and very high frequencies are attenuated; thus, it is a true **band-pass filter**. Graph the power function and compare with the power function of a simple resonance (see Chapter 5). Circuits with inductances and capacitances exhibit resonance, which is a filter effect. Consider the circuit of Figure 9.15. The output voltage is the voltage across the capacitor, or

$$\dot{W} = \frac{1}{C}I,$$

so

$$V = W + LC\ddot{W} \qquad\qquad (9.12)$$

or

$$\hat{V} = \hat{W} + LC\hat{W}(-\omega^2).$$

This gives the resonant low-pass filter. The differential equation (9.12) is the familiar forced oscillator of Chapter 5.

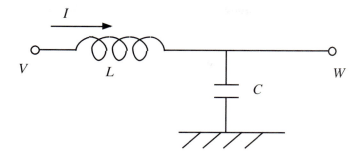

Figure 9.15 Resonant low-pass filter.

On the other hand, the circuit of Figure 9.16 is a high-pass resonant filter. In that circuit the differential equation relating input voltage to output involves time derivatives of the input:

$$\ddot{V} = \ddot{W} + \frac{1}{LC}W,$$

with the frequency-domain transfer function

$$\frac{LC\omega^2}{LC\omega^2 - 1}.$$

The graphs of these two resonant filters are shown in Figure 9.17.

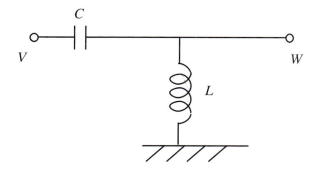

Figure 9.16 Resonant high-pass filter.

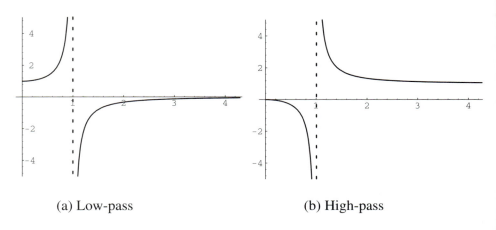

       (a) Low-pass   (b) High-pass

Figure 9.17 Resonant filters. $LC = 1$.

As with the RC circuits, the filtering capabilities of LC circuits can be improved by chaining them together into ladder networks. In fact, their use as filters is generally easier over long distances (see the discussion of transmission lines in Chapter 10). A general ladder network is represented in Figure 9.18 in terms of the component impedances.

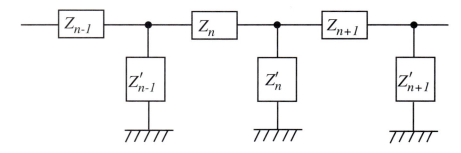

Figure 9.18 Ladder of impedances.

Recall that impedance gives the ratio of voltage to current through each component (see Chapter 5). The admittance is the reciprocal of the impedance. The impedances for simple components, resistors, capacitors, and inductors, are

$$Z_R = R, \qquad Z_C = \frac{1}{i\omega C}, \qquad Z_L = i\omega L.$$

The impedances of compound components can be made up from the impedances of their components by using the rules of circuit theory. Two impedances in series (Figure 9.19) can be replaced by the sum of the individual impedances,

$$Z_{\text{series}} = Z_1 + Z_2.$$

Two admittances in parallel sum to give the resulting impedance:

$$\frac{1}{Z_{\text{parallel}}} = \frac{1}{Z_1} + \frac{1}{Z_2}.$$

(a)   (b)

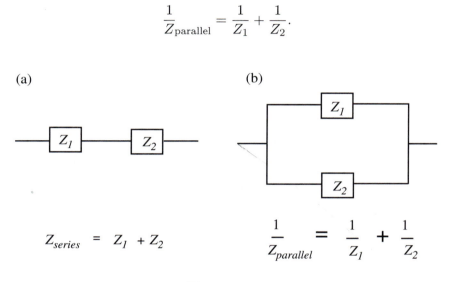

$$Z_{series} = Z_1 + Z_2 \qquad\qquad \frac{1}{Z_{parallel}} = \frac{1}{Z_1} + \frac{1}{Z_2}$$

Figure 9.19

As an example, the overall impedance of the circuit in Figure 9.15 can be found by eliminating W from the differential equation (9.12):

$$\dot{V} = \dot{W} + L\ddot{I}.$$

In the frequency domain, this is

$$\hat{V}/\hat{I} = (i\omega L + \frac{1}{i\omega C}).$$

Notice that the circuit of Figure 9.16 has the same impedance.

The impedance of a ladder can be found by combining impedances over and over again. In this way the impedances of all sections of the ladder combine to give the *characteristic impedance* of the ladder, Z_0 (Figure 9.20). Instead of computing this whole series, we can compute

Z_0 by noticing that the ladders of Figures 9.20 and 9.21 are equivalent. Thus,

$$Z_0 = Z_1 + \frac{1}{\frac{1}{Z_2} + \frac{1}{Z_0}}.$$

This is a quadratic equation for Z_0; solving, we get

$$Z_0 = \frac{Z_1}{2} + \sqrt{(Z_1^2/4) + Z_1 Z_2}.$$

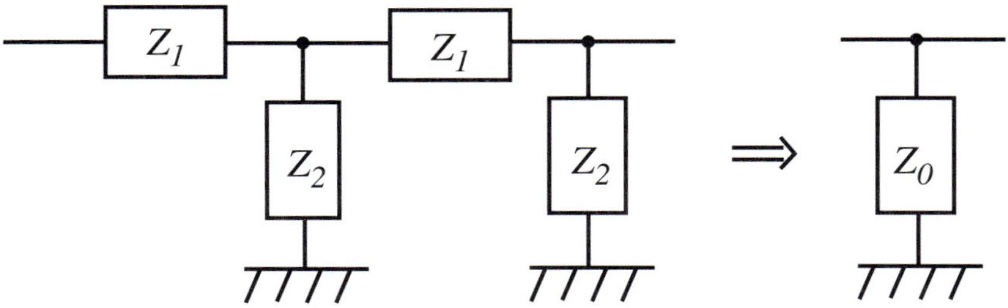

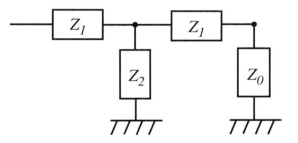

Figure 9.20

Figure 9.21

The characteristic impedance of any ladder of the form shown in Figure 9.18 can be found from this. For example, if sections of the (low-pass) circuit in Figure 9.15 are put together to form a ladder, with $Z_1 = i\omega L$, and $Z_2 = (1/i\omega C)$, we get

$$Z_0 = \frac{i\omega L}{2} + \sqrt{\frac{-\omega^2 L^2}{4} + \frac{L}{C}}$$

Notice that the impedance is purely imaginary if $\omega^2 > (4/LC)$. What does that have to do with the idea that this ladder should be a good low-pass filter? If the characteristic impedance is purely imaginary for some frequency, then signals of that frequency will not pass down the line. The reason is that in the ladder we need to continually supply a real current to each new section in turn. To get currents to flow down the line, we need a real part to the impedance. To see this more vividly, let us calculate the transfer function from one section to the next in the infinite ladder. Note that for any n, the voltage across the nth section, V_n, satisfies

$$V_n = I_n Z_0.$$

But the circuit gives

$$V_n - V_{n+1} = Z_1 I_n.$$

Putting these together gives

$$V_{n+1} = (1 - \frac{Z_1}{Z_0})V_n \tag{9.13}$$
$$= \Pi V_n \tag{9.14}$$

The number Π is known as the propagation factor. In general it is a complex number,

$$\Pi = \alpha e^{i\delta}.$$

The modulus α gives the factor by which the voltage is reduced at each section, and δ is the amount by which the phase is delayed.

EXERCISE 13 (i) For the LC ladder of Figure 9.22, show that $\alpha = 1$ if $\omega^2 < (4/LC)$; otherwise, $\alpha < 1$. If $\alpha < 1$, the relation (9.14) means that the signal is exponentially attenuated as it goes down the line.
(ii) For an RC ladder made up of sections such as those in Figure 9.9, show that $\alpha < 1$ for ω bigger than some value.

More examples and applications of filters are to be found in the problem section.

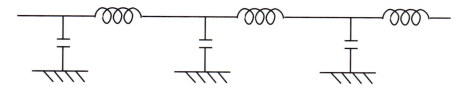

Figure 9.22

3   *Modeling the Ear

The hearing process is the transduction, or translation, of atmospheric pressure waves (sound) into electrical signals in the nervous system. These signals are then processed by the brain. How much processing must be left to the neural networks of the brain in this mechanical-to-electrical filtering process? Brain modeling is in its infancy and no doubt will eventually help us understand hearing, but perhaps some of the hearing functions can be explained through known physical properties of the ear. Hearing is a complex activity with many specific tasks. For example, the ears are necessary for determining the direction and distance of a sound, but most of the work necessary for these functions must be delegated to the higher processing centers. We are aware of manifold qualities of sound, such as timbre, clarity, color, but these are really functions of the two basic perceptions: loudness and pitch. *Loudness* perception refers to the transduction of the acoustic *energy* and amplitude in a sound wave, while pitch perception refers to the trapping of the dominant *frequency* (or frequencies) in a sound wave. We will concentrate on models that attempt to explain pitch perception, but it is a fact that the ear is nonlinear and these two properties cannot be fully separated. It is the hallmark of a linear system that the quality of the system is independent of quantity. In experiments on a variety of animals, it has been shown that frequency combinations (subharmonics) are produced by the hearing process. There are other nonlinearities that we will need to mention in due course, one common one being the experience that pitch is hard to detect when the sound intensity level is very high. (Even putting aside the questions raised by the form of the nonlinearities, the fact that the ear is a finite transduction mechanism means that its frequency resolution is affected by its time discrimination via the uncertainty principle, as we pointed out during the discussion of resonance in chapter 5.)

In referring to the *ear* or *hearing* we have in mind, certainly, the human ear (although it seems all mammal ears function similarly), but also envision general possibilities for the design of a hearing mechanism. In fact, the sensitivity of this biological signal processor has yet to be equaled by modern technology.

One possibility is that the ear should function like a microphone, whereby sound pressure fluctuations are converted directly into electrical signals. These signals would have to be analyzed to reveal the frequencies. The problem with that idea is that the nerves, because

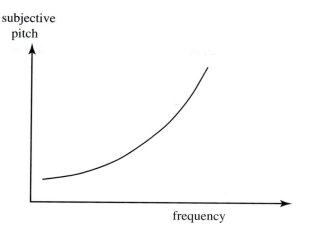

Figure 9.23 If a human subject would recognize the same musical interval (as supplied by an oscilloscope) no matter where it appeared on the musical scale, then this graph would be a straight line.

of inherent temporal limits due to saturation and latency, cannot completely process all the information in a signal fast enough to capture it all. It would be possible with the given quality of hardware to improve this situation with an extremely large bank of nerves, but this is not found in any anatomy. The result is that only time-averaged signals could be picked up for processing. But then all hope of fine frequency selectivity is lost. Apparently, insects have ears of this type; they do not possess good frequency resolution (sorry, no conversation for jiminy cricket), but they enjoy a quick reaction to instantaneous changes in sound pressure.

How then should we "tune in" to a particular frequency? The idea of resonance may jump to mind. For example, given an electrical signal, we can pass it through an LC circuit and start changing the capacitance until we notice a resonant response. Building on our analogies, we can likewise build a seismometer with a stiffness-variable mass–spring system to tune into mechanical vibrations. But the need to change parameters in this way is too slow to explain the actual response of the ear. Another possibility is to have a large number (say 10^5) of individual resonant oscillators all lined up, each with a different resonant frequency so that, when a sound wave of a particular frequency comes by, it should be easy to recognize the one that matches with the forcing frequency. This is essentially the model that Helmholtz proposed. The model has the virtue of suggesting a possible interface with the nervous system. If we suppose an all-or-nothing model for the firing of nerves, that is, a nerve will fire only when stimulated beyond a threshold, then, at moderate sound levels, only the nerve leading from the oscillator in resonance will fire and the others will be silent. At higher levels of intensity, vibrators near resonance will also fire, which leads to "fuzzy" pitch discrimination. (In essence, this is the *place principle* of hearing: if we imagine that the lined-up oscillators are indexed by k, then there is the obvious mapping

$$k \longleftrightarrow \omega_k,$$

that relates frequency and position.)

Any damping will tend to broaden the resonance peak so that at any finite "Q" there will be a definite relationship between possible nerve thresholds and pitch discrimination. And Q, it turns out, must be very high for this model to work.

EXERCISE 14 Assuming the all-or-nothing model for nerve firing, what must Q be to be able to distinguish semitones at 10 dB above the threshold of hearing (and that's not asking for much).

If this model is true, it means that the nervous system postprocessor must do most of the work.

How well does this model do in explaining the results of psychoacoustical experiments? One result, which we've already mentioned is that the pitch discrimination facility is extremely sensitive, especially at low sound levels and at higher frequencies. Even ignoring the fact that a linear system is inadequate, such a system would require a very high tuning parameter, Q. Another important result is that the loudness of one tone does not simply add to another tone unless separated by a critical frequency bandwidth. That is, two moderate sounds played together will make a loud sound if they are far apart in pitch. However, if their frequencies are close together, then the resulting sound is not much louder than either of the tones sounded individually.

EXERCISE 15 Can this be explained by the Helmholtz resonators-plus-nerve-threshold model?

Another important effect is that the least discriminable change in frequency increases with frequency. This is illustrated by Figure 9.23, where the abscissa is frequency and the ordinate plots a measure of "subjective pitch."[2] The graph would be a straight line if a particular

---

[2]It must be remarked that pitch discrimination is usually taken to mean the

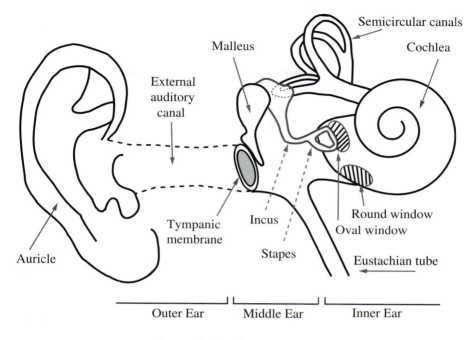

Figure 9.24 (a) Whole Ear.

musical interval was heard as the exact same proportion of frequencies wherever it was sounded on the musical scale.

EXERCISE 16 How would you need to arrange the distributions of resonant frequencies in the Helmhotz model in order to simulate this effect?

Since the time of Helmholtz, of course, a lot has been learned about the function and, especially, the anatomy of the ear. The ear is reminiscent of one of those Rube Goldberg contraptions. Air pressure waves enter a horn and pass through a tube to pound a drum to which is attached a linkage of bones; at the end of this is a stirrup that pulls on the end of a coiled tube (the cochlea) filled with liquid; along the length of the tube is a membrane that converts pressure waves to mechanical shear; hair cells respond to these shears with a change in electrical potential, innervating nerve cells that send these signals via the auditory nerve to the brain (Figure 9.24). noindent The basilar membrane sepa-

discrimination of two tones that are played at different times, as opposed to being able to distinguish two tones played simultaneously. Imagine an experiment where the two different tones are played alternately into headphones. The tones are distinguishable when a sirenlike sound is heard.

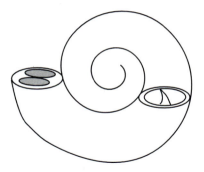

Figure 9.24 (b) Cochlea cut-away.

rates the cochlea into upper and lower fluid-filled chambers. The basilar membrane is very narrow near the forced end (base) and gets progressively wider along the longitudinal axis to the apex (Figure 9.25). The basilar membrane is actually the bottom piece of a complex structure known as the organ of Corti (Figure 9.26). Because of the different rotation centers of the basilar membrane and the tectorial membrane, relative shear is produced whenever the basilar membrane is displaced by fluid (Figure 9.27). The sensory end organs, which are the stereocilia of the hair cells, are bent by this shear strain. This mechanical bending alters the electrical resistance of the hair cells, leading to a change in current that causes a firing of a synapse, which depolarizes the axons at the next level of neurons, the beginning of the auditory nerve. There are two types of hair cells, the inner hair cells and the outer hair cells (named for their position). It is probable that only the inner hair cells are pickups and that the outer hair cells are involved in a feedback mechanism. The functioning of the cochlea according to the place principle can be pictured as in Figure 9.28.

Based on experiments done with cadavers, Békésy modeled the basilar membrane as an isotropic elastic plate that widened along the longitudinal axis. He pointed out that an elastic sheet under constant tension admits distensions that increase as the fourth power of the width.

EXERCISE 17 Show that this is the case. (*Hint:* the potential energy of bending is proportional to the curvature.)

According to this, he expected to see a wave on the basilar membrane growing as it moved down the longitudinal axis. This type of

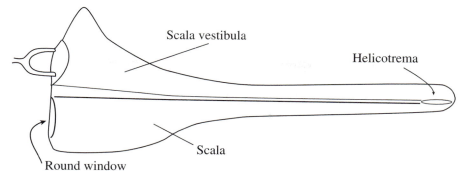

Scala vestibula

Helicotrema

Scala

Round window

(a) This cut-away of unrolled cochlea gives a side-view of the basilar membrane separating the cochlea into upper and lower fluid chambers.

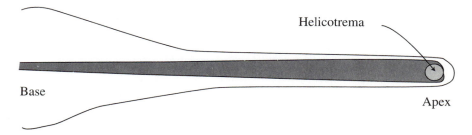

Helicotrema

Base

Apex

(b) This cut-away of unrolled cochlea gives a top-view of basilar membrane. The basilar membrane gets wider along the longitudinal axis from base to apex.

Figure 9.25

traveling wave will move at a speed much lower than the usual sound wave in a liquid (see Chapter 10). In fact, he carried out microscopic observations of cochlear vibrations and for a fixed frequency he plotted the amplitude versus distance (Figure 9.29a). He also plotted amplitude versus frequency for different positions along the membrane (Figure 9.29b). From the asymmetry of the last figure, Békésy concluded that the cochlea must function as a low-pass filter rather than a "tuner."

EXERCISE 18 Where have we seen a low-pass mechanical filter? Is there some connection?

The conclusion that must be drawn from the Békésy model, as

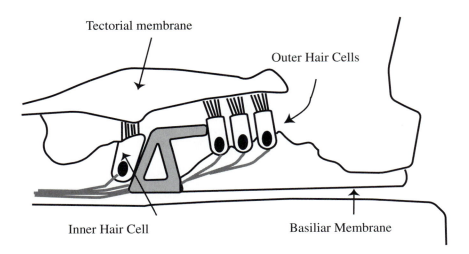

Figure 9.26 The area at the right (not covered by the organ of Corti) is negligible at the base at wide at the apex.

from the Helmhotz model, is that the mechanical cochlea displays poor frequency selectivity and thus obviates the need for highly developed neural tuning.

However, the story doesn't end there. Starting in the 1970s, more refined experiments have been carried out on live or freshly killed cochlea, using nondestructive techniques such as the Mössbauer effect, at moderate forcing levels. (The forcing level that allows a visual microscopic observation, such as those carried out by Békésy, corresponds to a sound level of 130 dB.) The results show that the cochlea itself acts

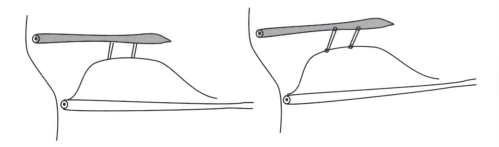

Figure 9.27 The basilar membrane and the tectorial membrane rotate about different centers of rotation which results in a relative shear between them.

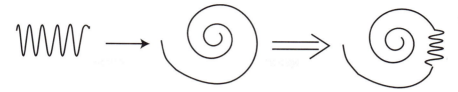

(a) Response to single high frequency.

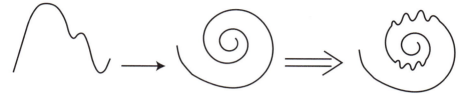

(b) Response to composite midrange frequency.

Figure 9.28 The "place principle".

as a highly selective band-pass filter. The response is especially highly peaked for low-intensity forcing (near the threshold of hearing). These experiments, typically done at a single *place* on the basilar membrane, come in two varieties: the *amplitude response curve* and the (threshold)*tuning curve*. Figure 9.30 shows amplitude response curves for different sound levels. The tuning curves of Figure 9.31 are labeled for displacement thresholds (yes, that is less than a nanometer; these are atomic length scales).

Meanwhile, careful measurements of fresh cochlea showed that the basilar membrane is not isotropic, but that it has negligible stiffness in the longitudinal direction. The effect is something like a row of piano strings (shades of Helmholtz again). To get the sharp cut-off from a row-of-strings model requires that the strings be coupled to each other in some way. The fluid in the tube can supply such a coupling. It does this through the added mass or virtual mass due to the movement of each string. If you have read about the modeling of flow-induced vibrations in Chapter 8, you know that the added mass due to motion through a fluid depends on the motion. For wavelike motions, long wavelengths correspond to large virtual mass because they are pushing on the fluid all at once over some longitudinal distance, while small-wavelength motions have many up and down motions over

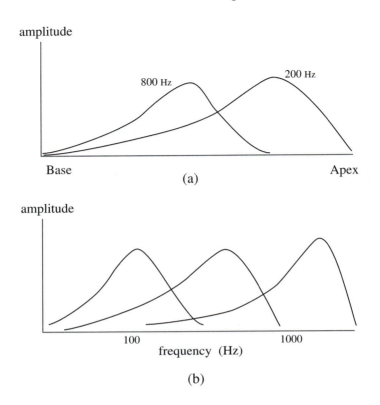

Figure 9.29 Békesy's curves.

that same distance that tend to cancel each other. The fluid that is
pushed up at some point can easily move over to the next point where
the displacement of the membrane is downward.

Let's see how the added mass modifies the Helmhotz resonator
model. In this way, we consider a row of oscillators, like piano strings.
There may be finitely many oscillators, but it turns out to be easier to
work with the situation where there is a continuum of such oscillation
lined up along the x-axis. They are uncoupled, but their stiffness varies
with x:
$$S(x) = S_0 e^{-x/\xi}.$$

We will assume that there is only a small amount of damping so that
Q is very large. If we now force all the oscillators at a frequency $\bar{\omega}$,
then on a logarithmic scale we see the usual resonance curve (Figure
9.32a), except that the abscissa is position instead of frequency. The

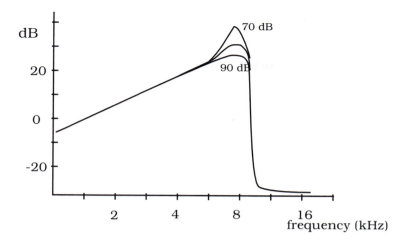

Figure 9.30 Basilar membrane displacement relative to malleus displacement at one place. Why must the cochlea be a nonlinear filter? (Adapted from Rhode.)

locus of resonance is the point x_c that satisfies

$$\bar{\omega}^2 = \frac{S(x_c)}{m} = \frac{S_0}{m}e^{-x_c/\xi},$$

or

$$x_c = \xi \ln(S_0/m\bar{\omega}^2).$$

The key to the whole discussion, though, is the phase delay curve. This also will be the usual resonance curve (Figure 9.32b) laid out along the x-axis.

The oscillators near $x = 0$ will be in phase with the forcing function; the phase will be delayed progressively until at the critical point, x_c, displacement will lag force by $\pi/2$ (phase quadrature). The oscillators down at the far end will lag the force by a half-cycle. The slope of this curve turns out to be very significant. Whenever there is a continuum (or very large number) of things vibrating in different phases relative to each other, then the gradient of the phase delay is called the **wave number** :

$$\frac{d\phi}{dx} = \text{wavenumber} = k.$$

(It is because the letter k is usually used for wave number that we used S for stiffness here.) The wave number is most easily interpreted

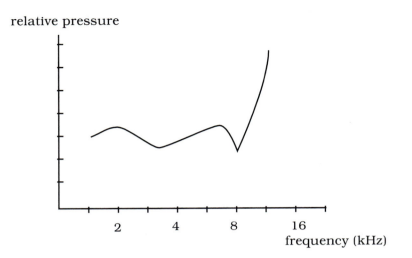

Figure 9.31 Sound pressure level needed to get a 1 nm displacement at a particular point on the basilar membrane (adapted from [57]).

in the case where it is constant. Then this means that as we move down the line each oscillator (assuming they are equally spaced) is delayed by a constant amount. But this is just the picture we had for waves moving around a ring of oscillators. In that case, $\frac{d\phi}{dx}$ is the amount of the cycle (in radians) that we go through per unit length. Thus, when k is constant, we have

$$\frac{2\pi}{k} = \text{wavelength} = \lambda.$$

(You can also think of the wave number as a wave *density*.)

The important thing to notice is that the resonant position is where the wave number has a maximum. In fact, if you do the mathematics, you will find that the functional form of k is similar to that of the amplitude (Figure 9.33).

It turns out, and you will show this in Chapter 10, that the added mass given to the array of piano strings by the ambient liquid depends on the wave number. The functional form is approximately

$$\mathcal{M} \sim k^{-\alpha}, \quad 1 < \alpha < 2,$$

where the approximation with $\alpha = 1$ is valid for small k and the approximation with $\alpha = 2$ is valid for large wave number. But the actual

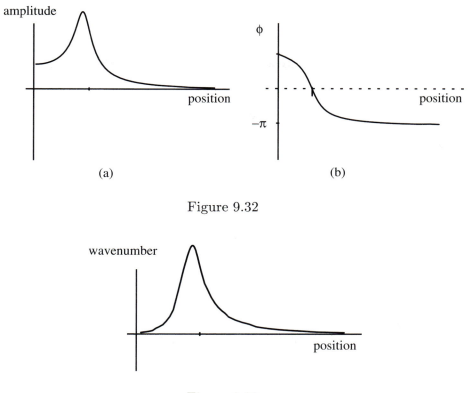

Figure 9.32

Figure 9.33

form is not so important; what is necessary is that $\mathcal{M}$ is a decreasing function of wave number. For the sake of simplicity, let us suppose that it varies as k^{-1}. This has the effect of changing the resonant frequency at any point:

$$\omega_k(x) = \sqrt{\frac{S(x)}{m + k^{-1}}}.$$

Now, frequency can be thought of as a function of either x or wave number. Let's fix x and see how it varies with k. This is shown in Figure 9.34. This is an example of a dispersion relation. The relations we found between frequency and mode wavelength in the first section of this chapter are also dispersion relations.

The graph of ω versus x is shown in Figure 9.35 for the case of no added mass (or, equivalently, as $k \to \infty$). Because Q is large, we can deduce from Figure 9.33, that there is only a small region about

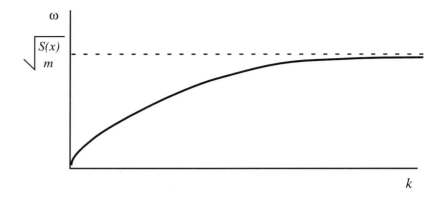

Figure 9.34

the critical point, x_c, that will be significantly affected. Since k is positive in that region, the graph of ω versus x will be correspondingly affected. To the left of x_c, the graph will be lowered. This means that some points to the left of x_c are brought closer to a resonant condition; they tend to act as though they are in resonance. This means that for such points the wave number is closer to a maximum. So if we look at the phase delay curve, it gets steeper earlier, and it gets steeper for all $x < x_c$. But it can't get just a little steeper; it has to get a lot steeper. The reason for this is topological: the oscillators at the far end haven't experienced any change due to the added mass, and so they must still be the equivalent of a half-cycle out of phase with the forcing. This means that the phase delay is some odd multiple of π. Since the phase delay curve starts getting steeper earlier, the effect is that in the small region near resonance the phase must lag far behind, at least $1^1/2$ cycles. Points in the region $x < x_c$ will have a correspondingly higher amplitude. Meanwhile, on the other side of the critical point the opposite is the case. So the amplitude graph has the form of Figure 9.36. Since, according to this resonance argument, a higher amplitude is associated with a higher wave number, we see that the amplitude curve is the envelope of a traveling wave that has a "pile-up". Another way of looking at it is that to the left of x_c we move up the dispersion curve, and so $k \to \infty$.

The above argument (or numerical computations) can only account for part of the experimental results. The left side of the curve is still too high. Furthermore, to get the steep cut-off on the right side, we

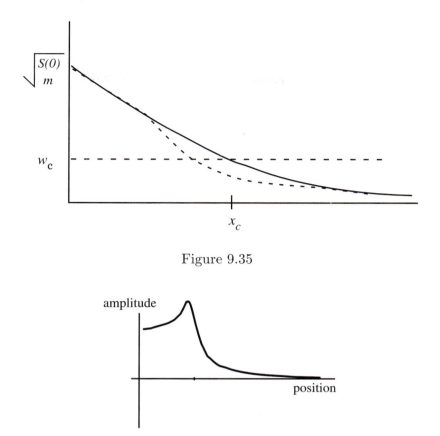

Figure 9.35

Figure 9.36

have had to assume that the quality of linear resonance Q is very high. It is possible to use dimensional analysis to estimate the Q of the ear. Other damping mechanisms in the ear notwithstanding, there is clearly present the viscous damping due to the viscosity of the fluid in the cochlea. But the order of magnitude of Q is given by the Reynolds number:

$$Q \sim \frac{\omega}{\lambda} \sim \frac{\omega^2}{|i\omega\lambda|} \sim \frac{\text{inertial energy}}{\text{viscous dissipation}} = \text{Reynolds number.}$$

Now, the Reynolds number is something that can be estimated. This is

$$R_e = \frac{\rho U d}{\mu},$$

where U and d are a characteristic velocity and length, and ρ and μ are the density and viscosity of the fluid. We can suppose that the fluid has properties very similar to seawater, so we can take as values for density and viscosity

$$\rho = 1 \text{ kg/m}^3, \qquad \mu = 10^{-3} \text{ Ns/m}^2.$$

For characteristic values of U and d, we can be generous and allow

$$d = 1 \text{ mm}, \qquad U = 10 \text{ cm/s}.$$

The result is
$$R_e = 10^{-1}.$$

This estimate for Q seems impossibly low, as it corresponds to a critically damped oscillator, but we are unlikely to get much of an improvement through better identification of the appropriate parameters. This means that in order to get the experimental amplitude and tuning curves, some mechanism must be working to effectively reduce the damping. The actual mechanism responsible is a topic of current research. Some possible approaches are outlined in the problems.

4   *Earthquakes

The tectonic plates of the earth's crust ride like surfboards on top of the slowly flowing mantle. Faults arise in the crust along lines where the mantle experiences the highest shear in the flow velocity. Compared to the mantle, the plates are quite stiff and move slowly relative to each other along the faults at the rate of about 1 to 5 cm per year. This movement is not steady, however. It tends to happen in fits and jerks. These are earthquakes. The stresses can build up over hundreds of years, leading to sudden relative motions of several meters. Actually, the whole side of the plate doesn't move at once, but the motion happens as a wave. Many models have been developed to describe earthquakes. An early model, modifications of which are still being studied, is due to Burridge and Knopoff. The idea is that we have two solids that are rubbing against each other on a line, subject to a (more or less constant) tectonic force. Stresses can build up and all of a sudden be released. We think of one side of the fault as a row of block masses attached by springs (Figure 9.37). The masses are all being driven by the tectonic force represented pictorially by the rig on top. Meanwhile, the rubbing against the other side of the fault is modeled by a frictional force experienced by the masses. This is *dry friction*, which, if you recall from Chapter 5, has two components: a *sticking friction*, which is a force that must be overcome before motion is possible, and a *dynamic friction*, which is some function of velocity, though usually taken to be a constant (less than the sticking force). Taking all masses and spring constants to be constant, the equations of motion are

$$m\ddot{u}_n + k(2u_n - u_{n-1} - u_{n+1}) = T_n - D_n,$$

as long as $\dot{u}_n$ is nonzero. It will stay nonzero until the sticking force is overcome. T_n is the tectonic driving force and D_n is the dynamic friction.

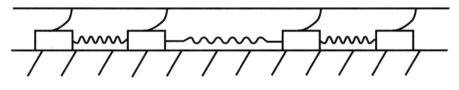

Figure 9.37

Now imagine what happens as we pull along the top: eventually one of the blocks starts to move; this starts pushing on the next one. It may be that the forcing is strong enough that the next one unsticks and then pushes on the one after that, and so on: a simple slow wave is generated. Another possibility is that the first block comes to a stop before unsticking the second block. But the spring between them is compressed. In this way uneven stresses can be built up. Can we ever have a catastrophic slippage in this model (as opposed to the slow wave)?

EXERCISE 19 Scale the right-hand forcing term properly and find the continuum-limit equation. This will be a linear nonhomogeneous wave equation.

It will be shown in Chapter 10 that the linear wave equation cannot exhibit the behavior that we are looking for. Besides, already we can see that the discrete problem is nonlinear, so the linear wave equation can't be a good continuum model. The hard part is how to deal with the sticking friction (which is the source of the nonlinearity in this model). The model should allow solutions that are discontinuous in x

to evolve. (In the presence of discontinuities, you can't justify the limit in the difference quotient.)

There are ways of approaching the problem without having to use a phenomenological friction law. One possibility is the other side of the fault presents a potential energy surface. This is a model that focuses in on a smaller length scale; a microscopic model, if you will, as opposed to the macroscopic model discussed above. As a simple example, suppose that the potential is sinusoidal, so that the equation of motion for the nth mass is

$$m\ddot{u}_n + k(2u_n - u_{n-1} - u_{n+1}) = A\sin u_n. \qquad (9.15)$$

This is equivalent to a line of coupled pendula.

EXERCISE 20   What is the continuum limit of equation (9.15)?

5   Snakes versus Chains

When should you use snakes to model a chain? That is, when is the continuum equation valid? The importance of this question will become more fully appreciated after you read Chapter 10 and see how much there is to be gained in terms of analysis by going to continuum models.

We have already seen that the continuum limit forms an appropriate model for phenomena that have some scale tending to zero. One way for this to happen is if our physical model is perceived as a smoothly varying distribution of parameters. In a mechanical model it may be that we are incapable of observing motion at small scales. If there is motion at smaller scales, it will have to be accounted for by the conservation of some macroscopic variable such as temperature. This would be the case in a chain model if the distance between particles is smaller than any length scale of observation.

An important length scale is the scale over which any mathematical solution to the model varies. This affords an important consistency check. If we find that a significant wavelength of our solution of the continuum equation is smaller than the interparticle distance, then we cannot really use the continuum equation, since the limiting procedure breaks down. On the other hand, this can be turned around: even if the interparticle distance is not small, we can still use the continuum limiting model if the only motions that we are interested in are long-wavelength motions. In this use of the continuum model the dependent variable does not represent the dependent variable of the discrete system *at a point*, but more correctly represents an average value (averaged over values of the independent value).

This leads to another way in which the continuum models are used and that is by postulating variables that represent from the very beginning some average or in some way have only a statistical meaning. Let us make a chart comparing the two viewpoints

chains	particle mechanics	systems of point masses
snakes	continuum mechanics	medium varies continuously

A route from particle mechanics to continuum mechanics can be found by computing averages. We represent point particles in the fashion of delta functions and then integrate over regions that are large enough to encompass many particles, but still much smaller than any length scale that we are interested in. This implies that there are certain scale changes that we deem irrelevant. Thus, there are two scales: the scale that is buried by averaging and the scale of significant variation of our averaged parameter. Let us look at a couple of examples.

Solid mechanics If we let go of the notion of trying to track every particle, we can represent mass in terms of a smoothly varying density by averaging over blocks representing many particles. This implies the existence of two scales, l and L, with $l \ll L$. In this case

$l \equiv$ intermolecular distance
$L \equiv$ scale of large density changes

This implies that we take our averages on length scales slightly greater than 10 Å, while L is on the order of something greater than $10\,\mu$m. Inherent motion on smaller scales would have to be taken care of by an average; temperature exemplifies this.

Galactic dynamics We can carry out the same type of averaging procedure, but over vastly different physical scales. It would be appropriate to use

$$l \equiv 0.1 \text{ parsec} \approx 3 \cdot 10^{13} \text{ km.}$$

The continuum model uses all the advantages subject to the averaging procedure; the model is much simpler because much detail is ignored. Once we have gone to averaged variables, we can derive equations for the variables on the assumption that they represent conserved quantities. We have avoided a discussion here of time scales, but it is clear that over small time scales behavior at small length scales will not have an effect on the behavior of the averaged quantities, but they could have a significant effect over longer times. How to incorporate this effect in the macroscopic equations is a very difficult problem.

6   Problems and Recommended Reading

1. *Bobs on a string* Consider a string with N point masses equally spaced along the string. Suppose that the masses are heavy enough that we can neglect the mass of the string, leaving us to model the motion of the N masses. Show that the equations of motion of this chain are the same as for the mass–spring chain, (9.4). [*Hint:* the restoring force of each oscillating mass reacting to its neighbor is proportional to the slope of the string joining them:

$$F_k = -T\left(\frac{y_k - y_{k-1}}{\Delta x}\right).$$

The constant of proportionality is the (constant) tension in the string.)

2. *A continuum string* Derive the equations for (small) displacements of a string under tension by finding the continuum limit of the equations in the previous problem. Compare to (9.10).

3. *A mechanical high-pass filter* Suppose that a flexible string under constant tension is attached to the ground by springs equally spaced along its length (Figure 10.41). The extra return force due to the springs is independent of curvature or slope. If k is the stiffness of a spring, then $\sigma = k/\Delta x$ is the stiffness per unit length. Derive the equations of motion and the dispersion relation, and show that there is a minimum frequency for vibrations.

4. *Traffic stability* Suppose that a stable solution to the equation (9.2) is known. When will small perturbations to this equation be unstable? The equation is linear, so the answer to this question can be found by examining the roots of the characteristic equation corresponding to (9.2). An unstable root is one with a positive real part. Since the equation incorporates a delay, the characteristic equation is not a polynomial equation, but a transcendental equation. However, the roots can easily be approximated by graphical means. Use the data in [1] (and references

therein) to compare the observed stability with the calculated stability criterion.

5. *Traffic flow* In a continuum model of traffic flow, the relevant macroscopic variables are the density of traffic (which has dimensions of *number of cars per unit length*) and the flux (the flux at a point is the number of cars passing that point in unit time). What is the relation between these variables?

(i) Show that the density, ρ, and the flux, q, are related via the flow velocity v by

$$q = \rho v.$$

This gives q as a function of ρ if v is a function of ρ. Based on observation, we should expect that v is a decreasing function of ρ and to actually become zero as the density reaches a large enough value, $\rho_{\max}$ (bumper-to-bumper). Also, it should be bounded:

$$v \leq v_{\max}.$$

The maximum value of flux, $q_{\max}$, is called the *carrying capacity* of the road. What is q_{max} when $v(\rho)$ is taken to be a linear function? For which value of ρ does it occur?

(ii) The relation between ρ and v is an example of what is called a *constitutive law*. Usually, such laws (friction force is given by such a law) are entirely empirical. It may be thought desirable to derive such a law from deductions made from a more refined model, but this is usually very difficult. In the case of traffic flow, even our microscopic equation, (9.2), contains a phenomenological constant, λ, but this is just a number and is estimable. Can we find a relation between v and ρ involving only this constant, λ? Note that

$$\frac{1}{\rho} = \text{``volume'' assigned to a car}$$

$$= \text{car length} + \text{space between cars}$$

Assume a steady state in (9.2); integrate the resulting equation, and use the relation given above to relate v to ρ. How should the resulting relation be modified to fit observation? Why doesn't it fit automatically?

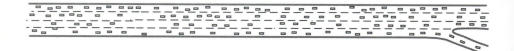

Figure 9.38 Snapshot of highway traffic.

6. *Traffic densities* In evaluating a continuum model of traffic flow, we need data on the macrovariables, density and velocity. The gathering of such data is not always so straightforward.

   (i) One possiblity is that, if we can take advantage of some technology that allows us to obtain snapshots of a large stretch of the highway taken over some time interval, then we can look at them and compute densities and, by comparing the photos from different times, compute velocities. An example of such a snapshot is Figure 9.38. Compute a density function based on this. Remember that the density should be a continuous function; you must decide on a scale Δx so that the total traffic on each interval $[x_c, x_c + \Delta x)$ is represented by $\rho(x_c)\Delta x$. Remember that to compute the traffic density, which is really an average, you must decide on an appropriate size "cell" over which to take this average.

   (ii) If these snapshots are not available, then to estimate these variables you will have to set up an observation post on the highway and count cars as they come by. It is easy to see that a single observation post is inadequate, so there will have to be at least two. In essence, what you have to do now is reconstruct the snapshots from these counts. The variable that is actually observed here is the flux q. Outline a procedure for computing densities and velocities from count data. Note that it is possible to compute the density, which is a space average, by taking a time average. Explain the difference in this context between a single-lane road and a multilane road. In the article by Gazis and Szeto in [42] a sophisticated averaging procedure is presented.

7. *Colliding galaxies* On the length scale at which two galaxies interact, the individual stars that make up the galaxies are irrelevant for the gross motion. Represent a spiral galaxy by a continuum disk of radius R. If the total mass of the galaxy is $\mu\sqrt{R}$, what is the mass density? When two galaxies are close enough to in-

teract, the important questions that arise are *how the exchange of mass occurs* and *how they change their shape*. Represent two galaxies in center-of-mass coordinates and model their interaction by focusing on the above questions.

8. *Ear modeling* A good place to learn about the physiology of hearing is [52]. That book provides many references to the experimental literature. Ground-breaking experiments on live cochlea using the Mossbauer technique are reported in [57]. Physical properties of the basilar membrane are discussed in [78]. A thorough account of the role of added mass in cochlear dynamics is given in [44]. There have been many attempts to account for the sharpness of experimental tuning curves. As can be imagined, mathematical models incorporating special negative damping mechanisms, such as the Van der Pol model, have been proposed and analyzed. Simulations done with these models can match closely the experimental data curves. An experimental verification of an active nonlinear damping mechanism seems to be the cochlear echo: sometimes, when stimulated, the cochlea responds with a signal of higher energy. However, it is more difficult to indentify the actual physical (feedback) mechanism that would give rise to the negative damping. One possibility is that the basilar membrane vibration is just part of a complex of coupled resonators that can feed into each other to cancel out damping at some points and to add to damping at other points. The tectorial membrane could supply a second resonant oscillating system. The outer hair cells are expected to play an important role in the feedback, as being the actual point of electrical-to-mechanical transfer. There has been some modeling of the function and electroelastic properties of these cells. Interesting accounts of these efforts are to be found in the two collections [13] and [84].

9. *Earthquake* In [37] the blocks and masses earthquake model is taken to a continuum limit by carefully accounting for *collective* motions in a conservation of momentum equation. They look at a small section of the chain that is moving and derive a condition for the lead mass (the crack tip). Suppose that the subchain consists of m masses starting at the nth block, i.e., $\{n, n+1, \ldots, n+m\}$. Suppose that the nth block moves at time t, and the $(n+m)$th block moves at time $t + \Delta t$. Newton's law says that the change in momentum of the subchain during that time is the time integral

of all the forces acting on the subchain. You have to sum up all the forces acting on the whole subsystem. These consist of three types. One is the force exerted by the $(n-1)$st spring on the whole system. Show that this is proportional to

$$u_x \Delta t.$$

Next there is the driving force $T_n - D_n$ acting on all the m masses. Show that this is proportional to

$$(\Delta t)^2,$$

and so can be ignored in the limit. Finally, there is the frictive force that has to be overcome as each successive block breaks free. Show that this is proportional to

$$(T_n - B_n)\Delta t,$$

where B_n is the sticking friction. Conservation of momentum then leads to a first-order partial differential equation valid at precisely the crack tip. This is an example of a jump condition conservation law (see Chapter 10). For details of this computation, see [37].

10. *Gutenberg–Richter law* According to the Richter scale, the *magnitude* of an earthquake is a logarithmic measure of the energy released,

$$m = a_1 - b_1 \log E.$$

The Gutenberg–Richter relationship gives the number of earthquakes, of magnitude m or greater, in a certain time period as

$$\log N = a_2 - b_2 m.$$

Thus there is a power-law relationship between energy and frequency of earthquakes. Do the block–spring models give this law? Perform many simulations of the block–spring model using some convenient measure for the energy of an "event." One possibility is to calculate the appropriate integrals in the last problem. Or you can try something less sophisticated, such as simply counting the number of blocks that move at once. You might want to throw out small events when they occur. Does your simulated frequency distribution match the Gutenberg–Richter law? What are your estimates for the parameters?

Chapter 10

Waves

*You know that something is a wave when it shows you its wavelength.
The order of magnitude of the wavelength can be deduced through observations of diffraction. For example, suppose that a boat wake produces waves with wavelength of 1 foot. This straight pattern of waves
is unaffected by a stick being placed in the wave of its progress, but a
dock piling will scatter the waves into a more complex pattern. The
wavelengths of visible light are too large for diffraction at the atomic
scale; there, x-rays are used. Diffraction patterns can be built up from
many examples of radiation of a single photon. But the same is true
for electrons. Thus, at the quantum level, everything has some wavey
characteristics, even single "particles."*

1   Waves Here and There

There are these things we call waves everywhere around us: water
waves in oceans, lakes, rivers, and kitchens; electromagnetic radio and
optical waves moving through us here at home through the far reaches
of space; sound waves whenever one thing knocks against another; even
the "human wave" in a stadium. What do these things have in common?

There are important differences.

We can distinguish between *traveling waves* and *periodic waves*. A
wave moving up a tidal river is an example of the former; there is a
momentary increase in the water level at some place (Figure 10.1). A
sharp electrical signal pulse is another example (Figure 10.2). For each
of these waves there is a functional form with a noticeable *front* that
is propagated without spatial variation (Figure 10.3). At the opposite
extreme are the modes of vibration in the chain of mass–spring oscillators (Chapter 9). Such *standing waves* do not have a distinguished

Figure 10.1

front. When we walk into a room where middle C is steadily generated
through a sound system, we walk into a standing wave (Figure 10.4).

We can distinguish between firm waves and dispersive waves. A
traveling wave can keep its shape as it moves along (Figure 10.3).
Thanks to such properties we can use waves to communicate. On the
other hand, a traveling wave can fall apart as it moves (Figure 10.5).
And yet such *dispersive* waves can be used for communication.

For flag signaling (Chapter 9), it is clear that the phase of the
signaler is a complete indication of the state of the wave. There is a
definite wavelength, λ, and a period of action, τ, and from these the
wave speed is

$$c = \lambda/\tau.$$

But in a chain of atoms, the state is a dynamic combination of phase
and amplitude. Each mode has a different wave speed, where wave
speed is defined as the linear velocity of a particular phase. This is
termed the *phase velocity*. Is this the speed of signals?

Suppose that forcing at one end of the chain is turned on then off
after a short time, but not too short. For example, instead of a steady
middle C through the sound system a middle C piano key is struck
(Figure 10.6). Because of the uncertainty principle, this signal is made

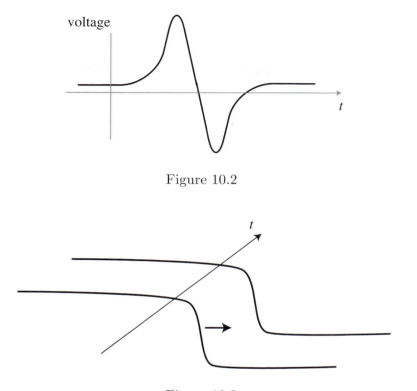

Figure 10.2

Figure 10.3

up of a broad band of frequencies (the more definite the signal, the broader the band). All these frequencies (being associated to different modes of vibration of the medium) have different wave speeds, or rather *phase speeds*. And yet the wavepacket in Figure 10.6 has a definite form and is transmitted at a definite speed, the local sound speed.

It is possible to see, in a group of waves generated by a storm at sea, the peaks of the waves moving faster than the group, each growing and then disappearing as it moves through the group.

Finally, we should mention that if we choose a frame of coordinates moving along a wave front then the wave is not travelling at all. This allows us to recognize as waves certain things that do not at first look like waves. An interesting example is the the hydraulic jump in a washtub illustrated in Figure 10.7, an easy home experiment.

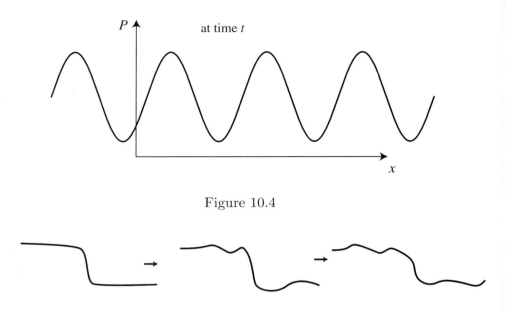

Figure 10.4

Figure 10.5 A dispersive wave. This shows wave profiles at successive times, the observer moving with the front.

2   Conservation Laws

A useful guiding principle toward a mathematicization of many phenomena, natural or otherwise, is to isolate a certain quantity or quantities whose distribution in space and time characterize to some extent the phenomenon at hand and whose behavior in some region of space can be accounted for by the fact that it will be conserved if not physically moved or otherwise acted upon by relevant agents. To take a broad view, these agents may be chemical, molecular, electromagnetic

Figure 10.6 A finite-duration wave (a wave packet).

Figure 10.7

gravitational, or nuclear forces for natural phenomena or relevant social forces for phenomenological models in the "soft" sciences.

The idea of conservation of mass as a fundamental physical principle is no doubt familiar; particle density and traffic flow density studied in Chapter 9 are also good examples; other examples that may be familiar correspond to the quantities energy and momentum. The student has probably worked out the equations for a collision of point particles by making use of the conservation of energy and momentum. In this chapter we are interested in applying the ideas of conservation to continuum models, where the relevant conserved quantity is a function — scalar or vector or tensor — of extent in the material. An example is the mass density in a fluid (the mass per unit volume of the fluid), keeping in mind that we think of the fluid as a continuum. This last assumption is only valid on certain length scales.[1] In the same way one may define a *concentration* of some chemical in a fluid in moles per unit volume. Seawater can be considered a fluid made up of two components, water and salt, the latter dissolved in the former; a quantity of interest would then be the concentration of salt.

---

[1]In the example of the flow of cars on a road, while knowing full well that cars come in discrete quantities, each car having some finite size, we define a *car density*, the number of cars per unit length, which varies pointwise as a function defined on a continuum.

In this introduction to conservation laws, $c(x, t)$, a function of space
and time, will denote a conserved quantity. Sometimes it is useful to
think of c as a function of material "particles," and *then* we must keep
track of the movement of these particles. The quantity, c, could be any
of the things we discussed above; we'll just call it the concentration of
"stuff." (It has dimensions of *amount per unit volume*.) If you need a
concrete example to have in mind, it is perhaps best to focus on a mass
density with dimensions of mass per unit volume.

To simplify matters, let us assume that all quantities vary only
in one direction; they are functions of the single variable x, and all
movement is restricted to the directions of increasing or decreasing x.
It is worth pointing out that we have made two distinct assumptions:

1. All variables are functions of a single variable, x.

2. If there are any vector quantities, they only have components in
   the x-direction.

Take an infinitesimal "box" (a cylinder) (Figure 10.8). The ends of the
box are at x and $x + \Delta x$.

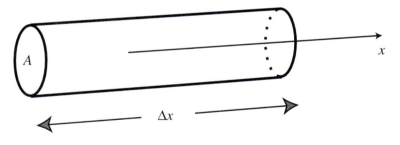

Figure 10.8

In our model, will allow the cross section area A to vary. We may
have to do consistency checks to evaluate assumptions 1 and 2 in light
of the variability of A; typically, we would only allow A to vary slightly,
and we are interested in the effects of this variability. Thus, although
A is a function of x, it is constant over distances of the order of Δx,
the box length.

The volume of the box is $A\Delta x$, and the total amount of "stuff" in
the box, at time t, is

$$c(\bar{x}, t)A\Delta x,$$

where $x \leq \bar{x} \leq x + \Delta x.$[2] Thus, the change of the amount in the box from time t to time $t + \Delta t$ is

$$\text{total change} = c(x, t + \Delta t)A\Delta x - c(x, t)A\Delta x. \tag{10.1}$$

What is responsible for any change during that time interval? Any of a number of things could be agents of change: there could be a local input of stuff; it may decay; it may be shared or equalized among different places in the material, and other such. We will assume that all change is due only to gross movement of the material (e.g., think of mass density) so that there is a flux of stuff,

$$q = cv, \tag{10.2}$$

where v is the velocity of the material. This velocity, which is positive for motion to the right and negative for motion to the left, is also defined pointwise on a continuum. So the only change during the time interval $(t, t + \Delta t)$ is due to this flux:

$$\text{total change} \equiv \left(\begin{array}{c} \text{flux "in" at left} \\ \text{during } \Delta t \end{array} \right) + \left(\begin{array}{c} \text{flux "out" at right} \\ \text{during } \Delta t \end{array} \right)$$
$$= q(x)A\Delta t - q(x + \Delta x)A A \Delta t$$
$$= c(x)v(x)A\Delta t - c(x + \Delta x)v(x + \Delta x)A\Delta t (10.3)$$

The negative sign must be introduced because, if the flux is inward at the right end, the movement must be to left, which is negative. Equate (10.1) to (10.3), divide by Δx and Δt, and take the limit as these quantities go to zero to get the differential equation

$$\frac{\partial}{\partial t}(cA) = -\frac{\partial}{\partial x}(cvA).$$

We cannot cancel the A from both sides because A may not be constant; if A is constant, then

$$\frac{\partial c}{\partial t} + \frac{\partial}{\partial x}(cv) = 0.$$

We have derived a *conservation law in differential form*. It cannot be a complete model for anything because it is only one equation relating two variables, c and v. To complete the model, we need to

---

[2] The concerned reader is invited to make rigorous the machinations of this paragraph by appeal to any of several mean-value theorems.

supply some other relation between c and v; this may consist of another conservation law for v or a law giving v as a function of c. In general, where we don't necessarily take $q = cv$, a *conservation law in differential form* is simply the relation

$$\frac{\partial c}{\partial t} + \frac{\partial}{\partial x} q = 0, \tag{10.4}$$

and we need to specify a relation between c and its flux q.

EXERCISE 1   For the equation to hold, we must assume certain differentiability properties of the relevant quantities. Make this precise by appeal to the mean value theorem.

EXERCISE 2   Another derivation of (10.4) proceeds by considering not an infinitesimal box, but a finite one, say $A \times [a, b]$, and a conservation law in integral form,

$$\frac{d}{dt} \int_a^b c(x, t) A \, dx = q(a) A(a) - q(b) A(b).$$

Obtain (10.4) by applying the fundamental theorem of calculus and the Dubois–Reymond lemma.

EXERCISE 3   Write down the conservation law in integral form in three dimensions using the conserved quantity, $c(\mathbf{x}, t)$ and its flux vector $\mathbf{q}(\mathbf{x}, t)$. By use of the divergence theorem, derive the conservation law in differential form.

Another way to come by a conservation law is to consider a box that moves with the stuff [it moves with the underlying material because of our assumption implicit in (10.2) that the stuff is attached to the material]. At time $t = 0$ the box is

$$A \times [a, b],$$

where $[a, b] = \{\xi : a \leq \xi \leq b\}$. As time evolves, the particle that was at ξ initially is now at

$$x = x(\xi, t)$$

(see Figure 10.9).

If, once again, we ignore sources or sinks and other such, then conservation of stuff means

$$0 = \frac{d}{dt} \int_{a(t)}^{b(t)} c(x, t) \, dx. \tag{10.5}$$

For simplicity, we have assumed that the cross-sectional area is constant. The fact that the domain of integration depends on t makes

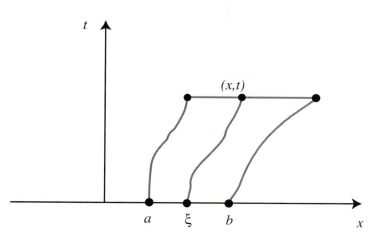

Figure 10.9

things difficult, but we can pull the integration back to the initial interval by use of the change of variable:

$$x = x(\xi, t).$$

This induces a change in measure:

$$dx = \frac{\partial x}{\partial \xi} \, d\xi = J(\xi, t) \, d\xi.$$

(We use this notation in anticipation of the higher-dimensional context, where J is called the *Jacobian* of the coordinate change.) In these variables, our conservation law is

$$0 = \frac{d}{dt} \int_a^b C(\xi, t) \, J(\xi, t) \, d\xi,$$

where

$$C(\xi, t) = c(x(\xi, t), t).$$

We have changed the name of the dependent variable when we made the change of independent variable to emphasize that it is a different function. The reader is warned, however, that this is rarely done in the literature.

The integral is fixed with respect to t, so now we can differentiate under the integral sign. This derivative is

$$\frac{\partial C}{\partial t} J + C \frac{\partial J}{\partial t}.$$

The derivative in the second term in this expression can be written

$$\frac{\partial J}{\partial t} = \frac{\partial}{\partial t}\frac{\partial x}{\partial \xi} = \frac{\partial}{\partial \xi}\frac{\partial x}{\partial t}.$$

Now,

$$\frac{\partial x}{\partial t}$$

is just the derivative of $x(\xi, t)$ with ξ held fixed. If we fix ξ, then

$$t \to x(\xi, t)$$

is the path of a material particle that started at ξ (see Figure 10.9). Thus, $\frac{\partial x}{\partial t}$ is just the velocity, v, and is usually written as an ordinary derivative:

$$V = \frac{dx}{dt}.$$

We have been careful and denoted it by a capital letter to indicate that we are thinking of it as a function of ξ. In this way,

$$v(x, t) = V(\xi, t).$$

Thus, by the chain rule;

$$\frac{\partial J}{\partial t} = \frac{\partial}{\partial \xi}V = \frac{\partial v}{\partial x}\frac{\partial x}{\partial \xi} = \frac{\partial v}{\partial x}J.$$

Our conservation law becomes

$$
\begin{aligned}
0 &= \int_a^b (\frac{\partial C}{\partial t} + C\frac{\partial v}{\partial x})J\,d\xi \\
&= \int_{a(t)}^{b(t)} \frac{\partial c}{\partial t} + v\frac{\partial c}{\partial x} + c\frac{\partial v}{\partial x}\,dx \\
&= \int_{a(t)}^{b(t)} \frac{\partial c}{\partial t} + \frac{\partial}{\partial x}(cv)\,dx.
\end{aligned}
$$

The time t is fixed, so we can apply the DuBois–Reymond lemma to conclude that the integrand is zero and thus

$$c_t + (cv)_x = 0. \tag{10.6}$$

The notation

$$\frac{D}{Dt}c = \frac{\partial c}{\partial t} + v\frac{\partial c}{\partial x}$$

is often used. This is the total derivative along the particle curves. Now, it may seem that we should expect our conservation law to be written

$$\frac{Dc}{Dt} = 0.$$

This is clearly not true—compare (10.6).

EXERCISE 4 Why not?

The fallacy is in thinking of the concentration as something attached to a particle. It is not: it is a continuum field; the concentration of a particle at x_0 at time t_0 is $c(x_0, t_0)$ and represents an *average* taken about that point (refer to the discussion in Chapter 9, section 5). Notice that $\frac{Dc}{Dt}$ is the total derivative of the quantity $c(x, t)$ along the curves in the (x, t) plane that are integral curves of the differential equation

$$\frac{dx}{dt} = v(x, t). \tag{10.7}$$

More significant than these particle paths is another set of curves. If

$$v = h(c),$$

then (10.6) is equivalent to

$$c_t + [v + h'(c)]c_x = 0.$$

Then

$$\frac{dc}{dt}\Big|_{\substack{\text{along} \\ \text{solu-} \\ \text{tions to} \\ 10.7}} = 0,$$

when the derivative is taken along the curves where

$$\frac{dx}{dt} = v + h'(c).$$

They are called *characteristic curves* for the partial differential equation (PDE) (10.6). If

$$\frac{\partial v}{\partial x} = 0,$$

then

$$\frac{Dc}{Dt} = 0$$

along these curves; thus c is constant along them; they are the level curves of c. We see in this way that any function whose levels coincide with these curves is a solution to the partial differential equation

$$\frac{\partial c}{\partial t} + v\frac{\partial c}{\partial x} = 0.$$

EXERCISE 5 Draw the figure (see Figure 10.9) that corresponds to this case. With that in mind, interpret the significance of the term $\frac{\partial v}{\partial x}$ (recall how it arose in the derivation).

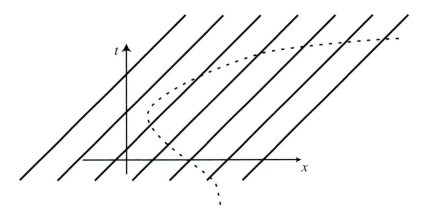

Figure 10.10

Let us consider the special case where the velocity is a constant,

$$v(x,t) = v_0.$$

Then the characteristic curves are just straight lines, all having slope $\frac{1}{v_0}$ (Figure 10.10).

Now, of course a set of level curves doesn't determine a function.

EXERCISE 6 Think of two different functions that have the level curves of Figure 10.10.

If any function having these lines as level curves solves our PDE, then what information must be supplied auxiliary to the PDE in order that we obtain a definite solution? If we give the value of c at any point then c is determined at every point on the line of slope $\frac{1}{v_0}$ passing through that point. So if we give the value of c along a curve that intersects each characteristic line in exactly 1 point, then we have specified c for all x and t and have in hand our definite solution. We are, in effect, solving an infinite number of "initial"-value problems of the form

$$\frac{dc}{dt} = 0, \qquad c(x(\tau, \tau)) = \bar{c}(\tau),$$

where the "initial" τ is a parameter that varies transversely to the family of characteristics; that is, it has a different value from line to line. Solving the equation is this way is known as the *method of characteristics*. (This is sometimes called the Cauchy problem.)

An Example of the Method of Characteristics

For any finite value of

$$v < \infty,$$

an acceptable data curve is

$$t = 0, \quad -\infty < x < \infty.$$

Supplying data on $t = 0$ amounts to an initial value problem (IVP) for the PDE: we know the distribution of "stuff" initially (at $t = 0$) and we wish to determine the temporal evolution of this distribution.

Let us look at a particular example. If $v_0 = 2$ and $f(x)$ is some function, our IVP is

$$\frac{\partial c}{\partial t} + 2 \frac{\partial c}{\partial x} = 0, \qquad u(x, 0) = f(x).$$

The characteristic curves, with

$$\frac{dx}{dt} = 2,$$

are lines $x = 2t + \xi$, and u is constant along these lines:

$$\begin{aligned} c(x, t) &= c(\xi, 0) \\ &= c(x - 2t, 0) \\ &= f(x - 2t). \end{aligned}$$

Such a function is called a *wave*. It is easy to see why when we look at successive time shots of $c(x, t)$ (see Figure 10.11): these are just the function f shifted over, moving to the right with speed 2. For this reason, v_0 is called the *wave speed*.

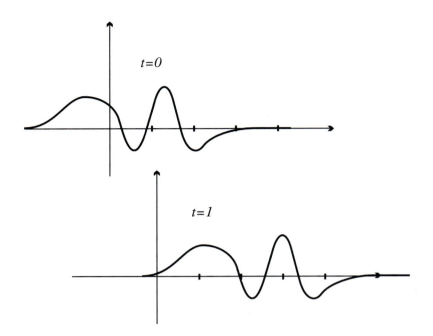

Figure 10.11 Successive time-shots.

It is obvious that the solution c is continuous for all x and all t if f is continuous; it is, indeed, as smooth in both its variables x, t as f is in its one variable.

EXERCISE 7 Justify the use of the term *wave* to mean any function of the form

$$u(x,t) = f(g(x,t)),$$

where g is a smooth real-valued function such that $\nabla g \cdot \mathbf{i} > 0$ for $t > 0$.

Figure 10.12 corresponds to the IVP; we say that the *domain of dependence* of the point (x,t) is the point ξ that lies on both the characteristic curve passing through (x,t) and the curve where data are specified. The value of $c(x,t)$ depends, among all initial values, only on the value of c at $(\xi, 0)$; once $c(\xi, 0)$ is given, we know the value of c at every point on the characteristic curve that passes through $(\xi, 0)$. This curve is sometimes referred to as the *range of influence* of the point $(\xi, 0)$.

In the Figure 10.10, if data are specified on the red curve, then we cannot associate an unambiguous domain of dependence for some values (x,t).

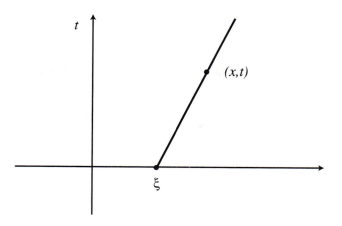

Figure 10.12

It may be relevant and important in your model of a continuum that the spatial domain be restricted somehow. For example, we may consider a tube filled with liquid that can flow indefinitely in one direction, but one end of the tube is distinguished by a specified inflow of liquid. When the domain is restricted in this way, it usually is the case that some auxiliary data are given on the boundary.[3] In particular, for the present case, instead of the IVP, one might wish to solve a boundary-value problem. Take as an example the equation

$$\frac{\partial c}{\partial t} + v_0 \frac{\partial c}{\partial x} = 0, \quad x > a, \quad t > 0$$

subject to the conditions

$$c(x,0) = f(x), \quad x > a$$
$$c(a,t) = g(t), \quad t > 0$$

Note that this problem can always be solved if $v_0 > 0$; if however, $v_0 < 0$, the problem cannot be solved (see Problem 1).

EXERCISE 8 Find $c(x,t)$ that solves

$$\frac{\partial c}{\partial t} + 2\frac{\partial c}{\partial x} = 0, \quad x > 0, \quad t > 0$$

with the data

$$c(x,0) = x, \quad x > 0,$$
$$c(0,t) = 0, \quad t > 0.$$

---

[3] In some problems extra information must be given even in situations modeled by PDE's on unbounded regions in order to guarantee a unique solution.

Constitutive Laws

The conservation laws that we have so far solved are *linear*. They are distinguished by having a linear constitutive law, that is, a linear relation between the conserved quantity and its flux:

$$q = cv_0.$$

EXERCISE 9 Show that this makes the partial differential equation linear in the sense that any multiple of a solution is a solution and the sum of two solutions is a solution.

The velocity of the particle flow is independent of the concentration. This corresponds to "no interaction between particles." This covers, for example, a line of traffic with each vehicle moving at the same speed: the relative distribution of vehicles looks the same at each time. Likewise for salt particles in water if all move according to an underlying uniform current, without any interaction. Clearly, linear conservation laws are not of much interest.

EXERCISE 10 Change coordinates in the (x, t)-plane so that the linear conservation law is equivalent to the ordinary differential equation

$$\frac{dc}{ds} = 0.$$

In Chapter 9 we justified the use of the flux–density relation for traffic flow,

$$q = cv,$$

but noted that the velocity at any point was necessarily dependent on the local density. We noted that v had a maximum and that it was zero for large traffic densities and was otherwise always decreasing. We pointed out that the simplest such function was linear (Figure 10.13). This gives a parabolic relation between flux and density (Figure 10.14).

The use of the linear velocity–density relation was based on a few gross observations. A more extensive empirical study of this relation, might display the relation shown in Figure 10.15. This leads to a modified flux–density relation (Figure 10.16). It turns out that qualitative features of the solutions to the resulting conservation law are similar

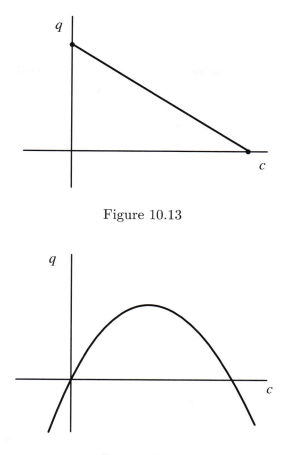

Figure 10.13

Figure 10.14

in either case—what is important is the negative curvature of the q versus c curve—but that there are some differences in the solution that may be important in a study for a traffic-light placement. You can investigate this in the problem section.

In the case of the flow of a dilute gase, we still use the relation

$$q = \rho v$$

between flux and density and velocity, but the empirical relation observed is different from that between velocity and density of traffic flow. In fact, it is noticed that the velocity of a gas *increases* with density: if you squeeze a gas, it flows faster. The simplest relation in this case

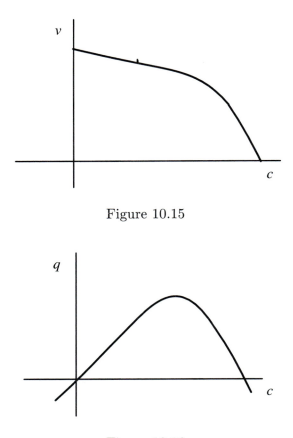

Figure 10.15

Figure 10.16

is to make the velocity proportional to density:

$$v \propto \rho.$$

This linear relation again gives a parabolic flux–density relation with *positive* curvature (Figure 10.17). This opposite curvature is responsible for the behavior of solutions of the mass conservation law to be opposite in some way to the corresponding solutions for the traffic flow problem.

A single conservation law for gas flow is an incomplete model. It is not generally acceptable to rely on an empirical rule to relate velocity to density. In fact, the velocity will also satisfy a conservation law (conservation of momentum). Then we have two equations relating the two variables. But not quite: the momentum conservation will

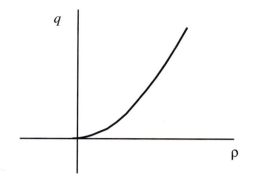

Figure 10.17

involve pressure forces; pressure is a third variable. So we still need another law. However, such laws (relating density, pressure, and thermodynamic variables) are more general and more easily obtained in experiments than would be a velocity–density law. Yet a single conservation law can still be used as a preliminary model to sketch the outline of a situation.

It is not always true that the flux is a pointwise function of the conserved quantity. The flux may depend on gradients. In that case we are led to consider diffusion equations, which is the subject of Chapter 11.

The Initial Value Problem for a Nonlinear Conservation Law. Shocks

Let's consider a simple initial-value problem for a conservation law. We will suppose that the initial condition is piecewise constant with only one jump:

$$c(x,0) = \begin{cases} c^- & x < 0 \\ c^+ & x > 0 \end{cases}.$$

Let us also suppose that in the conservation law

$$\frac{\partial c}{\partial t} + \frac{\partial q}{\partial x} = 0 \tag{10.8}$$

we set

$$q = f(c) = \frac{c^2}{2}.$$

Then, by the method of characteristics, the characteristic curves
are given by

$$\frac{dx}{dt} = f'(c) = c(x,t). \tag{10.9}$$

But the relation (10.8) is equivalent to the statement that c is constant along characteristic curves, and so these *curves are straight lines*.
In view of our initial data, we see that there are two families of would-
be characteristics: those straight lines emanating from the negative
x-axis with slope $1/c^-$ and those emanating from the positive x-axis
with slope $1/c^+$.

In a qualitative sense, there are two different possibilities at this
point, corresponding to whether

$$c^- < c^+$$

or

$$c^- > c^+.$$

In either case, the (x,t)-plane is split into three regions. If $c^- < c^+$,
these regions are the following (see Figure 10.18):

I. The region covered by the characteristics with slope $1/c^-$

II. The region covered by characteristics with slope $1/c^+$

III. A region covered by neither family of characteristics.

If $c^- > c^+$, these three regions are (see Figure 10.19) as follows:

I. The region covered by the characteristics with slope $1/c^-$

II. The region covered by characteristics with slope $1/c^+$

III. A region covered by *both* sets of characteristics.

We will look at each case in turn and figure out what type of solution
we can expect for each.

In the case where $c^- < c^+$, the method of characteristics will determine the solutions (in a unique way) in regions I and II, but will
leave the solution in region III undetermined. It may seem that one
is free to choose any values at all for c in region III, but this is not a
proper course on two accounts: (1) we want to obtain a solution that
actually satisifies the differential equation wherever possible, and (2)
the solution should be uniquely determined by the initial condition.

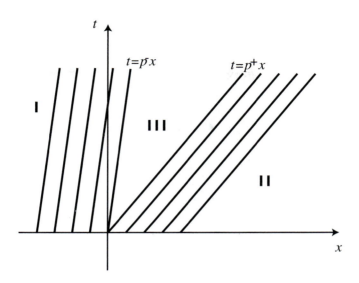

Figure 10.18

When $c^- < c^+$, it seems that the whole of region III is the range of influence of the single point 0; but this turns out not to be the case: the solution will actually be determined by the values of c on the lines

$$t = c^- x, \qquad t = c^+ x.$$

(This is an example of a *characteristic boundary-value problem*.)

EXERCISE 11 Show that for any $\lambda > 0$, $c(\lambda x, \lambda t)$ is a solution to the PDE whenever $c(x, t)$ is a solution; this prompts you to consider a solution of the form

$$c(x, t) = g(x/t).$$

What relation does this lead you to consider? What are the properties of your solution?

Since region III is formed by the discontinuity of the initial data at $x = 0$, let us see if we can back into a solution to our problem by considering a sequence of problems having continuous initial data. The sequence of initial data will converge to

$$f(x) = \begin{cases} c^- & x < 0 \\ c^+ & x > 0 \end{cases}.$$

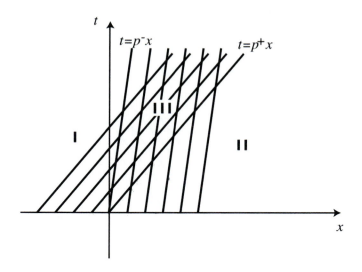

Figure 10.19

To be concrete, we will use the parametrized family of functions

$$c_\epsilon(x,0) = \begin{cases} c^- & x < 0 \\ \frac{c^+ - c^-}{\epsilon} x + c^- & 0 \le x \le \epsilon \\ c^+ & x > \epsilon \end{cases} , \qquad (10.10)$$

where ϵ is going to be sent to zero. With the initial data $c_\epsilon(x,0)$, we can solve the problem by the method of characteristics. The solution is the same as before in regions I and II (just think of region II as being shifted over). Suppose that we pick a point in region III,

$$(x,t) \in \ \text{III}.$$

Then it lies on the characteristic given by (see Figure 10.20)

$$\frac{x_0 - \bar{x}}{0 - \bar{t}} = \frac{c^+ - c^-}{\epsilon} x_0 + c^-,$$

which gives

$$x_0 = \frac{x - c^- t}{1 + \frac{c^+ - c^-}{\epsilon} t}$$

as the x-intercept of the relevant characteristic curve. Thus the solution is

$$c_\epsilon(x,t) \quad = \quad \frac{c^+ - c^-}{\epsilon} \frac{x - c^- t}{1 + \frac{c^+ - c^-}{\epsilon} t} + c^-$$

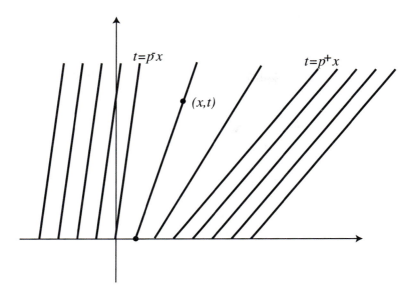

Figure 10.20

$$= \frac{(c^+ - c^-)(x - c^-t)}{\epsilon + (c^+ - c^-)t} + c^-.$$

We can let ϵ tend to zero and see what happens:

$$\lim_{\epsilon \to 0} c_\epsilon(x,t) = \frac{(x - c^-t)}{t} + c^-$$
$$= x/t.$$

The solution to the problem for $\epsilon = 0$ is

$$c(x,t) = x/t. \tag{10.11}$$

EXERCISE 12 Is the resulting solution continuous?

EXERCISE 13 The solution (in region III) is seemingly independent of c^+ and c^-. To what extent is that true?

EXERCISE 14 By considering initial data $c(x,0)$ as the limit of functions that are continuous, but not differentiable, we obtained a solution that is continuously differentiable. Argue that the results would have been the same even if we took $c(x,0)$ as a limit of smooth functions.

The solution, (10.11), is known as a *rarefaction wave* or a *fan*.

Now let us turn to the situation where

$$c^- > c^+,$$

which is illustrated in Figure 10.19. The problem is that, although we can find the solution uniquely in regions I and II by the method of characteristics, the solution is overdetermined in region III. Flush with our success in the previous case, we might want to apply the same technique. That is, we approximate the initial data by some smooth data and then solve. The difficulty with this is that any initial data that approximates our step function

$$f(x) = \begin{cases} c^- & x < 0 \\ c^+ & x > 0 \end{cases},$$

no matter how smooth it is, will have such a region (III) of overdeterminancy. To see this, consider any smooth $c_\epsilon(x, 0)$ whose graph has the form displayed in Figure 10.21a. Then, given the facts

1. characteristics are straight lines, and

2. their slopes are given by the intial data,

we see that there will always be the situation similar to that exhibited in Figure 10.21b. Note, however that if the intial data are smooth, the nasty region doesn't touch the origin; there is a small time, $0 < t < \bar{t}$, during which the solution is smooth.

EXERCISE 15 Verify that this is also true if the initial condition is the piecewise linear function (10.10).

A regular solution exists up to time $t = \bar{t}$, but we cannot get past that point with the method of characteristics.

There are two ways out of our impasse. One possibility is to extend the notion of the differential equation to allow solutions that are surfaces in three-dimensional space of the variables (x, t, c). These surfaces will not be of the form

$$c = c(x, t),$$

but will satisfy a more general relation of the form

$$G(x, t, c) = 0. \qquad (10.12)$$

Of course, the differential equation has to be satisfied in some sense. Solutions of this type, even when they can be produced, may not be

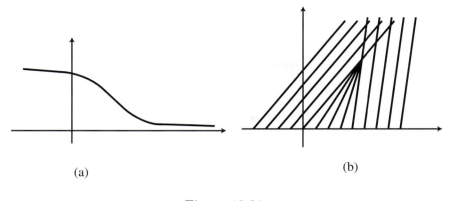

(a)   (b)

Figure 10.21

acceptable in some problems of physical interest where the density, c, has to be a single-valued function. This is the case when, for example, c represents a concentration of a substance or a mass density.

The other option open to us is to accept the fact that the integral form of the conservation law is more fundamental than the derivative form [the PDE (10.4)] and, while recalling that the solutions of the former satisfy the latter only when they are smooth, to use the integral form to suggest a (possibly discontinuous) solution that is uniquely related to the initial data. To see how this might work, at least in the present context, imagine a line

$$x = \alpha t,$$

across which the solution has a jump discontinuity. This line separates region I where $c = c^-$ from region II where $c = c^+$. In this way, region III is squeezed to this line. We might then check to see when this "solution" satisfies the integral form of the conservation law. It turns out that in the problem we have been considering the solution has just such a discontinuity across the line

$$x = \frac{c^- + c^+}{2} t. \qquad (10.13)$$

(See Figure 10.22.) Thus, the discontinuity propagates at the speed that is the average of the characteristic speed from the left and the characteristic speed from the right. Such a solution is sometimes called a *shock wave* or simply *a shock*.

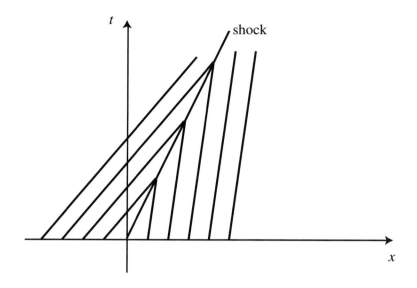

Figure 10.22

How do we calculate the speed of propagation of a discontinuity, that is, the slope of the line in Figure 10.22?

Suppose that the moving front is given by

$$x = \phi(t). \tag{10.14}$$

This curve separates regions in each of which the conserved quantity and flux are continuous, but they suffer a jump discontinuity across the interface. The quantities are still integrable, so the integral form of the conservation law still holds:

$$\frac{d}{dt} \int_a^b c(x, t) \, dx = q(a, t) - q(b, t).$$

We can no longer use any of our previous techniques to differentiate under the integral, but we can work piecemeal. In this way,

$$\frac{d}{dt} \int_a^b c(x, t) \, dx = \dot{\phi} c^- + \int_a^\phi \frac{\partial c}{\partial t} \, dx - \dot{\phi} c^+ + \int_\phi^b \frac{\partial c}{\partial t} \, dx. \tag{10.15}$$

On the other hand,

$$\begin{aligned} q(a, t) - q(b, t) &= q(a) - q(\phi^-) + q(\phi^-) - q(\phi^+) + q(\phi^+) - q(b) \\ &= -\int_a^{\phi^-} q_x \, dx - \int_{\phi^+}^b q_x \, dx + (q^- - q^+). \end{aligned}$$

Now apply the conservation law on each piece separately,[4] cancel the appropriate terms, and we are left with

$$[c]\dot{\phi} = [q], \tag{10.16}$$

where the bracket denotes the size of the jump across the discontinuity:

$$[c] = c^+ - c^-.$$

This relation relates the speed of propagation of the discontinuity to the jumps in concentration and flux across the curve of discontinuity. We can treat this as a supplementary differential equation. If the curve is given by (10.14), then (10.16) is

$$\frac{dx}{dt} = \frac{[q]}{[c]}.$$

In some simple cases, such a relation gives an immediate solution to the problem.

EXERCISE 16   For the problem considered directly above, show that (10.16) becomes (10.13), and that the shock is a straight line in the (x, t)-plane.

In more complicated situations the shock may have to found numerically.

Shock solutions are found to be useful in describing traffic flow at a signal light (see problems).

*Conservation Laws in Higher Dimensions

Suppose that c is a conserved quantity (mass density, or a concentration). To say that $\mathbf{q}$ is its flux vector means that for any piece of surface S the integral

$$\int_S \mathbf{q} \cdot d\mathbf{S}$$

gives the amount of c that passes through S in unit time, so

$$\frac{d}{dt} \int_B c \, d\mathbf{x} = -\int_{\partial B} \mathbf{q} \cdot d\mathbf{B}, \tag{10.17}$$

where B is any volume fixed in space and where the outward orientation is chosen for the bounding surface, ∂B.

---

[4]You can use the differential equation on each piece separately because the solutions are assumed continuously differentiable away from the curve $x = \phi(t)$.

Applying the divergence theorem to the right side and differentiating on the left gives

$$\int_B \left(\frac{\partial c}{\partial t} + \operatorname{div} \mathbf{q} \right) d\mathbf{x} = 0.$$

If the integrand is fixed, then, since the volume B is arbitrary, by the DuBois–Reymond lemma, we get

$$\frac{\partial c}{\partial t} + \operatorname{div} \mathbf{q} = 0,$$

the differential form of the conservation law. To close this differential equation, we must specify a relation between $\mathbf{q}$ and c. If c is convected by a flow, then the proper relation is

$$\mathbf{q} = c\mathbf{u},$$

where $\mathbf{u}$ is the velocity vector. If the flux is due to diffusion, then

$$\mathbf{q} = -k\nabla c.$$

3   Population Models

In the simple population models of Chapter 9, a single number, $p(t)$, represented the total size of the population at time t. The basic equations for the growth of the population, ignoring migration, were based on the use of a relative growth rate,

$$r = \dot{p}/p = \alpha - \omega,$$

where α is the *birth rate* and ω is the *death rate*. A balance between birth rate and death rate results in no net growth. It was found realistic to take r to depend on p; for example,

$$r = a - bp,$$

leading to the logistic equation

$$\dot{p} = ap - bp^2.$$

This equation has two stationary solutions (refer to Figure 4.9)

$$\begin{aligned} p_0 &= 0 && \text{unstable,} \\ p_1 &= a/b && \text{stable.} \end{aligned}$$

It is possible to get a better model for population evolution by considering the distribution among age groups in the population. With the use of a finite number of age groups, such as

$$
\begin{array}{ll}
A & 0\text{—}12 \\
B & 12\text{—}25 \\
C & 25\text{—}50 \\
D & 50\text{—}70 \\
E & 70\text{—}
\end{array}
$$

one can use a discrete population model. Or one can consider a continuum of ages. In this latter scheme, the population size is a function of two variables:

$$p = p(a, t).$$

Let's think of p as a conserved quantity and derive its conservation law. The change in the total population in an age interval, Δa,

$$[p(a, t + \Delta t) - p(a, t)]\Delta a,$$

is to be equated to the influx at the boundaries plus any sources or sinks. But the "flux" in this case is just p itself, since, at time t, the number entering the interval at a is just the population at a at that time, and likewise for the number leaving the interval at $a + \Delta a$. So the net influx is

$$[p(a, t) - p(a + \Delta a, t)]\Delta t + Q\Delta a\Delta t.$$

This leads to the linear equation

$$\frac{\partial p}{\partial t} + \frac{\partial p}{\partial a} = Q. \tag{10.18}$$

Since we are ignoring any migration, Q can only represents deaths in the population; births can only occur at $a = 0$. It is reasonable to write Q in terms of a *relative* death rate, ω, that depends only on age:

$$Q = -\omega(a)p.$$

Let's try to solve the initial-value problem for the equation

$$p_t + p_a = -\omega p.$$

The characteristic curves are given by

$$\frac{da}{dt} = 1,$$

and thus are lines of slope 1,

$$a = t + c.$$

Along these lines we have

$$\left.\frac{dp}{dt}\right|_{\substack{\text{along a} \\ \text{characteristic}}} = -\omega(a)p.$$

Integrate to get

$$p(a,t) = p(c,0)e^{-\int_0^t \omega(s+c)\,ds}. \tag{10.19}$$

[This illustrates the general method for solving any non-homogeneous linear equation of the form

$$a(x,t)u_t + b(x,t)u_x = c(x,t).]$$

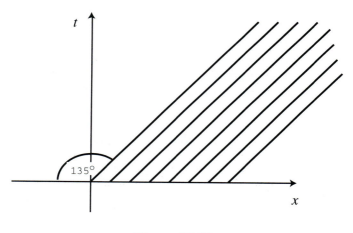

Figure 10.23

The function (10.19) shows us the decay along characteristics, but what is happening to the population? Look at the total population:

$$P(t) = \int_0^\infty p(a,t)\,da.$$

Compute the growth, using equation (10.18):

$$\frac{dP}{dt} = \int_0^\infty \frac{\partial p}{\partial t}\,da = \int_0^\infty \left(-\frac{\partial p}{\partial a} - \omega p\right)\,da.$$

Here we see the balance between growth,

$$-\frac{\partial p}{\partial a},$$

and decay,

$$-\omega p.$$

(The reason that $-p_a$ represents a growth term is that, if that term is positive at some age, a, it means that the distribution of ages is such that there are more in the population at ages just lower, and so by aging of the population that age group should grow in number.) Given some initial distribution of the population,

$$p(a, 0) = f(a), \qquad\qquad (10.20)$$

we can read the solution from (10.19). This is illustrated in Figure 10.23. The solution can only be determined in

$$a > t,$$

because otherwise the characteristic curves do not intersect the positive a-axis. In order to get a complete solution, we need to impose a boundary condition on the positive t-axis. This is the birth condition. This condition should be given in terms of a relative birth rate, but this rate should not be simply proportional to the population, since not all age groups contribute equally to the birth rate. We expect a low birth rate in a mostly senile population. Thus,

$$p(0, t) = -\int_0^\infty \alpha(a, t)p(a, t)\, da.$$

The weighting distribution, α, is expected to be insignificant outside some interval of fertile age groups. We can write

$$p(0, t) = \beta(t) \int_{a_1}^{a_2} h(a)p(a, t)\, da. \qquad\qquad (10.21)$$

The complete initial-value boundary-value problem is given by (10.18, 10.20, and 10.21) .

EXERCISE 17 Show that the solution can now be obtained by the method of characteristics. (*Hint:* see Figure 10.23.)

*Birth–Death Processes

In Chapter 7 we found that we could calculate all the moments of a birth–death process if we could calculate its probability generating function,

$$G(t, s) = \sum_{k=0}^{\infty} P_k(t)s^k.$$

For some simple processes, such as the Poisson process, the generating function could be found easily since it is the solution of an ordinary differential equation. It was remarked that, in general, the generating function will satisfy a partial differential equation. Let us derive the generating function for the birth–death process:

$$\dot{p}_n = -n(\alpha + \omega)p_n + (n-1)\alpha p_{n-1} + (n+1)\omega p_{n+1}, \quad n = 1, 2, 3, \ldots,$$
$$(10.22)$$

$$\dot{p}_0 = \omega p_1.$$

Differentiating with respect to t, we get

$$
\begin{aligned}
\frac{\partial G}{\partial t} &= \sum_{k=0}^{\infty} \dot{P}_k(t)s^k \\
&= -(\alpha + \omega)\sum_{0}^{\infty} kp_k s^k + \alpha \sum_{1}^{\infty}(k-1)p_{k-1}s^k + \omega\sum_{0}^{\infty}(k+1)p_{k+1}s^k \\
&= -(\alpha + \omega)s\sum_{0}^{\infty} kp_k s^{k-1} + \alpha \sum_{0}^{\infty}(k)p_k s^{k+1} + \omega\sum_{0}^{\infty}(k+1)p_{k+1}s^k \\
&= -(\alpha + \omega)s\frac{\partial G}{\partial s} + \alpha s^2 \frac{\partial G}{\partial s} + \omega\frac{\partial G}{\partial s}.
\end{aligned}
$$

So that the equation for G is

$$\frac{\partial G}{\partial t} + \{(\alpha + \omega)s - \alpha s^2 - \omega\}\frac{\partial G}{\partial s} = 0.$$

That is, we must integrate

$$\frac{dG}{dt} = 0$$

along the curves

$$\frac{ds}{dt} = (\alpha + \omega)s - \alpha s^2 - \omega.$$

EXERCISE 18   Carry this out and compute the moments of $\{p_n\}$.

EXERCISE 19   Compute the higher moments of the $\{p_n\}$ from Problem 5 in Chapter 7.

4 *First-order Quasilinear Equations: General Theory

If

$$z = f(x, y)$$

is a function that satisfies

$$P(x, y, z)\frac{\partial z}{\partial x} + Q(x, y, z)\frac{\partial z}{\partial y} = R(x, y, z), \qquad (10.23)$$

then a relation between the vector field

$$\mathbf{V} = (P, Q, R)$$

and the graph of f,

$$\Gamma(f) = \{(x, y, z) : z = f(x, y)\},$$

can be found by noting that a normal vector field on $\Gamma(f)$ is

$$N = (\frac{\partial f}{\partial x}, \frac{\partial f}{\partial y}, -1),$$

and (10.23) expresses the fact that

$$\mathbf{V} \cdot \mathbf{N} = 0. \qquad (10.24)$$

In other words, $\mathbf{V}$ is tangent to the surface $\Gamma(f)$. This suggests a method for solving (10.23): Extend the notion of a solution in the following way. A *generalized solution* of (10.23) is a surface in R^3 whose normal vector at each point satisfies (10.24). Having in hand such a surface, we can then verify whether it is the graph of a function of x and y, that is, whether it is a classical solution of (10.23). Such a surface is called an *integral surface* of the vector field $\mathbf{V}$.

There are many ways to express the condition (10.24). For example, if the surface is given by a relation

$$u(x, y, z) = 0,$$

then $\mathbf{N}$ can be taken to be the gradient of u,

$$\nabla u = (u_x, u_y, u_z).$$

(This is because the surface is a level surface for u, so u cannot have any tangential variation.) The condition (10.24) can thus be written

$$Pu_x + Qu_y + Ru_z = 0.$$

Now, the vector field $\mathbf{V}$ gives a tangential direction field everywhere on an integral surface, so it seems reasonable to build up an integral surface from *integral curves* of $\mathbf{V}$. Integral curves thread the vector field $\mathbf{V}$, that is, they are curves

$$s \to (x(s), y(s), z(s))$$

that solve

$$\frac{dx}{ds} = P, \quad \frac{dy}{ds} = Q, \quad \frac{dz}{ds} = R. \tag{10.25}$$

The solutions of (10.25), the integral curves of $\mathbf{V}$, will be called the *characteristics* of the partial differential equation.

If u_1 and u_2 are solutions of (10.23) ($u_1 = 0, u_2 = 0$ are integral surfaces of $\mathbf{V}$), that satisfy

$$\nabla u_1 \times \nabla u_2 \neq 0,$$

then the simultaneous solution of the equations

$$u_1 = 0, \quad u_2 = 0,$$

specifies an intergral curve of $\mathbf{V}$. Thus, integral surfaces of $\mathbf{V}$ can intersect, and they always intersect in an integral curve. Moreover, there are infinitely many integral surfaces that can intersect on an integral curve.

*Cauchy Problem

The *Cauchy problem* for (10.23) is to find an integral surface S of the vector field $\mathbf{V}$ that contains a given curve C.

Remark 1: We consider generalized solutions.

The first thing to do is to examine the curve in hand, C, to see whether it is *characteristic*. If it is, then our case is hopeless, because a characteristic curve has infinitely many integral surfaces passing through it. If it is *not* characteristic, then our plan of action is very simple:

At each point $(x_0, y_0, z_0) \in C$ construct the characteristic (or portion thereof) that passes through (x_0, y_0, z_0). This is done by solving

(10.25) with (x_0, y_0, z_0) as the initial vector. This always has a unique solution, by the existence and uniqueness theorem for ordinary differential equations (assuming that $\mathbf{V}$ is smooth); the result is illustrated in Figure 10.24a.

Repeat this for every point on C, obtaining Figure 10.24b; we thereby construct an integral surface passing through C.

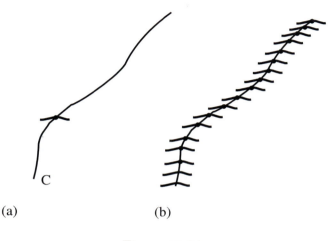

(a)   (b)

Figure 10.24

Thus, in principle, the Cauchy problem for a noncharacteristic curve can always be solved. Now, the solution so obtained may not be a *classical solution*. Is there any way to determine before we begin our procedure whether the solution *will* turn out to be a classical solution? The answer is yes: simply determine whether the characteristics that pass through C have the property of extending C in the (x, y)-direction, or only in the z-direction, that is, whether the projection onto the (x, y)-plane of the respective tangents of C and the characteristics are parallel.

Let us solve some initial-value problems.

1. Nonlinear conservation law

$$u_t + uu_x = 0, \qquad u(x, 0) = f(x). \qquad (10.26)$$

Here the inital curve in 3-space is the curve

$$C : \tau \to (\tau, 0, f(\tau)).$$

The equations for the characteristics are

$$\frac{dt}{ds} = 1,$$
$$\frac{dx}{ds} = u,$$
$$\frac{du}{ds} = 0.$$

Note that the tangent to C is

$$(1, 0, f'(x)),$$

and the tangents to the characteristics that stick out along C are

$$(1, f(x), 0).$$

C is *not* a characteristic; thus, we *can* find a solution. However, this cannot be a classical solution anywhere that

$$f(x) = 0.$$

EXERCISE 20 Intepret the last statement in light of our previous handling of this problem (shocks).

The solution to the characteristic equations is

$$t = s, \quad x = x_0 + \int_0^s u\, ds, \quad u = u_0$$

or

$$t = s, \quad x = x_0 + su, \quad u = u_0.$$

The last equality tells us that the characteristics lie in horizontal planes. Thus, the third dimension is not important, and we can work with the projections onto (x, t)-plane of the characteristics. This is what we did in section 10.2, calling these projections the characteristic curves. We can parametrize our surface by (s, t):

$$t = s, \quad u = f(\tau), \quad x = \tau + f(\tau)s.$$

If f is one–one, we can eliminate s and τ in order to bag a relation between x, t, u:

$$u = f(x - ut).$$

2. Linear equation

$$a(x,t)u_t + b(x,t)u_x = c(x,t), \tag{10.27}$$

$$u(x,0) = f(x).$$

Again, the initial curve is

$$C : \tau \to (\tau, 0, f(\tau)),$$

and the characteristics are given by

$$(1) \qquad \frac{dt}{ds} = a(x,t), \tag{10.28}$$

$$(2) \qquad \frac{dx}{ds} = b(x,t), \tag{10.29}$$

$$(3) \qquad \frac{du}{ds} = c(x,t)u. \tag{10.30}$$

Notice that, even though the partial differential equation (10.27) is *linear*, the ordinary differential equations for the characteristics are *nonlinear* (in x and t). (This is also seen in optics, where the characteristics of the *linear* equation of electromagnetics are the solutions to the *nonlinear* eikonal equations.)

Now, (10.28) and (10.29) give the planar equation

$$\frac{dx}{dt} = b(x,t)/a(x,t).$$

Generally, for arbitrary a and b there will be no way to solve this analytically. We must then compute the solution numerically. Actually, in that case one might as well compute $(10.28) - -(10.30)$ numerically.

A particular linear equation is the population growth with a linear age-dependent death rate:

$$\frac{\partial p}{\partial t} + \frac{\partial p}{\partial a} = -\omega_0 a p.$$

The characteristics are

$$\frac{dt}{ds} = 1, \tag{10.31}$$

$$\frac{dx}{ds} = 1, \tag{10.32}$$

$$\frac{dp}{ds} = -\omega_0 a p, \tag{10.33}$$

As is easily seen, this reduces to an ordinary differential equation for p:

$$t = t_0 + s,$$
$$a = a_0 + s,$$
$$\frac{dp}{ds} = -\omega_0(a_0 + s)p,$$

or

$$p(s) = p(0)e^{-\omega_0 \int_0^s a_0 + \sigma \, d\sigma},$$

or

$$p(a,t) = p(0)e^{-a_0\omega_0 s - \omega_0 \frac{s^2}{2}} = p(a-t,0)e^{-a_0\omega_0 t + \omega_0 \frac{t^2}{2}}.$$

The Cauchy problem can be solved analytically in this case; this is a special case of (10.19). (Of course, the Cauchy problem is not a realistic model for population dynamics; but we already know how to deal with a realistic boundary-value model.)

3. An interesting illustration is afforded by a Related but Nonlinear partial differential equation, for which we want to solve the initial-value problem:

$$\frac{\partial \rho}{\partial t} + \rho \frac{\partial \rho}{\partial x} = -x\rho,$$
$$\rho(x,0) = f(x).$$

The equation for the characteristics [the integral curves of $(1, \rho, -x\rho)$] are

$$\frac{dt}{ds} = 1, \qquad\qquad (10.34)$$

$$\frac{dx}{ds} = \rho, \qquad\qquad (10.35)$$

$$\frac{d\rho}{ds} = -x\rho, \qquad\qquad (10.36)$$

from which we get the relation

$$t = s.$$

The last two equations reduce to

$$\frac{d\rho}{dx} = -x.$$

Thus,

$$\rho = -\frac{x^2}{2} + C = -\frac{x^2}{2} + \frac{x_0^2}{2} + f(x_0).$$

The first thing to notice is that for a very special function

$$f(x) = -\frac{x^2}{2},$$

the solution is a parabolic cylinder

$$\rho(x, t) = -\frac{x^2}{2}. \tag{10.37}$$

The solution does not depend on t!

Now, such surfaces can be generated very easily with straight lines parallel to the generator of the cylinder (here, the t-axis). However, the lines

$$\rho = -\frac{x^2}{2} = \text{const.}$$

are *not* characteristics. To see this (and to compute the correct characteristics), look at the equation for the (x, t)-characteristic curves:

$$dt/dx = -2/x^2.$$

This gives

$$t = \frac{2}{x} + C = 2(\frac{1}{x} - \frac{1}{x_0}),$$

or

$$x = \frac{2}{t + 2/x_0}.$$

The characteristic curves (projections of the characteristics) are shown in Figure 10.25. Note that, although the integral surface (the solution to the partial differential) is independent of t, and thus *a fortiori* exists for all t, integral curves (characteristics) starting at $x < 0$ do not exist for all t. There is a "hidden blow-up." This may indicate a deficiency in a model. When $x < 0$, the forcing term is positive for ρ positive.

The surface, (10.37), forms a qualitative dividing point: if x_0 is such that

$$f(x_0) > -x_0^2/2,$$

then

$$2f(x_0) + x_0^2 = a^2,$$

and we have

$$\frac{dt}{dx} = \frac{-2}{x^2 - a^2}.$$

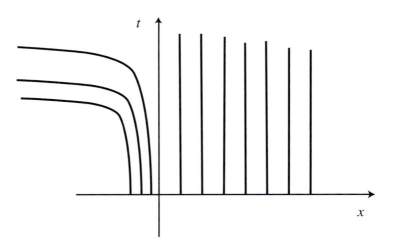

Figure 10.25

This can be integrated by partial fractions:

$$\frac{dt}{dx} = \frac{1}{a}\left[\frac{1}{x+a} - \frac{1}{x-a}\right]$$

to get

$$at = \ln\left(\frac{x+a}{x-a}\right) + C = \ln\left(\frac{x+a}{x-a}\right) - \ln\left(\frac{x_0+a}{x_0-a}\right),$$

or

$$e^{at} = \frac{x+a}{x_0+a}\frac{x_0-a}{x-a}.$$

We can solve this for x:

$$x = a\left[\frac{e^{at} + \frac{x_0-a}{x_0+a}}{e^{at} - \frac{x_0-a}{x_0+a}}\right].$$

Note that

$$a > x_0 \quad \leftrightarrow \quad f(x_0) > 0,$$
$$a < x_0 \quad \leftrightarrow \quad -x_0^2/2 < f(x_0) < 0.$$

If x_0 is such that

$$f(x_0) < -x_0^2/2,$$

then

$$2f(x_0) + x_0^2 = -a^2,$$

and the equation is

$$\frac{dt}{dx} = \frac{-2}{x^2 + a^2}.$$

The solution is

$$t = -\frac{2}{a}[\arctan\frac{x}{a} - \arctan\frac{x_0}{a}],$$

or

$$x = -\tan(\frac{at}{2} + \arctan x_0 a).$$

In principle, a solution of the partial differential equation can be put together based on these analyses.

5 Linear Systems

Blood Flow. Acoustics

Blood is an incompressible liquid flowing in flexible tubes. The waves that appear wouldn't seem to call for the same model as a compressible gas flowing in a rigid tube, or waves in elastic media, or the surface waves in a river. Yet all these can be modeled using the linear system of acoustic equations. The term *acoustic* here refers to an approximation, where we assume that the medium is characterized by small variations about a constant density, a constant flow speed, and a constant hydrostatic pressure. This approximation means that, in the course of deriving equations, we ignore any products of variables that appear. We will see later that these acoustic equations can be found as a linearization of a nonlinear system.

We can assume, by choice of coordinates, that the velocity and pressure are small. Thus, we are looking for equations where the variables satisfy

$$\rho \approx \text{constant,} \tag{10.38}$$

$$\mathbf{u} \approx 0, \tag{10.39}$$

$$p \approx \text{constant.} \tag{10.40}$$

The assumption of small velocity has to be dealt with carefully. There is a mean current of flow, but this velocity is small compared to the speed of waves that appear (as a consistency check should show). Although the local speed of particles may be high in the passage of the peaks of waves, they move over very small distances and, when averaged over

distances on the order of the wavelength, should be expected to satisfy
the approximation (10.39).

We are going to assume that friction forces are negligible so that the
only stresses are pressure forces. We also will assume a one-dimensional
flow. The material is constrained in the directions orthogonal to the
x-direction. So we will derive the equations by looking at conservation
of material and momentum (force balance) in a small "box" (cylinder)
(Figure 10.26). The length of the box is to be small enough so that

Figure 10.26

we don't have any problems in using a constant cross-sectional area,
A. We will take material volume as the conserved quantity. Thus,
variables of interest are

$$V \;=\; \text{volume},$$
$$Q \;=\; \text{volume flow rate}.$$

The volume flow rate is the amount of volume flowing per unit time.

EXERCISE 21 What are the dimensions of Q?

Over a small time period, Δt, the change in volume,

$$V(t + \Delta t) - V(t),$$

is due to an influx of material during the time Δt:

$$[Q(x) - Q(x + \Delta x)]\Delta t.$$

Since the cross-sectional area, A, is constant over the distance, Δx, we
can assume that
$$V(t) = A(t)\Delta x, \tag{10.41}$$

giving
$$V(t + \Delta t) - V(t) = [A(t + \Delta t) - A(t)]\Delta x.$$

The relation, (10.41), is only valid if the fluid is incompressible. In that case the volume can only increase if the sides of the container expand or contract. Upon expansion at some point, we expect a flow to be induced as material moves in to fill the new space.

If a fluid is compressible, the volume can change even when the cross sectional area, A, doesn't. In that case the flow can increase if the fluid is compressed so as to be able to fill more space. In the acoustic approximation, a relative change in volume is balanced by a change in the pressure:

$$\frac{\Delta V}{V} = K\Delta p. \tag{10.42}$$

K is called the *compressiblity*. Thus, for compressible fluids a change in volume over time corresponds to a change in pressure:

$$V(t + \Delta t) - V(t) = KV(p(t + \Delta) - p(t)) = KA(p(t + \Delta) - p(t))\Delta x, \tag{10.43}$$

and thus the conservation equation is

$$[Q(x) - Q(x + \Delta x)]\Delta t = KA(p(t + \Delta) - p(t))\Delta x. \tag{10.44}$$

The relative change in the cross-sectional area can also be related to a change in pressure. It is sufficient for the acoustic approximation to have

$$\frac{\Delta A}{A} = D\Delta p. \tag{10.45}$$

D is called the *distensibility* and is related to the elasticity constant for the material that bounds the fluid. Thus,

$$V(t + \Delta t) - V(t) = DA(p(t + \Delta) - p(t))\Delta x, \tag{10.46}$$

and in this case we have

$$[Q(x) - Q(x + \Delta x)]\Delta t = DA(p(t + \Delta) - p(t))\Delta x. \tag{10.47}$$

To review: we have two material conservation equations, (10.47) or (10.44), depending on whether the fluid is incompressible or compressible. Of course, if a compressible fluid is in a flexible channel, both equations would have to be used; in the acoustic approximation, the dynamical equation is additive.

The momentum equation is comparatively straightforward. The momentum of fluid in the box is (check the dimensions !)

$$\rho_0 Q \Delta x.$$

The change in momentum during the time Δt is

$$[Q(t + \Delta t) - Q(t)]\rho_0 \Delta x. \tag{10.48}$$

This must be balanced by net forces acting on the box during the time interval Δt. The only forces *acting on the box* are the pressure forces at either end:

$$p(x)A(x)\Delta t = \text{ pressure force on the left end}, \tag{10.49}$$

$$-p(x + \Delta x)A(x + \Delta x)\Delta t = \text{ pressure force at the right end}. \tag{10.50}$$

[We have taken the inward pointing orientation for the box, so that the right side of the box is negatively oriented (with respect to the x-axis), and thus the negative sign.]

We balance the change in momentum (10.48) with the forces acting, (10.50) and (10.49), to get

$$[Q(t+\Delta t) - Q(t)]\rho_0 \Delta x = [p(x)A(x) - p(x+\Delta x)A(x+\Delta x)]\Delta t. \tag{10.51}$$

Taking the limit, $\Delta t \to 0$, $\Delta x \to 0$, gives the differential equation

$$\rho_0 \frac{\partial Q}{\partial t} = -\frac{\partial}{\partial x}(pA). \tag{10.52}$$

Because of our assumptions, we end up with

$$\rho_0 \frac{\partial Q}{\partial t} = -A_0 \frac{\partial p}{\partial x}. \tag{10.53}$$

Likewise, we can take the limit in (10.44) and (10.47) to get

$$\frac{\partial Q}{\partial x} = -KA_0 \frac{\partial p}{\partial t} \tag{10.54}$$

and

$$\frac{\partial Q}{\partial x} = -DA_0 \frac{\partial p}{\partial t}, \tag{10.55}$$

respectively. Thus, the equations for sound propagation in a rigid pipe are (10.53) and (10.54), while the equations to describe the waves of an incompressible fluid in a flexible tube are (10.53) and (10.55). Notice that the equations are exactly the same except that one involves a constant that represents the elasticity of the fluid medium, while the other involves a constant that represents the elasticity of the flexible tube.

In Chapter 9 we saw an example, during the investigation of the inner ear, of a wave in an incompressible fluid with a flexible boundary. In that case, we were interested in the vibrations of the bounding membrane that were induced by this fluid wave. Clearly, the assumption of slowly varying cross-sectional area would not be appropriate for that case, and the equations of the model would have to be different (see problems).

By differentiating (10.53) with respect to x and (10.54) with respect to t,

$$\rho_0 Q_{tx} = -A_0 p_{xx},$$
$$Q_{xt} = -A_0 p_{tt},$$

and using the equality of mixed partials, we can eliminate Q from the equations and get a single second-order partial differential equation for the pressure:

$$p_{xx} = \rho_0 K p_{tt}. \tag{10.56}$$

Compare this to equation (9.8). Remark on the difference. This is what we called the second-order linear wave equation. At that time it was found as the continuum limit of a microscopic model for vibrations of a crystal and thus for sound waves in a solid. We see here that sound waves in a fluid satisfy the same equations. If you go back through our derivation in this section, you will find that, because we neglected the fluid velocity, we used nothing that would distinguish a fluid; thus, the same equations are valid for a solid, using the appropriate material parameters.

EXERCISE 22 Show that Q satisfies the same second-order wave equation, (10.56).

Transmission Lines

A simple transmission line consists of a cable stretched across the country side (Figure 10.27). When grounded, this forms a circuit, and if we ignore, for the time being, any losses, can be approximately represented as an LC line (Figure 9.22), where an inductance, L, and a capacitance, C, appear every Δx along the line. Consider a single one of these sections (Figure 10.28). Since we are going to treat the line as a continuum by letting

$$\Delta x \to 0,$$

Figure 10.27

we will introduce inductance and capacitance per unit length:

$$L = l\Delta x, \quad C = c\Delta x.$$

The voltage drop along each section is due to the current oscillating in the inductance:

$$V(x) - V(x + \Delta x) = L\dot{I} = l\dot{I}\Delta x. \tag{10.57}$$

In the limit, we get the equation

$$-\frac{\partial V}{\partial x} = l\frac{\partial I}{\partial t} \tag{10.58}$$

The branching of the current at each section gives

$$I(x) = I(x + \Delta x) + C\dot{V} = I(x + \Delta x) + c\dot{V}\Delta x. \tag{10.59}$$

The resulting differential equation is

$$-\frac{\partial I}{\partial x} = c\frac{\partial V}{\partial t}. \tag{10.60}$$

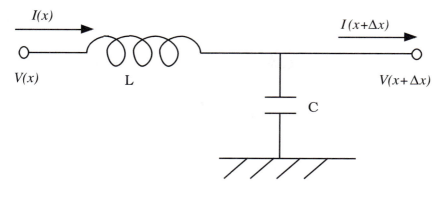

Figure 10.28

Table 10.1 Analogy between Mechanical and Electrical Waves

Mechanical	Electrical
Q/A_0	I
p	V
ρ_0	l
K	c
$\frac{1}{2}(Kp^2 + \rho_0(Q/A_0)^2$	$\frac{1}{2}(cV^2 + LI^2)$

Notice that the equations (10.58, 10.60) are the same equations as (10.53, 10.54). This inspires the analogy of Table 10.1, which should be compared to Table 5.1. One should refer also to the derivation of the wave equation in Chapter 9.

We did not explicitly treat the voltage and current as conserved quantities, but the relation to conservation laws is clear, and one *could* now rederive them as conservation laws.

EXERCISE 23 Show that
(i) V/l is a current flux, and
(ii) I/c is a voltage flux
Hint: check dimensions.

EXERCISE 24 Show that V and I satisfy the second-order linear wave equation.

Dissipative losses are very important in transmission line technology. Small leakages and resistances can add up over a long line to attenuate and distort the signal. For example, the first attempts at laying the transatlantic cable met with difficulties until it was understood how to deal with losses. This basically amounted to an analysis of the transmission line equations with loss terms included.

Irreversible drop in voltage across the section Δx will be due to a resistance

$$\Delta V_{\text{irrev}} = -RI. \tag{10.61}$$

Likewise, leakage of current is assumed proportional to voltage:

$$\Delta I_{\text{irrev}} = -GV. \tag{10.62}$$

These are empirical relations [to be contrasted with the more fundamental laws (10.57, 10.59)], in fact only valid in restricted circumstances, for example when V and I are not too large (refer to the discussion in Chapter 5). Combining (10.57–10.61) as well as (10.59–10.62), we now have

$$V(x) - V(x + \Delta x) = rI\Delta x + l\dot{I}\Delta x, \tag{10.63}$$
$$I(x) - I(x + \Delta x) = gV\Delta x + c\dot{V}\Delta x. \tag{10.64}$$

The resulting equations are

$$l\frac{\partial I}{\partial t} + \frac{\partial V}{\partial x} + rI = 0, \tag{10.65}$$
$$c\frac{\partial V}{\partial t} + \frac{\partial I}{\partial x} + gV = 0. \tag{10.66}$$

EXERCISE 25 *Telegraph equation* Show that I and V each satisfy a second-order partial differential equation. *Hint:* If we eliminate I from the system (10.65,10.66), then we get

$$lcv_{tt} - v_{xx} + (lg + rc)v_t + rgv = 0.$$

Solving Linear Hyperbolic Systems

The transmission line equations (10.65, 10.66) are a special case of linear systems of the form

$$\mathbf{u}_t + A(x,t)\mathbf{u}_x + B(x,t)\mathbf{u} = 0, \tag{10.67}$$

where **u** is a vector and A, B are square matrices. For the equations (10.65, 10.66), we have

$$\mathbf{u} = \begin{pmatrix} I \\ V \end{pmatrix}, \quad A = \begin{pmatrix} 0 & 1/l \\ 1/c & 0 \end{pmatrix}, \quad B = \begin{pmatrix} r/l & 0 \\ 0 & g/c \end{pmatrix}.$$

We will distinguish a special type of system of the form (10.67) whose solution is particularly easy. A system is *hyperbolic* if the $n \times n$ matrix A has n real distinct eigenvalues, $\{\lambda_1, \ldots, \lambda_n\}$. Recall that a number λ is an *eigenvalue* for A if it satisfies

$$Ap = \lambda p$$

for some nonzero vector p; such vectors are called *eigenvectors*. For such matrices, the set of eigenvectors, $\{p_k\}$, forms a basis. If

$$\Lambda = \begin{pmatrix} \lambda_1 & \cdots & 0 \\ 0 & \ddots & 0 \\ 0 & \cdots & \lambda_n \end{pmatrix}$$

and $P = \begin{pmatrix} p_1 & p_2 & \cdots & p_n \end{pmatrix}$, then we can write

$$AP = \Lambda P = P\Lambda. \tag{10.68}$$

Since the $\{p_k\}$ are independent, P is invertible; thus, we can make the change of variable

$$\mathbf{u} = P\mathbf{v}.$$

The system (10.67) transforms to

$$0 = \mathbf{u}_t + A(x,t)\mathbf{u}_x + B(x,t)\mathbf{u} = P\mathbf{v}_t + \frac{\partial P}{\partial t}\mathbf{v} + AP\mathbf{v}_x + A\frac{\partial P}{\partial x}\mathbf{v} + BP\mathbf{v},$$

or

$$\mathbf{v}_t + \Lambda\mathbf{v}_x + C\mathbf{v} = 0, \tag{10.69}$$

where

$$C = P^{-1}\left(\frac{\partial P}{\partial t} + A\frac{\partial P}{\partial x} + BP\right).$$

The system (10.69) is said to be in *canonical form*. We have n equations of the form

$$v_t^{(k)} + \lambda_k v_x^{(k)} + \sum c^{kj} v^{(j)} = 0. \tag{10.70}$$

Since there is no coupling among the $v^{(k)}$ in the derivatives in (10.70), it is a straightforward matter to define n independent families of characteristic curves,

$$\frac{dx}{dt} = \lambda_k. \tag{10.71}$$

The numbers λ_k thus represent speeds of propagation, since information is transmitted along characteristics at that speed.

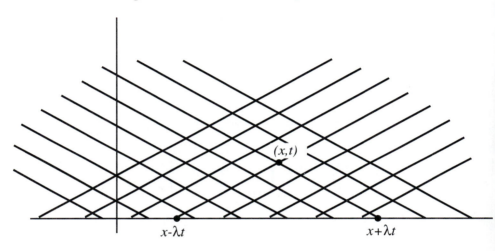

Figure 10.29

Solving the Transmission Line Equations

We'll start by putting the transmission line equations in canonical form. The eigenvalues satisfy

$$0 = \det(A - \lambda I) = \begin{vmatrix} -\lambda & l^{-1} \\ c^{-1} & -\lambda \end{vmatrix} = \lambda^2 - (lc)^{-1}.$$

These are

$$\lambda_\pm = \pm \frac{1}{\sqrt{lc}}. \tag{10.72}$$

The number $|\lambda_\pm| = \frac{1}{\sqrt{lc}}$ is the transmission speed of the line. The two families of characteristic equations are illustrated in Figure 10.29. One family carries waves forward and the other backward. To put it another way, for every point on the x-axis there are two characteristics emanating from that point in the positive t-direction, one carrying information forward and the other backward.

To find the eigenvectors, we must solve

$$(A - \lambda I)p = 0.$$

This equation,

$$Ap = \pm \frac{1}{\sqrt{lc}} p,$$

has for components

$$
\begin{aligned}
\frac{1}{l} p_2 &= \pm (lc)^{-1/2} p_1, \\
\frac{1}{c} p_1 &= \pm (lc)^{-1/2} p_2,
\end{aligned}
$$

and is equivalent to the relation

$$\sqrt{l} p_1 \pm \sqrt{c} p_2 = 0. \tag{10.73}$$

Let us set $p_1 = \sqrt{c}$; then, for the choice of the plus sign, we get

$$(+) \qquad p_1 = \sqrt{c}, \quad p_2 = -\sqrt{l}.$$

If we choose the minus sign, we get

$$(-) \qquad p_1 = \sqrt{c}, \quad p_2 = \sqrt{l}.$$

To put the system in canonical form, we will need

$$P^{-1} = \begin{pmatrix} \sqrt{c} & \sqrt{c} \\ \sqrt{l} & -\sqrt{l} \end{pmatrix}^{-1} = \frac{1}{2\sqrt{lc}} \begin{pmatrix} \sqrt{l} & \sqrt{c} \\ \sqrt{l} & -\sqrt{c} \end{pmatrix}$$

The system is

$$\frac{\partial}{\partial t} \begin{pmatrix} v_1 \\ v_2 \end{pmatrix} + \begin{pmatrix} \frac{1}{\sqrt{lc}} & 0 \\ 0 & -\frac{1}{sqrtlc} \end{pmatrix} \frac{\partial}{\partial x} \begin{pmatrix} v_1 \\ v_2 \end{pmatrix} + C \begin{pmatrix} v_1 \\ v_2 \end{pmatrix} = 0, \tag{10.74}$$

where

$$
\begin{aligned}
C &= \frac{1}{2\sqrt{lc}} \begin{pmatrix} r\sqrt{\frac{c}{l}} + g\sqrt{\frac{l}{c}} & r\sqrt{\frac{c}{l}} - g\sqrt{\frac{l}{c}} \\ r\sqrt{\frac{c}{l}} - g\sqrt{\frac{l}{c}} & r\sqrt{\frac{c}{l}} + g\sqrt{\frac{l}{c}} \end{pmatrix} \\
&= \frac{1}{2} \begin{pmatrix} \frac{r}{l} + \frac{g}{c} & \frac{r}{l} - \frac{g}{c} \\ \frac{r}{l} - \frac{g}{c} & \frac{r}{l} + \frac{g}{c} \end{pmatrix}
\end{aligned}
$$

The system of equations, (10.74), is equivalent to equations of the form

$$\frac{dv^k}{dt} = -c^{11}v^1 - c^{12}v^2,$$

where

$$\frac{d}{dt}$$

denotes the total derivative along the appropriate characteristic curve. The solution method for the initial-value problem is simple and straightforward: to find the solution at (x, t), we must find the two characteristic curves that pass through that point and follow them back to $t = 0$ (Figure 10.29). Then the value is found by integrating the right side over the characteristic curve. For v^1, this is

$$v^1(x,t) \;=\; v^1(x - \lambda t, 0) - \int_0^t c^{11}v^1(x - \lambda(t - xs), s)\, ds$$
$$- \int_0^t c^{12}v^2(x - \lambda(t - xs), s)\, ds.$$

Notice that the integration of v^1 requires that, except for a couple of special cases, we know v^2 everywhere along the characteristc curve. One special case is when there is no dissipation:

$$r = g = 0.$$

This results in the system (10.58, 10.60). There is another special case. If

$$c^{12} = \frac{r}{l} - \frac{g}{c} = 0, \qquad (10.75)$$

then the equations are decoupled and can be easily integrated:

$$v^1(x,t) = v^1(x - \lambda t, 0)e^{-(\frac{r}{l}+\frac{g}{c})t}.$$

The value of $v^1(x, t)$ is simply the value of the function at $v^1(x - \lambda t, 0)$ multiplied by an attenuation factor. This is just exponential relaxation, and the relaxation time

$$\tau = \frac{lc}{rc + gl}$$

is the same as that for the RLC circuit when there is no leakage from the line (see Chapters 4 and 5). Actually, the discrepancy depends only on the size of the ratio, g/c, which is small because of the large capacitance in most practical cases.

EXERCISE 26 Change variables to show that, under the condition (10.75), the current and voltage are

$$I(x,t) = I(x - \lambda t, 0)e^{-(\frac{r}{l} + \frac{g}{c})t}$$
$$V(x,t) = V(x - \lambda t, 0)e^{-(\frac{r}{l} + \frac{g}{c})t}$$

The significance of the condition (10.75) can be seen by considering a short signal that is transmitted from one end of a line, $x = 0$, at time $t = 0$. For simplicity, we will signal with

$$v^1 = \frac{1}{2}(I/\sqrt{c} + V/\sqrt{l}).$$

If the signal lasts for 2 seconds then the signal will be received at $x = x_r$ after a time of exactly x_r/λ seconds and will last for exactly 2 seconds; most importantly, the signal will have exactly the same functional form as the transmitted signal. The only difference is that it has been reduced in amplitude by the appropriate relaxation factor. In other words, it can be reproduced by a constant amplification. On the other hand, if c^{12} is nontrivial, then the signal received, v^1, will involve the integration of v^2 over its whole journey. This will lead to distortion of the signal. Thus, the particular combination, (10.75), is precisely what is needed for distortionless signaling. When this condition was understood, it was realized that it could be achieved by compensating for the large capacitances with large inductances: coils were placed along the line. The interesting thing is that Pupin came at this idea by playing with a mechanical analogy, which he realized had the same mathematical structure.

6 Systems of Nonlinear Conservation Laws

A fluid can be modeled via some continuum fields, the mass density, ρ, and the velocity, **u**. We have already encountered equations for mass density. Conservation of mass in integral form is either [see 10.17]

$$\frac{d}{dt}\int_B \rho \, d\mathbf{x} = -\int_{\partial B} \rho \mathbf{u} \cdot d\mathbf{B}, \tag{10.76}$$

where B is fixed in space, or

$$\frac{d}{dt}\int_{B(t)} \rho \, dx = 0, \tag{10.77}$$

where $\mathcal{B}(t)$ "moves with the flow" [see 10.5]. If the integrands are continuous, we can write the equation in differential form

$$\frac{\partial \rho}{\partial t} + \text{div}\,(\rho \mathbf{u}) = 0. \tag{10.78}$$

One way to close this equation is to obtain a direct relation (through experiment) between $\mathbf{u}$ and ρ. This results in the scalar conservation law.

However, it is hard to justify an algebraic relation between $\mathbf{u}$ and ρ. In fact, the velocity will have to satisfy Newton's law, which states that the rate of change of the momentum is due to the net force acting:

$$\frac{dP}{dt} = F.$$

Among the forces acting on a particle of fluid are *body forces*, which act on the whole particle and are proportional to the volume (or mass) of the particle, and *surface forces*, which act across surfaces. Examples of the latter type of force are the usual mechanical forces of pressure and traction. Gravity is an example of a body force.

To simplify matters, we will derive the equations for one-dimensional flow (extensions to higher dimensions can be found in the problem section). We will consider, once again, a "box" and treat momentum as a conserved quantity. We will assume that the cross-sectional area of the box is constant, and all variables will be reduced by that area. This is easily done by choosing units in space perpendicular to x so that cross-sectional area equals 1. The total momentum in the box is

$$\rho v \Delta x.$$

The momentum density is

$$\rho v.$$

Because of Newton's law, the momentum will change not only due to its flux by material transport, but also because of any forces acting. Thus, over a small time interval, Δt, the change in momentum is

$$\rho v(t + \Delta t)\Delta x - \rho v(t)\Delta x = \left(\begin{array}{c} \text{momentum flux due to} \\ \text{inertial transport} \end{array} \right)$$

$$+ \left(\begin{array}{c} \text{force acting} \\ \text{during}\Delta t \end{array} \right)$$

$$= [((\rho v)v)(x) - (\rho v)v(x + \Delta x)]\Delta t$$

$$+ f\Delta x \Delta t + (\sigma(x) - \sigma(x + \Delta x))\Delta t,$$

where f is a body force density, and σ is a "surface" force acting across
the end points of the box and is taken to be positive when acting to
the right. After taking appropriate limits, we arrive at

$$\frac{\partial}{\partial t}(\rho v) + \frac{\partial}{\partial x}(\rho v^2) + \frac{\partial \sigma}{\partial x} = f. \tag{10.79}$$

Carrying out the derivatives and making use of the mass conservation
equation (10.78) gives us

$$\rho\left(\frac{\partial v}{\partial t} + v\frac{\partial v}{\partial x}\right) + \frac{\partial \sigma}{\partial x} = f, \tag{10.80}$$

the momentum balance equation. We will set out to close the mass
conservation equation by deriving another equation for these variables.
We have to determine the relation that f and σ bear to these variables.
The body force density, f, is usually a very simple function of ρ, v, but
σ may depend on ρ, v or involve other variables. A simple case arises
when the fluid is assumed to have no viscosity; then $\sigma = p$, a pressure,
which is related to the work done on the fluid. This means, in general,
that to close the equations we must also conserve energy. However,
in the adiabatic or isentropic case, p is given as a function of ρ. For
inviscid flow, the momentum balance equation is

$$\rho(v_t + vv_x) + p_x = f.$$

EXERCISE 27 Show that in higher dimensions the momentum balance for
inviscid flow is
$$\rho(\mathbf{u}_t + \mathbf{u} \cdot \nabla\mathbf{u}) + \nabla p = \mathbf{f}.$$

(The assumption is that the force $\boldsymbol{\sigma}$ acting across the surface is *normal* to
the surface and has magnitude p: $\boldsymbol{\sigma} = -p\mathbf{n}$.)

*Momentum Equation Using Pull-back Method

The integral form of the momentum conservation law is

$$\frac{d}{dt}\int_{\mathcal{B}(t)} \rho\mathbf{u}\,d\mathbf{x} = \int_{\mathcal{B}(t)} \mathbf{f}\,d\mathbf{x} + \int_{\partial\mathcal{B}(t)} \boldsymbol{\sigma} \cdot d\mathbf{S},$$

where $\mathcal{B}(t)$ moves with the flow. In the one-dimensional form, this is

$$\frac{d}{dt}\int_{a(t)}^{b(t)} \rho v\,dx = \int_{a(t)}^{b(t)} f\,dx + \sigma(a(t)) - \sigma(b(t)).$$

Pull back to $t = 0$ and differentiate, using the derivative of the Jacobian as we did in section 2 (the notation is similar):

$$\frac{d}{dt}\int_{a(t)}^{b(t)} \rho v\, dx = \frac{d}{dt}\int_{a(0)}^{b(0)} RUJ\, d\xi = \int_{a(0)}^{b(0)} \frac{\partial}{\partial t}(RU)J + RU\frac{\partial v}{\partial x}J\, d\xi.$$

By the chain rule,

$$\frac{\partial}{\partial t}(RU) = \frac{D}{Dt}(\rho v),$$

so when we change back to the moving domain, $\mathcal{B}(t)$, we get

$$\begin{aligned}
\frac{d}{dt}\int_{a(t)}^{b(t)} \rho v\, dx &= \int_{a(t)}^{b(t)} \frac{D}{Dt}(\rho v)\, dx \\
&= \int_{a(t)}^{b(t)} \frac{\partial}{\partial t}(\rho v) + v\frac{\partial}{\partial x}(\rho v)\, dx \\
&= \int_{a(t)}^{b(t)} \rho\frac{\partial v}{\partial t} + \rho v\frac{\partial v}{\partial x}\, dx,
\end{aligned}$$

where the equation for mass conservation has been used. Putting everyting under the same integral results in

$$\int_{a(t)}^{b(t)} \rho\frac{\partial v}{\partial t} + \rho v\frac{\partial v}{\partial x} - f + \frac{\partial \sigma}{\partial x}\, dx = 0.$$

Application of the DuBois–Reymond lemma gives the same equation we had before.

Sound Speed in Gases

Under the assumption that the compression of a gas is adiabatic (isentropic), the condition that relates the pressure in that gas to its mass density is

$$p = C\rho^{\gamma}, \tag{10.81}$$

where γ is a ratio of specific heats. Sound waves in a gas are linear compression waves. Let us suppse that a sharp loud sound is produced in a gas, say by snapping fingers. As such a disturbance passes, the local values of the density and pressure rise to high values and then drop back to normal (Figure 10.30). Let us choose coordinates that move with the wave, and let us assume that the velocity in the laboratory frame can be ignored. Thus, in the new system of coordinates the velocity of a particle is the wave speed referred to the laboratory frame.

With respect to the new coordinates, the wave is stationary; thus, the equations of mass conservation and momentum balance become

$$\frac{\partial}{\partial x}(\rho v) = 0, \tag{10.82}$$

$$\rho v \frac{\partial v}{\partial x} + \frac{\partial p}{\partial x} = 0. \tag{10.83}$$

We can eliminate the term involving $\frac{\partial v}{\partial x}$ from these equations:

$$\frac{\partial p}{\partial x} = v^2 \frac{\partial \rho}{\partial x},$$

or, in other words,

$$v^2 = \frac{dp}{d\rho}.$$

Using (10.81), we get

$$v^2 = \gamma p / \rho.$$

If the gas obeys the ideal gas law;

$$p = \frac{\rho R T}{M_w},$$

where M_w is the molecular weight, then we obtain the sound speed in terms of fundamental parameters:

$$v = \sqrt{\frac{\gamma R T}{M_w}}. \tag{10.84}$$

Actually, strong disturbances such as shocks can move through a gas at speeds quite different from the sound speed, (10.84). The previous derivation does not apply to such waves because (1) we cannot use the differential form of the conservation laws, (2) we cannot assume that the velocity of the gas is small, and (3) the thermodynamics are not isentropic: entropy increases across a shock. The fact that shocks can be an avenue of dissipation will appear when we study tidal bores below, but it is simply illustrated by the following mathematical example.

Shock as a Dissipative Structure

Consider the momentum equation under conditions of constant pressure:

$$u_t + u u_x = 0.$$

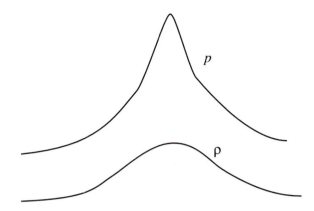

Figure 10.30 The rise of pressure and density in a solitary sound wave.

Suppose that the initial condition is of a small packet of gas moving through a stationary background:

$$u(x,0) = \begin{cases} 0 & \text{if } x > 1 \text{ or } x < 0 \\ 1 & \text{if } 0 < x < 1 \end{cases}.$$

Figure 10.31 shows the resulting solution. There is a "fan" solution at $x = 0$ and a shock of speed $1/2$ emanating from $x = 1$. At $t = 2$ the shock hits the fan. Then a new shock forms with the new speed

$$\dot{\phi} = \frac{x}{2t}.$$

To find the equation for this shock, we must solve the initial-value problem

$$\frac{dx}{dt} = \frac{x}{2t}, \qquad x(2) = 2.$$

This solution is

$$x = \sqrt{2t}.$$

Look at the jump in the solution

$$[u(x,t)] = u_l - u_r = \frac{x}{t} - 0$$

across the shock:

$$[u]|_{x=\sqrt{2t}} = \frac{\sqrt{2}}{\sqrt{t}}.$$

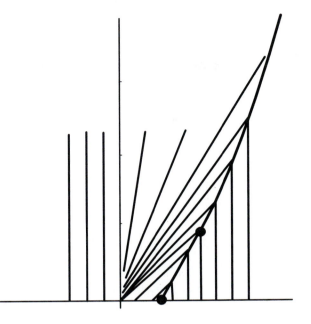

Figure 10.31 Two shocks. Each shock forms at a dot.

Thus, the velocity jump decays like

$$t^{-1/2}$$

as $t \to \infty$. That is, the shock decreases in strength; but meanwhile it has been doing its part to dissipate energy. To see this, let us compute the total kinetic energy per mass in the gas:

$$\int_{-\infty}^{\infty} u^2 \, dx = \int_{0}^{\sqrt{2t}} (\frac{x}{t})^2 \, dx = \frac{\sqrt{8}}{3} t^{-1/2}.$$

But wait a minute! This is supposed to be a conservation law. But, yes, total momentum (per unit mass) is conserved:

$$\int_{0}^{\sqrt{2t}} \frac{x}{t} \, dx = 1.$$

The energy that is dissipated can be considered to be a momentum flux. Waves can also cause stresses in the medium through which they move, which can be quantitatively related to a momentum flux.

Conservation of Energy

Recall that the thermodynamic state variables are: energy, E; volume, V; pressure, P; entropy, S; and temperature, T. It is known that if any two of these are specified then the rest are determined.

The first law of equilibrium thermodynamics states that, for systems in equilibrium,

$$dE = đQ - đW,$$

which is short for " if the system is always in equilibrium, then any change in energy is due to an addition of heat, Q, minus any work, W, done by the system." The $đ$ notation is to remind us that these are not true differentials since Q and W are not state variables. However, they have integrating factors

$$T^{-1}đQ = dS, \qquad P^{-1}đW = dV.$$

Thus, the equilibrium condition is

$$dE = T\,dS - P\,dV. \tag{10.85}$$

In fluids (and solids) we have distributed variables (per unit volume or per unit mass; the qualifier "specific" is for quantities that are referred to unit mass). These variables are p, pressure; θ, temperature; τ, specific volume; ρ, mass density; s, specific entropy; e, specific energy; and specific enthalpy, $h = e + p\tau$.

If we assume that thermodynamic equilibrium holds along the path of a particle, then (10.85) give

$$\frac{De}{Dt} = \theta\frac{Ds}{Dt} - p\frac{D}{Dt}\Big(\frac{1}{\rho}\Big).$$

If we assume that a fluid particle doesn't change its entropy during its evolution, then

$$\frac{Ds}{Dt} = 0. \tag{10.86}$$

(This means that we ignore heat generation due to viscosity, as well as conduction in the fluid. Such fluids are called *ideal.*) Then, using the equation for mass conservation, the equilibrium condition becomes

$$
\begin{aligned}
\rho\frac{De}{Dt} &= \frac{p}{\rho}\frac{D\rho}{Dt}\\[2mm]
&= \frac{p}{\rho}(-\operatorname{div}\mathbf{v})\\[2mm]
&= \frac{1}{\rho}(\mathbf{v}\cdot\nabla p - \operatorname{div}(\rho\mathbf{v})).
\end{aligned}
$$

Notice that the last term is the rate of working by the pressure forces. We can get an energy conservation equation by relating total energy change to rate of working by the applied forces (we consider only pressure forces here).

Total   specific energy

$$= \text{(internal) specific energy} + \text{specific kinetic energy}$$

$$= e + \frac{1}{2}\mathbf{v} \cdot \mathbf{v}.$$

The rate of working by the pressure forces across a surface with normal **n** is

$$-p\mathbf{v} \cdot \mathbf{n}.$$

Our equation then is

$$\frac{d}{dt}\int_{\mathcal{B}(t)} \rho(e + \frac{1}{2}\mathbf{v} \cdot \mathbf{v})\, d\mathbf{x} = -\int_{\partial\mathcal{B}(t)} p\mathbf{v} \cdot d\mathbf{S}.$$

In one dimension this reduces to

$$\frac{d}{dt}\int_{a(t)}^{b(t)} \rho(e + \frac{1}{2}v^2)\, dx = p(a,t)v(a,t) - p(b,t)v(b,t).$$

EXERCISE 28 If the integrands in this integral are continuous, then it follows that

$$\frac{d}{dt}\int_{a(t)}^{b(t)} \rho s\, dx = 0.$$

Relate to the use of the condition (10.86).

In the nonisentropic situation, the temperature can be used as a variable. The heat capacity, C, of a material is the amount of energy needed to raise the temperature (of 1 mole) 1 degree:

$$\Delta E = C\Delta T.$$

This is either C_V if the heat is supplied under conditions of constant volume or C_P if the heat is supplied under conditions of constant pressure. This can be written in terms of specific variables:

$$e(x,t) = c\theta(x,t),$$

where c is called the specific heat. Usually, the energy density is given per unit volume, and then this relation takes the form

$$e = c\rho\theta,$$

where ρ is the mass density.

*Quasilinear Hyperbolic Systems. Simple Waves

The systems of conservation laws that we have seen can be put in the form of a hyperbolic system

$$\mathbf{u}_t + A\mathbf{u}_x = \mathbf{B},$$

where, in contrast to the linear systems, the matrix A and the vector function B may depend on the components of the state vector, $u_1, \ldots, u_m$, as well as on (x, t). We cannot directly extend the analysis of the linear case, because that relied on the fact that the change of dependent variable that lead to a diagonal form of the system was brought about by a linear transformation of (right) eigenvectors,

$$\mathbf{u} = P\mathbf{v}.$$

If P depends on $\mathbf{u}$, this won't work. We can, however, get a partial decoupling of the system by multiplying by *left* eigenvectors of A, which in some cases, notably in two dimensions, will suggest a more complete diagonalization.

Let $\{\lambda_1, \ldots, \lambda_n\}$ be the eigenvalues of A; generally, they are functions of $x, t, u_1, \ldots, u_n$. Suppose that they are real and distinct, so that we can find n *left* eigenvectors of the matrix A:

$$\mathbf{l}_1, \ldots, \mathbf{l}_n.$$

Then

$$
\begin{aligned}
0 &= \mathbf{l}_k \cdot (\mathbf{u}_t + A\mathbf{u}_x - \mathbf{B}) \\
  &= \mathbf{l}_k \cdot \mathbf{u}_t + \lambda_k \mathbf{l}_k \cdot \mathbf{u}_x - \mathbf{l}_k \cdot \mathbf{B}
\end{aligned}
$$

is a scalar equation, and we have n such scalar equations. Each of these equations is an ordinary differential equation along a characteristic curve: that is,

$$\mathbf{l}_k \cdot \frac{D\mathbf{u}}{Dt} = \mathbf{l}_k \cdot \mathbf{B}$$

holds along

$$\frac{dx}{dt} = \lambda_k.$$

We have not derived the method of characteristics because the equations are formed with the characteristics.

As an example, consider the equations of one-dimensional isentropic gas flow

$$u_t + uu_x + \frac{c^2}{\rho}\rho_x = 0 \tag{10.87}$$

$$\rho_t + u\rho_x + \rho u_x = 0. \tag{10.88}$$

This is a hyperbolic system with

$$A = \begin{pmatrix} u & c^2/\rho \\ \rho & u \end{pmatrix}, \qquad \mathbf{B} = 0.$$

The eigenvalues are $\lambda_\pm = u \pm c$. The left eigenvectors are

$$\mathbf{l}_+ = (1, \frac{c}{\rho}), \qquad \mathbf{l}_- = (1, -\frac{c}{\rho}).$$

EXERCISE 29 Verify this.

The transformed equations are

$$\frac{Du}{Dt} + \frac{c}{\rho}\frac{D\rho}{Dt} = 0, \quad \text{along } \frac{dx}{dt} = u + c,$$

and

$$\frac{Du}{Dt} - \frac{c}{\rho}\frac{D\rho}{Dt} = 0, \quad \text{along } \frac{dx}{dt} = u - c.$$

The form of these equations suggests that we integrate the 1-forms

$$du \pm \frac{c}{\rho}\,d\rho.$$

Doing this gives that the functions

$$r = \int \frac{c}{\rho}\,d\rho + u,$$

$$s = \int \frac{c}{\rho}\,d\rho - u,$$

are constant along

$$\frac{dx}{dt} = u + c$$

and

$$\frac{dx}{dt} = u - c,$$

respectively. Thus,

$$\frac{Dr}{Dt} = 0,$$

$$\frac{Ds}{Dt} = 0,$$

are equations to which we can apply the usual method of characteristics. The functions r and s are called *Riemann invariants*.

A solution is called a *simple wave* in a region of the (x, t)-plane if one of r or s is constant throughout that region.

As an example, suppose that $s = 0$. Then

$$du = \frac{c}{\rho} \, d\rho.$$

Also,

$$\frac{Dr}{Dt} = \frac{D}{Dt}(2 \int \frac{c}{\rho} \, d\rho)$$

$$= 2\frac{c}{\rho}\frac{\partial \rho}{\partial t} + 2(u + c)\frac{c}{\rho}\frac{\partial \rho}{\partial x}.$$

Thus, the equation reads

$$\frac{\partial \rho}{\partial t} + (u + c)\frac{\partial \rho}{\partial x} = 0.$$

This is a scalar equation. Suppose that on the initial curve, $f(\rho) = u+c$; then

$$\rho = F(x - f(\rho)t)$$

is a parametrization of the solution surface.

7   Water Waves

It seems that wherever there is water in motion, there are waves. From the large ocean-going surges, to the breakers at the shore, to the small ripples in a stream. It is unlikely that we will get anywhere with a comprehensive model for water waves, so let us start with a more specific situation. Consider the surface waves in a shallow river. We want to know the speed and height of waves as well as the wavelengths. What are the relevant variables? The prominent one here is the depth of the water, h, that is, the height of the surface above the bottom (Figure

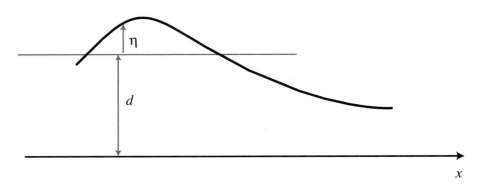

Figure 10.32

10.32). In addition, there are the usual fluid variables of velocity, density, pressure, viscosity, and surface tension. But we cannot have large waves without moving large amounts in the vertical direction. Because of water's matter, we have to do work against the gravitational force. Let us make a list of working assumptions:

1. We neglect viscosity and surface tension.

2. We neglect all variations in the z-direction (perpendicular to plane of Figure 10.32).

3. All flow is directed in the x-direction.

4. Water is incompressible.

5. Variables are small; the surface is close to flat.

EXERCISE 30 (i) What are the conditions that allow assumption 1?
(ii) Note that assumption 3 means that h should be small; but small compared to what?
(iii) The assumption 5 entails a linearization; what is the flow situation that we are linearizing about?

Since we are ignoring dissipation, we have the usual momentum equations. Because of assumptions 2 and 3, we can neglect the z-momentum equation. The other equations are

$$x\text{-momentum}: \quad \rho\left(\frac{\partial u}{\partial t} + u\frac{\partial u}{\partial x} + v\frac{\partial u}{\partial y}\right) = -\frac{\partial p}{\partial x}, \quad (10.89)$$

$$y\text{-momentum}: \quad \rho\left(\frac{\partial v}{\partial t} + u\frac{\partial v}{\partial x} + v\frac{\partial v}{\partial y}\right) = -\frac{\partial p}{\partial y} - \rho g. \quad (10.90)$$

Because of assumption 3, the y-momentum equation reduces to

$$\frac{\partial p}{\partial y} = -\rho g. \tag{10.91}$$

Integrating this, we get

$$p(y) = C - \rho g y.$$

We will set atmospheric pressure level as the zero level for p (see Figure 10.32), so

$$p(y) = \rho g(\eta - y). \tag{10.92}$$

Because of assumptions 3 and 5, the x-momentum equation reduces to

$$\rho \frac{\partial u}{\partial t} = -\frac{\partial p}{\partial x}. \tag{10.93}$$

We then use (10.92) to write

$$\frac{\partial u}{\partial t} = -g \frac{\partial \eta}{\partial x}. \tag{10.94}$$

Let us stop to interpret (10.94). It simply states that the horizontal acceleration is due to gravity, the gravitational force being proportional to the slope of the water surface. In other words, water flows downhill to obtain horizontal velocity.

EXERCISE 31 Is the equation (10.94) consistent with our assumptions?

The assumption of incompressibity (4) means that we must take care of the mass balance as we did for the flow through a flexible tube. That is, the mass flow balance is

$$[(\rho A)(t + \Delta t) - (\rho A)(t)]\Delta x = [(\rho A u)(x) - (\rho A u)(x + \Delta x)]\Delta t,$$

which gives the differential equation

$$\frac{\partial}{\partial t}(\rho A) + \frac{\partial}{\partial x}(\rho A u) = 0. \tag{10.95}$$

Since we ignore variation in z (assumption 2), we can set

$$A = h \cdot 1 = (d + \eta) \cdot 1,$$

so the equation reduces to

$$\frac{\partial \eta}{\partial t} + \frac{\partial}{\partial x}(du) + \frac{\partial}{\partial x}(\eta u) = 0.$$

Assumption 5 allows us neglect the last term so that, finally, mass conservation takes on the form of the equation

$$\frac{\partial \eta}{\partial t} = -d\frac{\partial u}{\partial x}. \tag{10.96}$$

The system of equations that we have derived, (10.94, 10.96), is identical to the system of acoustic equations derived earlier.

EXERCISE 32 Put the equations (10.94, 10.96) in dimensionless form.

Building on this analogy, we know what the solutions to this system look like: waves without distortion all moving at the speed,

$$v_{\text{wave}} = \sqrt{gd}.$$

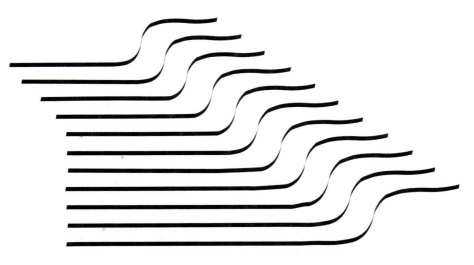

Figure 10.33

A Big Bore

These equations are clearly inadequate for a description of large amplitude waves let along the hydraulic jump and the big bores in tidal rivers (see Figure 10.33). Which assumptions can we keep and which must we flow out? We must get rid of number 5 because the surface is not gently varying in a tidal bore, but we can retain number 3 as long as the depth, though not varying gently is still small. The y-momentum equation is the same as before; thus, the pressure satisfies (10.92). The new equations are

$$x\text{-momentum} : \quad \frac{\partial u}{\partial t} + u\frac{\partial u}{\partial x} = -g\frac{\partial h}{\partial x}, \quad (10.97)$$

$$\text{mass conservation} : \quad \frac{\partial h}{\partial t} + \frac{\partial}{\partial x}(uh) = 0. \quad (10.98)$$

These same equations describe the one-dimensional compressible isentropic flow of a gas [see (10.87), (10.88), and Problem 27].

EXERCISE 33 Linearize these equations about $u = 0, h = d$, to recover the acoustic equations.

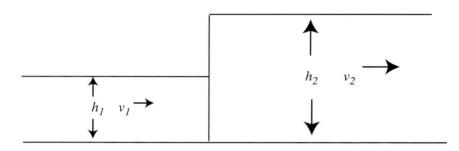

Figure 10.34

With regard to Figure 10.34, what do we want to predict? If we assume that the bore is a dissipative structure (comletely analogous to a shock), we will try to show that the water jumps in level. It is easy to see from the conservation of mass that this jump in water level is equivalent to a jump down in velocity.

Let's get quantitative.

To simplify matters, we'll suppose constant conditions on either side of the discontinuity. Look at the integral forms of the conservation laws. We will conserve mass flux and momentum flux per unit width of the channel. We get these by integration of the flux densities over the height of the water. The mass flux density is

$$\int_0^h \rho v\, dy = \rho vh.$$

The momentum flux is

$$\int_0^h p + \rho v^2 \, dy = \int_0^h \rho g y + \rho v^2 \, dy = \frac{1}{2}\rho g h^2 + \rho v^2 h.$$

The continuity of fluxes across the discontinuity gives

$$v_1 h_1 = v_2 h_2, \tag{10.99}$$

$$\frac{1}{2}gh_1^2 + v_1^2 h_1 = \frac{1}{2}gh_2^2 + v_2^2 h_2. \tag{10.100}$$

We can solve for v_1 and v_2:

$$v_1^2 = \frac{1}{2}g\frac{h_2}{h_1}(h_1 + h_2) \tag{10.101}$$

$$v_2^2 = \frac{1}{2}g\frac{h_1}{h_2}(h_1 + h_2) \tag{10.102}$$

Meanwhile, the energy flux density is

$$\int_0^h \left(\frac{p}{\rho} + \frac{1}{2}v^2\right)\rho v \, dy.$$

This integrates to

$$\int_0^h (gy + \frac{1}{2}v^2)\rho v \, dy = \frac{1}{2}\rho g v h^2 + \frac{1}{2}\rho v^3 h \tag{10.103}$$

$$= \frac{1}{2}(\rho v h)(gh + v^2) \tag{10.104}$$

The jump in energy flux density is

$$
\begin{aligned}
q_1 - q_2 &= Q[gh_1 + v_1^2 - gh_2 - v_2^2] \\
&= gQ[h_1 + \frac{h_2}{2h_1}(h_1 + h_2) - h_2 - \frac{h_1}{2h_2}(h_1 + h_2)] \\
&= gQ[(h_1 - h_2) + \frac{(h_2^2 - h_1^2)(h_1 + h_2)}{2h_1 h_2}] \\
&= gQ[(h_1 - h_2)(1 - \frac{(h_1 + h_2)^2}{2h_1 h_2}] \\
&= gQ(h_2 - h_1)\frac{h_1^2 + h_2^2}{2h_1 h_2}.
\end{aligned}
$$

Thus, $q_1 > q_2$ implies that $h_2 > h_1$; in other words, the water "jumps up".

The local wave speed for small-amplitude waves is $c_1 = \sqrt{gh_1}$ to the left of the bore and $c_2 = \sqrt{gh_2}$ to the right.

EXERCISE 34 Show that

$$v_1 > c_1 \;\; = \;\; \sqrt{gh_1}, \tag{10.105}$$
$$v_2 < c_2 \;\; = \;\; \sqrt{gh_2}. \tag{10.106}$$

Energetically, what happens is that kinetic energy, carried by flow velocity, is converted catastrophically to potential energy, represented by the height.

This is a decent model for the river bore, but the assumption of a sharp step in height is not true for the hydraulic jump in the washtub. In fact, the latter is a low Reynolds number flow, not because of low speed but because of the small characteristic length scale (h). In this regime, viscous effects are very important; energy is dissipated mainly through turbulence, and the profile of the jump is complicated. An interesting question is where does the jump form. For possibilities and ideas about setting up experiments consult [51] and [5].

Surge on Deep Water

We have found a characteristic wave speed for water waves to be

$$c = \sqrt{gd},$$

which is independent of wavelength. But in Chapter 3 we found through dimensional analysis that the wave speed of gravitational waves depended on the wavelength as

$$c \propto \sqrt{g\lambda}. \tag{10.107}$$

A little thought will convince you that if $\mathcal{L}$ is any characteristic length then

$$c \propto \sqrt{g\mathcal{L}}.$$

It is seen that our derivation of the linear shallow water equations above would only be valid for

$$\lambda \gg h.$$

In general, there should be some parameter $\mu = \lambda/h$, and the two regions that can be distinguished are

$$\text{shallow water} \;\; \mu \gg 1,$$

and

deep water $\mu \ll 1$.

In which of these do we classify the tsunami? Maybe both: out at sea, $\mu \ll 1$, but close to shore, $\mu \gg 1$. Of course, the smallest waves never enter the region $\mu \gg 1$, because surface tension is an important consideration.

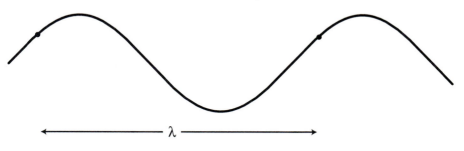

Figure 10.35

How can we find the constant in the relation (10.107)? In a wave (see Figure 10.35), we have the relation

$$\lambda = cT,$$

where T is the time between identical states for a fluid particle. We need an expression for T. We can get an idea of how to proceed by noting that fluid particles in this wave system move in near-circular motion (Figure 10.36a). For the purposes of our model, let us assume that it is precisely circular. We need to find the frequency of this circular motion. Using the force balance for a fluid particle moving at velocity v gives that the acceleration along its path of motion is approximately

$$a = \left(\frac{v}{c}\right)g.$$

EXERCISE 35 When is this valid? (Refer to Figure 10.36b.)

Equating this to the expression for centripetal acceleration gives

$$\omega = g/c,$$

and thus the formula (10.107) with its precise constant

$$c = \frac{1}{2\pi}\sqrt{g\lambda}. \tag{10.108}$$

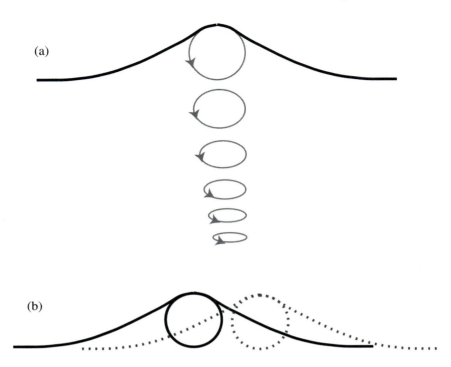

Figure 10.36 Pictures of particle trajectories (a) can be found in [17].

This is another example of a dispersion relation. It says that waves of different wavelengths move at different speeds. This is reminiscent of the modes of vibration of the chain of masses of Chapter 9. In fact, the dispersion relation in each of these examples is qualitatively similar, at least for long wavelengths. Dispersion laws are usually written as relations between angular frequency and wave number:

$$\omega = \omega(k).$$

The relation (10.108) can be written

$$\omega = \sqrt{2\pi g k}.$$

Another, qualitatively similar, dispersion relation was encountered in the cochlea model of Chapter 9. On the other hand, some systems have completely different dispersion relations (see Problem 29). More on waves can be found in [65].

8   Problems and Recommended Reading

1. Which of the following problems involving partial differential equations can be solved?

$$\frac{\partial u}{\partial t} + 2\frac{\partial u}{\partial x} = 0, \quad x > 0, \ u(0,t) = 1, \quad -\infty < t < \infty$$

$$\frac{\partial u}{\partial t} - 2\frac{\partial u}{\partial x} = 0, \quad x > 0, \quad t > 0, \quad \begin{aligned} u(x,0) &= \sin x, \quad x > 0 \\ u(0,t) &= \cos t, \quad t > 0 \end{aligned}$$

$$\frac{\partial u}{\partial t} - 2\frac{\partial u}{\partial x} = 0, \quad x < 0, \quad t > 0, \quad \begin{aligned} u(x,0) &= \sin x, \quad x < 0 \\ u(0,t) &= \cos t, \quad t > 0 \end{aligned}$$

2. *Numerical methods* Develop numerical methods to compute

$$\frac{\partial u}{\partial t} + v_0 \frac{\partial u}{\partial x} = 0, \quad v_0 > 0,$$

based on the following:

(i) The method of characteristics: The basic equation here is

$$u(x + v_0 t, t) = u(x, t).$$

You simply have to compute the position on the characteristic line at the time steps $t_0, t_0 + \Delta t, \ldots$. There is *no* truncation error.

(ii) Finite differences: The implementation using forward differences,

$$\frac{\partial u}{\partial t} \approx \frac{u(x, t + \Delta t) - u(x, t)}{\Delta t}, \quad \frac{\partial u}{\partial x} \approx \frac{u(x + \Delta x, t) - u(x, t)}{\Delta x},$$

is straightforward. The only problem is you won't get the right answer. Show that this is the case by computing the solution to the Riemann problem. What is wrong? How can you modify this?

3. *Energy flow in the wave equation* Show that for the equation

$$y_{tt} - c^2 y_{xx} = 0$$

the quantity

$$\mathcal{E} = \frac{1}{2}(y_t^2 + c^2 y_x^2)$$

satisfies a conservation law

$$\frac{\partial \mathcal{E}}{\partial t} + \frac{\partial J}{\partial x} = 0.$$

Show that the time average of energy flow for complex periodic solutions is

$$\frac{1}{T} \int_0^T J \, dt = \frac{1}{2} c^2 \omega \Im < y, y_x > .$$

4. *d'Alembert formula* Solve the initial-value problem for the wave equation

$$y_{tt} - c^2 y_{xx} = 0, \qquad t > 0, \qquad -\infty < x < \infty, \qquad (10.109)$$

$$y(x, 0) = f(x), \qquad y_t(x, 0) = g(x),$$

by the method of characteristics. (*Hint:* You can convert the equation to a system using the variables y_t and y_x.)

5. (i) Show how to solve the boundary-value problem for the wave equation using the method of characteristics:

$$y_{tt} - c^2 y_{xx} = 0, \qquad t > 0, \quad a < x < b, \qquad (10.110)$$

$$y(x, 0) = f(x), \qquad y_t(x, 0) = g(x),$$

$$y(a, t) = 0 = y(b, t).$$

(ii) Change to dimensionless variables in (10.110):

$$\xi = \frac{x - a}{b - a}, \quad \tau = \frac{ct}{b - a}.$$

6. *Absorbing boundary conditions* Can you find boundary conditions for the wave equation (10.110) so that the solution is zero after a certain time?

7. Choose coordinates ξ and η in the (x, t)-plane so that the wave equation (10.109) transforms to

$$Y_{\xi\eta} = 0.$$

8. A traffic light is to be placed at a certain intersection. Preliminary data from an empirical investigation have indicated that the velocity–density relation is as in Figure 10.37a. This will mean a flux–density curve with a high curvature (Figure 10.16).

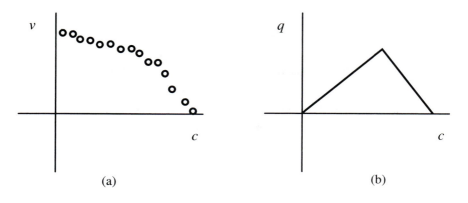

Figure 10.37

(a) For analytic work, it is easier to use the piecewise linear relation of Figure 10.37b. Using this relation, how does the traffic light behave? (What are the variables that characterize a controlled intersection? Pick a particular situation, say the green light "go." Can you reduce this to an initial-value problem?)

(b) Compare your results with those that follow from using a parabolic flux–density relation. What do you conclude?

(c) Now do the red light problem.

9. Having modeled the green light and the red light, how would you model the effect of a cop car entering highway traffic?

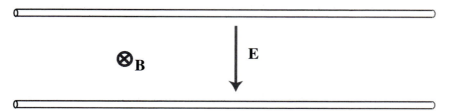

Figure 10.38

10. Figure 10.38 shows the electric and magnetic fields between the cables of a transmission line. Show that in this case Maxwell's

equations reduce to

$$-\frac{\partial H}{\partial z} = \epsilon_0 \frac{\partial E}{\partial t},$$
$$\frac{\partial E}{\partial z} = -\mu \frac{\partial H}{\partial t}.$$

Relate to the transmission line equations that we have derived.

11. *Plane EM waves* (i) Find the equations satisfied by plane electromagnetic waves. (*Hint:* In Maxwell's equations try solutions of a special form, or first derive a system similar to that in Problem 10.) (ii) Find the standing waves in a linear waveguide.

12. *One-dimensional cochlea* The equations (10.53, 10.54) for waves in a flexible tube are not suitable for the traveling wave through the cochlea (see Chapter 9), because they do not account for the inertia of the membrane. Modify the momentum balance equation to include a membrane inertial force. Derive a fourth-order partial differential for the membrane displacement. Show that you get the dispersion relation of Figure 9.27.

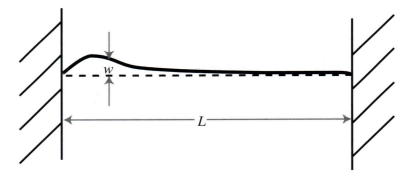

Figure 10.39

13. *Elastic string* A string is stretched along a length L (Figure 10.39). The elastic energy is represented by a *tension* (force/length), T, directed along the string.

(i) Show that the equation of motion for small transverse displacements of the string is

$$\mu w_{tt} = (T \sin \theta)_x,$$

where $\tan\theta$ is the slope of the string, and μ is a mass density (mass/length).

(ii) If the tension is uniform in the string, show that the equation becomes

$$\mu w_{tt} = T w_{xx}.$$

(iii) When a real string is plucked, the tension is not uniform, even for small displacements. The tension can be written as

$$T = T_0 + EA\left(\frac{\Lambda - L}{L}\right),$$

where T_0 is the tension in the flat position, E is Young's modulus of elasticity, A is the cross-sectional area, and Λ is the length of the plucked string:

$$\Lambda = \int_0^L \sqrt{1 + w_x^2}\, dx.$$

Show that

$$zL \approx \frac{1}{2}\int_0^L w_x^2\, dx,$$

and derive an integrodifferential equation for the string.

(iv) Can a string be "pumped up" by varying the tension?

14. *Bobs on a string* Derive the (linearized) equation of motion for N point masses equidistributed along an elastic string. Show that this reduces to the equation for the chain of springs and masses when we neglect the mass of the string.

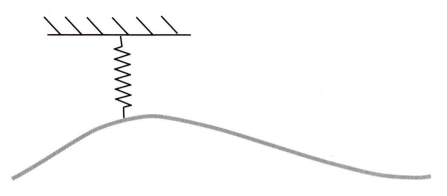

Figure 10.40

15. *Scattering by an oscillator* For standing waves on the linear string, the relation between wavenumber and frequency is

$$k = \frac{\omega}{c}, \tag{10.111}$$

where $c^2 = T/\rho$. Suppose that a linear harmonic oscillator is attached at some point along the string (Figure 10.40). Then the standing wave pattern will be changed. You can investigate this by imagining that a wave incident from the left is partially reflected and partially transmitted. Suppose that the oscillator is at $x = 0$. Then the standing wave solution is

$$y_l(x, k) = e^{ikx} + R(k)e^{-ikx}, \quad x \le 0, \tag{10.112}$$
$$y_r(x, k) = T(k)e^{ikx}, \quad x \ge 0. \tag{10.113}$$

By continuity of the string we have

$$1 + R(k) = T(k)$$

for all k. Write down the equation for the mass–spring oscillator (remember to include the force of the string). Derive a relation among k, R, T, and the spring stiffness.

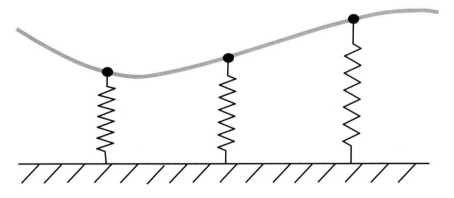

Figure 10.41

16. *A high-pass filter* In the arrangement of Figure 10.41, the string supplies mass and tension to the line of springs.
(i) Derive a continuum equation for the motion of this system. (Define κ, stiffness/length). Find the dispersion relation. Show

that it is high-pass filter. At high wave numbers it is approximately the usual wave equation. Does that make sense?

(ii) Suppose there is a significant increase in the mass density, μ, at $\bar{x}$ (or a decrease in κ). Show that waves are reflected at $\bar{x}$.

(iii) Show that waves can tunnel through a region, $\bar{x} < x < \bar{x} + \delta$, of high mass density.

17. *Characteristic wave impedance* (i) Recall that the characteristic impedance for an LC circuit is

$$Z_0 = V/I = \sqrt{L/C}.$$

Show that the characteristic impedance for a free transmission line is $\sqrt{l/c}$.

(ii) Now suppose that the line is terminated at a load with

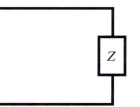

Figure 10.42

impedance Z (Figure 10.42). Then some of the wave will be absorbed and some fraction, r, will be reflected. The reflected wave may be phase-delayed with respect to the incident wave; this can be accounted for by allowing r to be complex. Suppose that the incident voltage wave is unit amplitude, then the incident current wave is Z_0^{-1}, and the reflected current is rZ_0^{-1}. Use continuity of current at the load to get a formula for r in terms of Z and Z_0. What happens if the line is terminated in a short circuit ($Z = 0$)? In an open circuit ($Z = \infty$)?

(iii) Suppose that a series load is on the line (Figure 10.43a). As in the scattering by an oscillator on a string, an incident wave is partially reflected and partially transmitted. Since impedances in series add, we can reduce this to the situation in (ii) by pretending that the line is terminated at the load (Figure 10.43b). Calculate the reflection and transmission coefficients, r and t. If Z is purely imaginary, show that r and t are 90° out of phase.

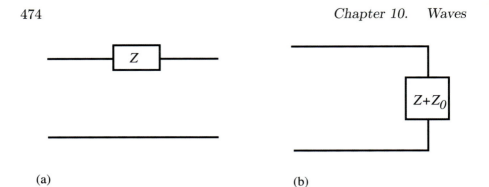

(a)   (b)

Figure 10.43

(iv) Calculate r and t for a parallel load. (*Hint:* admittances in parallel add.)

(v) How is the standing wave pattern affected by a load with impedance Z? Notice that as you move back down the line away from the load incident waves are *advanced* in phase by kx, while reflected waves are *delayed* by kx.

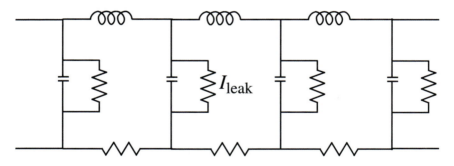

Figure 10.44

18. *Nerve propagation* A simple model for the propagation of signals along nerve fibers can be arrived at by assuming the form of a transmission line. In Chapter 1, we have seen the source of the capacitance and the conductance. Is there a possible source of inductance? Lieberstein ([43]) thinks that this could be due to toroidal currents aroung the nerve fiber, but this seems to be far from acceptable to most workers in the field. Consider a transmission line as in Figure 10.44 where the leakage current, I_{leak}, depends on voltage in the same way as the solid-state device

of Figure 5.15. Capacitance should also depend on voltage. How?
Write down a system of equations. In the continuum limit you get
a nonlinear telegraph equation. In [50] some data from Hodgin-
Huxley are matched to the I–V curve.

19. Show that the mass conservation equations derived for incompressible fluids can be found by integrating the equation

$$\text{div } \mathbf{v} = 0.$$

20. *Heat flow in a tube reactor* Some flows are clearly not isentropic;
such is the case when chemical reactions are taking place. Some
reactions are highly exothermic; they produce much heat. The
temperature is measured. The heat generated during a reaction is determined by a reaction rate, R which has dimensions
of (mole/volume)/time and a heat coefficient, Q:

$$\text{heat generated per volume per unit time } = QR.$$

Heat is usually transferred out the side of a tube reactor; the rate
of this transfer can be considered to be proportional to the jump
in temperature across the side. If the gas is being pushed through
a reactor at a steady velocity, v, show that the equation for the
temperature is

$$\rho c(\theta_t + v\theta_x) = QR + H(\theta - \theta_c).$$

21. *Heterogeneous reactions* In the last problem the chemical reaction
was occurring at interior points of the reactor. Some reactions
take place on surfaces, so that they show up in the boundary
conditions of continuum models. In that case, the heat generated
is given by

$$\text{heat generated per unit surface area per unit time} = Qr,$$

where the reaction rate, r, has dimensions of (mole/area)/time.
Suppose that such a reaction takes place at one end of a reactor.
How would you modify the model of the previous problem? In
the case of the long reactor, it would be more likely that such a
reaction would occur along (part of) the sides. Can you model
this with a one-dimensional model?

22. *Growing small crystals* For some applications, such as liquid crystal display, it is desired to have small crystals all of about the same particular size. It is thus interesting to be able to control the mean and variance of the size distribution of crystals. Suppose that crystals are grown in a continuous-stirred tank reactor, where small seed crystals are supplied to the growing tank in a steady stream. Let

$$
\begin{aligned}
n(x,t) &= \text{fraction of particles of size } \leq x \text{ at time } t, \\
\mu(t) &= \text{crystal growth rate}, \\
Q &= \text{steady seed crystal feed rate}.
\end{aligned}
$$

Derive equations for the size distribution based on conservation. How does the growth rate depend on n? Calculate the mean and variance. For variations, refer to [1].

23. *Species concentration in a chemical reactor* The second step in the process that converts coal to gasoline is the burning of carbon monoxide to create hydrocarbons in a highly exothermic reaction:

$$
\alpha_1 CO + \alpha_2 H_2 \longrightarrow \alpha_3 HC_1 + \alpha_4(\text{wax}) + \alpha_5(\text{methane}) + \text{HEAT}.
$$

There is the usual mass conservation for the whole gas, as well as conservation equations for each chemical species:

$$
\frac{\partial \rho^i}{\partial t} + \frac{\partial \rho^i v}{\partial x} = \alpha^i \beta^i R,
$$

where β^i is the molecular weight of the ith species. Set $\alpha_3 = 1$. Derive the equations

$$
\left(\frac{\rho^i}{\rho} - \frac{\alpha^i \beta^i}{\beta^3} \frac{\rho^3}{\rho} \right)_t + v \left(\frac{\rho^i}{\rho} - \frac{\alpha^i \beta^i}{\beta^3} \frac{\rho^3}{\rho} \right)_x = 0.
$$

Use the method of characteristics to show that

$$
\frac{C^i}{\rho} - \alpha^i X = \frac{C_0^i}{\rho_0},
$$

where C^i is the concentration of the ith species, and C_0^i and ρ_0 are constants such that $C_0^3 = 0$. Use the equation of state for an ideal gas to show that the set of equations for all species can be replaced by the single equation for X,

$$
(\rho X)_t + (\rho v X)_x = R.
$$

Notice that, because of this, the reaction rate is just a function of X (and T):

$$R = e^{-E/RT}\phi(X).$$

The actual function ϕ is nontrivial (see [1]).

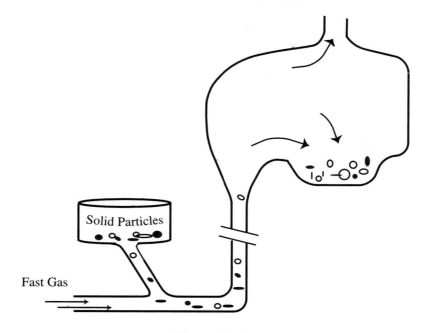

Figure 10.45

24. *Two-phase flow* A pneumatic lifter is a system used to raise solid particles by entrainment into a fast gas flow (Figure 10.45). They find applications in the agricultural, mining, and chemical industries. The question you would like to answer is when such a system would be feasible. Using conservation of mass and momentum, derive the equations to model the pneumatic lifter, assuming a one-dimensional flow. (You will need separate variables for the gas and the solid particles, ρ^G, ρ^S, v^G, v^S. The mass density is written as

$$\rho = \epsilon\rho^G + (1 - \epsilon)\rho^S.$$

For the momentum balance, you can assume inviscid flow for the gas and that ϵ is large enough so that the solid particles don't interact, but there is still an effective viscosity between the solid particles and the gas, giving rise to a drag force:

$$D(v^S - v^G),$$

where the drag coefficient can be written as

$$D = \frac{9\mu}{2} \frac{1 - \epsilon}{r^2} (\frac{2}{3\epsilon})^{2/3},$$

where μ is the viscosity of the gas, and r is the size (radius) of the solid particles (see [1]).

25. Discuss the Riemann problem for a scalar conservation when the flux, q, is a nonconvex function of the conserved quantity [i.e., $q''(c) = 0$ has roots]. Under what conditions will a smooth solution break down?

26. *Ion etching* is considered superior to chemical etching for the fabrication of semiconductor devices. In this technique the semiconductor is masked with a photoresist and then bombarded with an ion beam (see Figure 10.46). The rate at which the surface is eroded by the ion beam is a function of the slope of the surface, $f(p)$, where $p = \frac{\partial y}{\partial x}$, when $y = y(x, t)$ is the equation of the surface. Derive the conservation law

$$\frac{\partial p}{\partial t} + \frac{\partial}{\partial x} f(p) = 0.$$

The function, f, is nonconvex; it is even and has a local minimum at zero. It can have two or more maxima. Thus, this is an example of a conservation law with a nonconvex flux. For more details and other problems see [60].

27. *Aerodynamic shocks* This shock is completely analogous to the river bore. The dissipation energy condition is that the conditions analogous to (10.105, 10.106) should hold. That is, the speed of the flow is greater than the local sound speed to the left of the shock and less than it to the left. Thus the shock separates a supersonic region from a subsonic region. Kinetic energy is converted catastrophically into potential energy. This should correspond to an increase in some variable; what is the aerodynamic analogue of "height"?

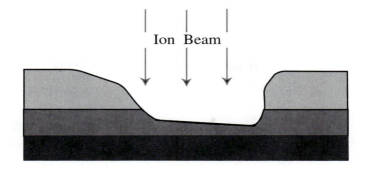

Figure 10.46

28. *Remarks on 3D versions of conservation laws*
    (i) Let ϕ be any quantity defined on $R^3 \times R$, and let $\boldsymbol{\xi}$ be the material coordinates. Then

    $$\phi(\mathbf{x}(\boldsymbol{\xi}, t), t) = \Phi(\boldsymbol{\xi}, t)$$

    and, applying the chain rule,

    $$
    \begin{aligned}
    \frac{\partial}{\partial t}\Phi &= \frac{\partial\phi}{\partial t} + \frac{\partial\phi}{\partial x_1}\frac{\partial x_1}{\partial t} + \frac{\partial\phi}{\partial x_2}\frac{\partial x_2}{\partial t} + \frac{\partial\phi}{\partial x_3}\frac{\partial x_3}{\partial t} \\
    &= \frac{\partial\phi}{\partial t} + \frac{\partial\phi}{\partial x_1}v_1 + \frac{\partial\phi}{\partial x_2}v_2 + \frac{\partial\phi}{\partial x_3}v_3 \\
    &= \phi_t + (\mathbf{v} \cdot \nabla)\phi \\
    &= \frac{D\phi}{Dt}, \text{ the convected derivative.}
    \end{aligned}
    $$

    [This defines the quantity $(\mathbf{v} \cdot \nabla)\phi$.]
    (ii) It can be show that, for 3D, the surface forces are

    $$\int_{\partial\mathcal{B}} \mathbf{T} \cdot d\mathbf{S},$$

    where $\mathbf{T}$ is a symmetric tensor. (For details, refer to [45] and [63].)

29. *Water waves* Show that for one-dimensional flow the material conservation equation can be written

    $$\frac{\partial}{\partial t}A + \frac{\partial}{\partial x}(Au) = 0. \tag{10.114}$$

[This can follow either from 10.78 or 10.95.] Write $A = S_0 + S$, where S_0 is an equilibrium cross-section, and $S = b\eta$, where b is the (constant) channel width. Then (10.114) becomes

$$b\frac{\partial \eta}{\partial t} + \frac{\partial}{\partial x}(S_0 u) + \frac{\partial}{\partial x}(ub\eta) = 0.$$

Notice that you can rid this equation of b by setting $S_0 = bh$. When can you neglect the last term? Suppose that the water is in a pond. Can you derive the following equation for η?

$$\eta_{tt} - gh(\eta_{xx} + \eta_{yy}) = 0.$$

This same equation describes the vibrations of a membrane. Is this reasonable? Use this equation to compute the standing wave pattern in a pond.

30. Can the energy in ocean waves be harnessed to generate electricity? This is a hope of many, although it has never been successful on a large scale. The biggest problem is that the ocean surge is too unpredictable; where there is a good surf, it is sometimes catastrophically strong. Figure 10.47 shows some of the devices that have been tried. The idea behind the Salter duck is to use the ocean as a periodic forcing for an oscillator. What should be the dimensions of the Salter duck? Under which conditions could this be made practical. The Cockerell raft is meant to absorb waves. What should be the dimensions of the rafts? How many should be coupled together? How should the motion of the rafts be converted? How about a piezoelectric conversion with a single flexible raft? Evaluate the other devices. The central problem is to predict the wave-induced forces and then to match the impedance properly. Recall from Chapter 8 that the equation for a buoyant fluid oscillator has the form

$$L\ddot{x} + B(x) = F_{\text{wave}}(t) + U(t).$$

31. *Group velocity* In a nondispersive wave propagation system, where there is one wave speed for all waves, if an impulsive signal is sent from a point at $t = 0$, then at all succeeding times the same signal will be received without distortion. On the other hand, in a dispersive system, where there is a nonlinear relation between frequency and wavenumber, there is a different speed for every

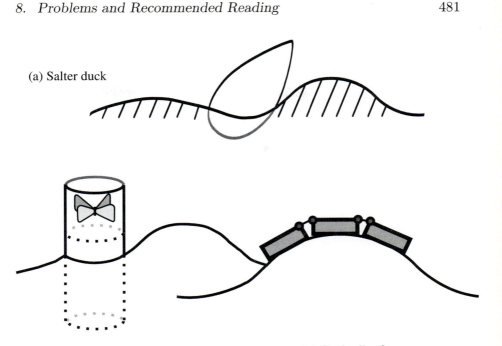

(a) Salter duck

(b) Vortex tube

(c) Cockrell raft

Figure 10.47

wavelength. Because an impulsive signal (or a sharp signal) contains all frequencies, it cannot keep its shape but must spread out. In terms of a Fourier integral, the signal at time t is

$$g(x,t) = \int_{-\infty}^{\infty} e^{i(kx-\omega t)}\, dk$$

The exponent

$$\phi = kx - \omega t$$

is the phase. The points of constant phase in the wave move along the lines

$$x = \frac{\omega}{k} t = v_p t.$$

This defines the phase velocity, v_p. This is not the speed at which information travels.

(i) Show the spread in a wave packet.

(ii) *Beats* Consider a signal made up of two frequencies, ω_1, and ω_2. Show that at time t the signal has the form

$$A\cos(\Delta k x - \Delta \omega t)\cos(\bar{k}x - \bar{\omega}t)$$

of a high-frequency, $\bar{\omega}$ carrier wave being amplitude modulated. If there is no modulation, then the speed of propagation is

$$v = \bar{\omega}/\bar{k},$$

the phase velocity. But what is important is the velocity of the envelope. This is

$$v_g = \frac{\Delta\omega}{\Delta k}.$$

In general, if

$$\omega = \omega(k)$$

is the dispersion relation,

$$v_g = \frac{d\omega}{dk}$$

is the *group velocity*.

(iii) Use the dispersion relation for long gravitational waves to explain the phenomenon mentioned at the beginning of the chapter, that in trains of oceans waves the peaks move through the train, disappearing at the front end.

(iv) Show that the opposite behavior is true for the string–spring system of Problem 16.

Chapter 11

Diffusion

Mr. Science wants to demonstrate diffusion to his class. He opens a bottle of perfume at one end of the room. A few moments later the students on the other end of the room can smell the perfume. Later, in the school cafeteria, Mr. Science is explaining to a colleague: "If you want the coffee to cool fastest, you should wait to add the cream."

1   How It Goes at Small Scales

The earth's magnetic field has reversed orientation many times in the past several million years.[1] The record of these reversals can be read in the geological magnetization. The three main geological contexts of this reading are the following:

- Cooled volcanic lava

- Magma forming the ocean floor

- Iron oxides in marine sediments

One of the outstanding questions is whether the field stays approximately dipolar during an *inversion*, the time period during which it is changing orientation. If one is interested in studying this period using evidence from the magnetization record, then the first two methods above are inconvenient because the reversals take place relatively quickly ($\sim 10^3$ yr), making it possible for volcanoes, which can stay dormant for long periods, to miss all the action. On the other hand, the sedimentation of the ocean floor is a continual process. It should be able to supply a picture of the magnetic field during any particular epoch. The question arises concerning the accuracy of the evidence. For the sedimentation record to be a fingerprint of the magnetic reversal, there

---

[1]See chapter 8, section 4

483

must be a one–one mapping between the timeline and the position of particles in a sedimentary layer. There are two difficulties: First, there may not be a regular enough sedimentation in the deep ocean basins. The magnetic particles typically come from airborne dusts or smokes. They are mixed with many other kinds of particles while they settle to the ocean bottom. A millenium is mapped to a layer of sediment that has a thickness on the order of 10^{-1} cm to 10^2 cm. We will assume that there exist places where the layer is thick enough to make its magnetization reading easy. The second difficulty with the reading is reflected by the small sizes of the particle. These range over about 10^{-3}cm – 10^{-1}cm; considering the extremely low Reynolds number of the fluid environment and the different possible disturbances and range of sedimentation rates, the result is at best a slightly smeared fingerprint. Is the resultant smearing negligible compared to the size of the time intervals that are recorded in the geology that we need to study?

To answer this in a qualitative manner, we need a model.

Let's start by looking at a small iron oxide particle. In order to make measurements statistically valid, we will have to examine many particles at once, but we can start with the simplest situation. The zeroth-order model considers just the gravitational force on the particle:

$$ma = -mg,$$

so the downward speed is

$$v = gt.$$

This is the Galilean result that the velocity increases linearly and is independent of the weight of the object. All objects would descend at the same rate; this is fine, but what really happens in a fluid? From Chapter 3, we know that this is a low Reynolds-number situation, so there will be a drag force proportional to velocity,

$$F_D = \eta v.$$

But what is the velocity? Gravity will accelerate the particle until its force is balanced by the drag force,

$$mg = \eta v.$$

This velocity,

$$v = \frac{mg}{\eta}, \tag{11.1}$$

is known as the *terminal velocity*.

EXERCISE 1 We know from Chapter 3 that

$$F_D \propto \mu v d,$$

where μ is the fluid viscosity, and d is a characteristic size of the particle. The exact formula for a sphere of radius a is

$$F_D = 6\pi \mu v d.$$

Use this to estimate the terminal velocity (11.1).

EXERCISE 2 Show that v increases with particle size. Why do we hope that the iron oxide particles do not have a wide range of sizes?

This last result shows that, in principle, sedimentation can be used to separate particles according to their size. In practice, gravity is too weak to be useful for separating small particles, particularly if their density is of the order of the ambient fluid. But the only thing we used in deriving the constant velocity, (11.1), was the existence of the constant driving force. Other such forces come to mind. A particle that is being spun at constant angular velocity experiences a constant centrifugal force; high-speed spinners known as ultracentrifuges are used to separate biomolecules according to size. The electrical field between two electrodes is approximately constant; this is the basis for separating charged macromolecules such as proteins. This technique, electrophoresis, is widely used in DNA fingerprinting and the human genome project.

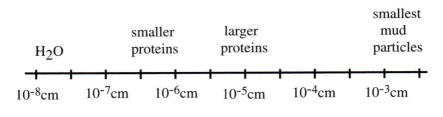

Figure 11.1

The range of particle sizes in all these different examples (Figure 11.1) would seem to indicate that divergent models would be needed to deal with, say, the movement of a foreign atom in a crystal or a silt particle in a fluid. On the other hand, since we are interested in all cases with the behavior of large numbers of these objects, it might be expected that the individual microforces would converge in the form of similar macrostatistics. The particles' microdynamics are determined through the nature of particle–particle, particle–water, water–water forces, and these should give a more or less random disruption to the the steady force responsible for the terminal velocity (11.1). These interactions are commonly ascribed to *thermal* agitation. Entropy is at work here.

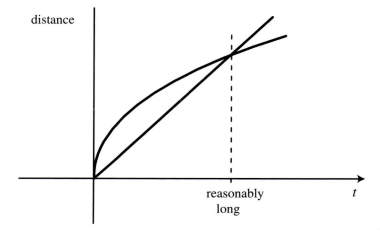

Figure 11.2 It takes a while for drift to overcome diffusion.

The principle of insufficient reason impels us to consider a one-dimensional model where the particle, in the absence of the driving force, undergoes a random walk with equal probabilities of moving either up or down. The driving force can be incorporated by considering an anisotropic random walk with the probability of moving in the direction of the force to be $p > q = 1 - p$. We know that what is interesting in large ensembles are the statistical moments, such as the mean and the variance. From Chapter 6, we know that the variance in the random walk is proportional to the time of passage. Thus, on average, the displacement and its spread (standard deviation) satisfy

$$\bar{x} \sim t, \quad \overline{\Delta x} \sim t^{1/2}. \tag{11.2}$$

This means that, if we can wait reasonably long, the deterministic drift should become more important than the random influence. Thus, as long as the random walk is a good model, we can expect to be able to use sedimentation, centrifugation, and electrophoresis as separating techniques. Of course, to identify what is *reasonably long*, we must be able to compare the constants of proportionality that appear in (11.2) (see Figure 11.2).

EXERCISE 3 Explain why it is precisely the resolution of what constitutes a reasonably long time that allows us to evaluate the possibility of using sediments to study the inversion of the geological field.

As is our practice in dealing with large ensembles, we pass from a microscopic model to a probabilistic model to a continuum model. We expect this continuum model to involve a density. The simplest microsopic model reduces the micro-interactions to the idea of a "collision" that changes the momentum of a particle. The simplest model is that these collisions are independent and infrequent; that is, they have Poisson statistics. Let τ be the *average* time between collisions; then the chance that a particle will collide in the next small time interval is proportional to that interval:

$$P_{\Delta t} = \frac{1}{\tau}\Delta t + o(\Delta t).$$

In other words, the collisions form a Poisson process with density $\lambda = \frac{1}{\tau}$.

EXERCISE 4 What is the probability that a particle will go a time t without a collision?

Meanwhile, what is the probability that the particle will go a distance Δx without suffering a collision? Under the same assumptions, the chance that the particle will collide before going the distance Δx is

$$P_{\Delta x} = \mu\Delta x + o(\Delta x).$$

EXERCISE 5 What is the relation between μ and τ? Argue that $\mu = n\sigma$, where n = particle density, and σ = particle cross section.

The quantity $l = \frac{1}{\mu}$ is called the *mean free path*. (In the language of Chapter 6, $l = \bar{\Lambda}_i$.)

Let us assume that the random walk is carried out on a fixed grid in one dimension with

$$\Delta x = l, \quad \Delta t = \tau.$$

From Chapter 6 we know that, if p is the probability of taking a step to the right, then the probability of taking a total of k right steps in N total steps is

$$P_N(k) = \binom{N}{k}p^k q^{N-k}.$$

For very large N (and $pq \approx 1$; see Problem 15 in chapter 6), we have the approximation

$$P_N(k) \approx \frac{1}{\sqrt{2\pi Npq}} e^{-(k-Np)^2/2Npq}.$$

Let us write this in terms of the coordinates (x, t) of the random walk:

$$t = N\Delta t = N\tau,$$

$$x = k\Delta x - (N - k)\Delta x = (2k - 1)l,$$

so

$$t/\tau = N, \quad x/l = 2k - N,$$

which implies that

$$k = \frac{1}{2}\left(\frac{x}{\Delta x} + \frac{t}{\Delta t}\right).$$

Then we can write

$$(k - pN) = \frac{1}{2l}\left(x - (2p - 1)\frac{l}{\tau}t\right),$$

so if

$$\tilde{p}(x, t)$$

denotes the probability of being at position x at time t, we have

$$\tilde{p}(x, t) \quad = \quad P_{t/\tau}\left(\frac{1}{2}\left(\frac{x}{l} + \frac{t}{\tau}\right)\right) \tag{11.3}$$

$$= \quad \frac{le^{-(x-v_0 t)^2/4\tilde{D}t}}{\sqrt{4\pi\tilde{D}t}}, \tag{11.4}$$

where

$$v_0 \quad = \quad (2p - 1)\frac{l}{\tau}, \tag{11.5}$$

$$\tilde{D} \quad = \quad \frac{l^2}{2\tau}p(1 - p). \tag{11.6}$$

We can go from the single-particle probability to a particle concentration (density) defined here by

$$c(x, t) = \tilde{p}(x, t)/l.$$

The particle concentration evolution thus follows a normal distribution (see problems in chapter 6):

$$c(x, t) \approx N(v_0 t, 2\tilde{D}t).$$

Recall the continuum approximation: we are assuming that quantities or properties that appear in the model do not vary significantly over distances of the order of l, so that c is a well-defined density. Thus,

Prob{particle started at 0 at time 0 is found at time t

between x and $x + dx$}

$$
\begin{aligned}
&= \quad c(x, t)\, dx \\
&= \quad \frac{e^{-(x-v_0 t)^2/4Dt}}{\sqrt{4\pi Dt}}\, dx \\
&= \quad \text{expected fraction of particles} \\
&\qquad \text{between } x \text{ and} x + dx.
\end{aligned}
$$

If N_0 is the original number of particles, then the expected number of particles in any interval (a, b) is

$$
N(t) = \int_a^b c(x, t) N_0\, dx.
$$

What are the parameters v_0 and $\tilde{D}$? The parameter (11.5) has dimensions of speed, and if we express it as

$$
v_0 = (p - q)\frac{l}{\tau} = \left(\begin{array}{c} \text{net probability of} \\ \text{moving to the right} \end{array} \right) \left(\begin{array}{c} \text{particle} \\ \text{velocity} \end{array} \right),
$$

we see that v_0 is a mean, or *drift*, velocity and can be associated with the terminal velocity discussed above. If $p = 1$, then $v_0 = v = l/\tau$ is the particle velocity. Notice that if we choose coordinate

$$
\xi = x - v_0 t
$$

then

$$
c(\xi, t) = \frac{e^{-\xi^2/4\tilde{D}t}}{\sqrt{4\pi \tilde{D}t}}, \tag{11.7}
$$

and we see that $\tilde{D}$ is the constant of proportionality in the time spreading of the variance; that is, $\tilde{D}$ is associated with the rate of the spread of the density.

Attached to each of these variables is a length scale, L_v and L_D. If T is a timescale of interest and associated with the particular technology or physical process, then

$$
L_v = v_0 T
$$

is a length scale associated to drift, and

$$L_D = \sqrt{2\tilde{D}T}$$

is a length scale associated to dispersion, and we can define a dimensionless number

$$\sigma = \frac{2\tilde{D}}{v_0 T}.$$

If

$$\sigma \ll 1,$$

then the spreading over time is not onerous and the separation technique will work. Clearly, we need some way of evaluating $\tilde{D}$. We need to relate $\tilde{D}$ directly to physical properties. Let's start by deriving an equation for the concentration.

2   The Diffusion Equation

We start with the random walk executed by an ensemble of particles, treating it as a continuous-time Markov process. Each time step corresponds to a collision, and at each collision a particle has probability p to go right and probability q to go left. These jumps (collisions) take place at the mean rate $1/\tau$. Let $R(t)$ be the random variable that represents the position of a particle at time t. Then, if Δt is a small time interval,

$$
\begin{aligned}
p_k(t + \Delta t) &= P(R(t + \Delta t) = x_k) \\
&= P(R(t + \Delta t) = x_k | R(t) = x_{k-1})p_{k-1}(t) \\
&\quad + P(R(t + \Delta t) = x_k | R(t) = x_{k+1})p_{k+1}(t) \\
&\quad\quad + P(R(t + \Delta t) = x_k | R(t) = x_k)p_k(t) \\
&= p/\tau \Delta t + o(\Delta t) \\
&\quad + q/\tau \Delta t + o(\Delta t) \\
&\quad\quad + (1 - p/\tau - q/\tau)\Delta t + o(\Delta t).
\end{aligned}
$$

This leads to the differential equations

$$\dot{p}_k = -\frac{1}{\tau}p_k + \frac{p}{\tau}p_{k-1} + \frac{q}{\tau}p_{k+1}. \tag{11.8}$$

Now, according to our continuum approximation assumption, we expand in powers of the mean free path and assume that the coefficients of powers of l greater than 2 are negligible:

$$p_k(t) = \tilde{p}(x_k, t),$$

$$p_{k-1} = \tilde{p}(x_k, t) - \frac{\partial \tilde{p}}{\partial x}(x_k, t)l + \frac{1}{2}\frac{\partial^2 \tilde{p}}{\partial x^2}l^2 + \cdots$$

$$p_{k+1} = \tilde{p}(x_k, t) + \frac{\partial \tilde{p}}{\partial x}(x_k, t)l + \frac{1}{2}\frac{\partial^2 \tilde{p}}{\partial x^2}l^2 + \cdots.$$

Then

$$\frac{\partial \tilde{p}}{\partial t}(x_k, t) = \dot{p}_k(t) = -\frac{1}{\tau}p_k + \frac{p}{\tau}\left(\tilde{p}(x_k, t) - \frac{\partial \tilde{p}}{\partial x}(x_k, t)l + \frac{1}{2}\frac{\partial^2 \tilde{p}}{\partial x^2}l^2 + \cdots\right)$$

$$+ \frac{q}{\tau}\left(\tilde{p}(x_k, t) + \frac{\partial \tilde{p}}{\partial x}(x_k, t)l + \frac{1}{2}\frac{\partial^2 \tilde{p}}{\partial x^2}l^2 + \cdots\right)$$

$$= -(2p-1)\frac{l}{\tau}\frac{\partial \tilde{p}}{\partial t} + \frac{l^2}{2\tau}\frac{\partial^2 \tilde{p}}{\partial x^2} + \cdots.$$

Our continuum model is the partial differential equation

$$\frac{\partial \tilde{p}}{\partial t} + (2p-1)\frac{l}{\tau}\frac{\partial \tilde{p}}{\partial t} = \frac{l^2}{2\tau}\frac{\partial^2 \tilde{p}}{\partial x^2} \tag{11.9}$$

EXERCISE 6 *Particle onservation* Consider a large number of particles undergoing random walks. If $n_k(t)$ is the number at site x_k at time t, and a particle concentration is defined by $n_k(t) = c(x_k, t)l$, show that c satisfies

$$\frac{\partial c}{\partial t} + \frac{\partial \tilde{q}}{\partial x} = 0$$

in the limit of small l. (*Hint:* Set up the usual conservation box and account for fluxes at both ends.)

If we define a particle concentration $\tilde{p} = cl$, then c satisfies the same differential equation (11.9):

$$\frac{\partial c}{\partial t} + v_0\frac{\partial c}{\partial x} = D\frac{\partial^2 c}{\partial x^2}, \tag{11.10}$$

where

$$v_0 = (2p-1)\frac{l}{\tau}, \tag{11.11}$$

$$D = \frac{l^2}{2\tau}. \tag{11.12}$$

EXERCISE 7 Carry out the random walk, with steps Δx and Δt, while $p = 1/2$. Define c appropriately. Take the limit $\Delta x \to 0$, $\Delta t \to 0$ under the condition that

$$\lim_{\substack{\Delta x \to 0 \\ \Delta t \to 0}} \frac{\Delta x^2}{2\Delta t} = D.$$

Show that you get the diffusion equation, that is, the equation (11.10) with $v_0 = 0$.

The velocity (11.11) is simply the drift speed (11.5) of Section 11.1. The *diffusivity*, D, is related to the corresponding quantity in (11.6) by

$$\tilde{D} = pqD.$$

That one depends on the probability and the other doesn't is an artifact of the discretization; in the former case a particle had to make a jump at each time step, but here it has a certain probability of resting. The $\tilde{D}$ was defined via an approximation that made use of the normal approximaion to the binomial distribution, which is only valid for

$$Npq \gg 1,$$

that is, for

$$pq \approx 1.$$

This suggests the extent to which the two paths in Figure 11.3 are commutative. This will be developed later.

The Einstein–Smoluchowski Relation

Suppose that the constant force acting on our particle induces an acceleration, a. Then the average velocity of the particle between collisions is

$$v = \frac{1}{2}a\tau = \frac{F}{2m}\tau.$$

This can also be interpreted as the drift speed so that

$$v = F/\eta.$$

Equating these two expressions, we get an expression for the frictional term: η,

$$\eta = \frac{2m}{\tau} = \frac{2m(l/\tau)^2}{l^2/\tau} = \frac{mv_p}{D}, \tag{11.13}$$

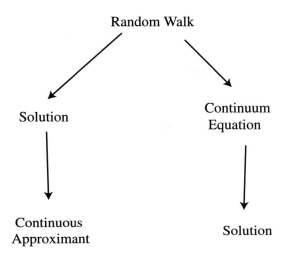

Random Walk

Solution

Continuum
Equation

Continuous
Approximant

Solution

Figure 11.3

where v_p is the particle velocity. According to statistical mechanics,

$$\frac{1}{2}mv_p = \frac{1}{2}kT;$$

thus, (11.13) is usually written

$$D = kT/\eta. \qquad (11.14)$$

There is another way to look at this that helps highlight the relation to equilibrium thermodynamics. Suppose that the concentration of particles is moving under the influence of the diffusive flux,

$$Q_D = -D\nabla c,$$

and convective transport by the drift velocity,

$$Q_d = cv.$$

But the drift velocity is given in terms of a retarding force,

$$v = -\frac{F}{\eta}.$$

If we suppose that this force is given in terms of a potential, then

$$Q_d = -\frac{c}{\eta}\nabla\phi.$$

At equilibrium the total flux is zero, so

$$0 = Q_D + Q_d = -D\nabla c - \frac{c}{\eta}\nabla\phi,$$

which leads to

$$\frac{dc}{d\phi} = -\frac{c}{\eta D},$$

so

$$c \propto e^{-\frac{\phi}{\eta D}}.$$

But at equilibrium, the concentration must be proportional to

$$e^{-\frac{\phi}{kT}}.$$

Once again, we get the relation, (11.14).

Conservation Laws. Heat Conduction

Notice that (11.10) has the form of a conservation law,

$$\frac{\partial c}{\partial t} + \frac{\partial q}{\partial x} = 0,$$

where

$$q = cv_0 - D\frac{\partial c}{\partial x}.$$

The constitutive law,

$$q = -D\frac{\partial c}{\partial x},$$

or, in higher dimensions,

$$\mathbf{q} = -D\nabla c,$$

is known as a *diffusion* law, and the resulting conservation equation is the *diffusion equation*. From the way it was derived we see that the diffusion flux takes the form of a "force" that averages or "smooths" a local distribution of the conserved quantity.

An example of this arises in heat flow. According to the second law of thermodynamics, a system will tend to the arrangement with the highest entropy; this is equivalent to the statement that heat must flow from a body with higher temperature to one with lower temperature. Thus the heat flux depends on temperature gradients,

$$\mathbf{q}_{\text{heat}} = -\kappa\nabla\theta, \tag{11.15}$$

where θ is the continuum temperature field, and κ is a material parameter known as the *conductivity*.[2] From the definition, we can see that the conductivity is the amount of heat flowing per unit time from one side of a unit cube to the other under a temperature difference of one degreee. The values of the heat conductivity for some substances are given in Appendix C. The heat conductivity depends on the temperature.

If the energy density is denoted by e, then the total energy in some region is found by integrating this density. The conservation of energy takes the form

$$\frac{dE}{dt} = \frac{d}{dt} \int_B e(\mathbf{x}, t) \, d\mathbf{x} = - \int_{\nabla B} \mathbf{q} \cdot d\mathbf{S},$$

where the flux is given by (11.15) when the flow of heat is due to heat conduction only. The use of the divergence theorem, assuming enough regularity of the composants, gives the conservation differential equation,

$$\frac{\partial e}{\partial t} = \text{div} \left(\kappa \nabla \theta \right). \tag{11.16}$$

We clearly need a way to relate the heat density to the temperature. The relation is found by measuring the temperature increase due to a step input of heat energy,

$$\Delta E = C \Delta \theta,$$

where C is the *heat capacity* of the material, which is a function of the temperature. In terms of the density we can write this as

$$e = c \rho \theta,$$

where ρ is the mass density, and c, the *specific heat*, is the heat capacity per unit mass. The equation (11.16) then becomes the heat conduction equation for temperature (the *heat equation*):

$$c \rho \frac{\partial \theta}{\partial t} = \text{div} \left(\kappa \nabla \theta \right). \tag{11.17}$$

Since c and κ depend on θ, this is a nonlinear partial differential equation. But in many instances these functions will be so slowly varying

---

[2]More properly, the *thermal conductivity*. Although there is a definite relation between them, this quantity should not be confused with the electrical conductivity.

that they can be taken to be constant. Then the heat equation takes the form of a diffusion equation,

$$\frac{\partial \theta}{\partial t} = K \nabla^2 \theta, \tag{11.18}$$

where the *heat diffusivity* is

$$K = \frac{\kappa}{c\rho}.$$

Newton's Law of Cooling

There can be a discontinuity in temperature at the interface between two materials or phases. If the boundary has conductivity, then there will be a flow of heat from the warmer side to the colder. Often this flow is seen to be proportional to the difference in temperature:

$$q = H(\theta_1 - \theta_2).$$

EXERCISE 8   What are the dimensions of H? Does H depend on any geometrical parameters?

*Reactions

Chemical reactions are common sources or sinks of heat. If the reaction occurs in the interior (homogeneous reaction), then the heat generated is given by a reaction rate R (see Chapter 10),

heat generated per volume per unit time $= QR$.

If the reaction occurs along some surface then the heat generated is given by a heterogeneous reaction rate, r,

heat generated per unit surface area per unit time $= Qr$,

where the reaction rate, r, has dimensions of (mole/area)/time.

Notice that reactions can change the density as well as redistribute the relative concentrations of composants.

Solving the Diffusion Equation

From the derivation and the form of the coefficients, we recognize a relation between the diffusion equation and the normal distribution. In other words, we expect that a solution to (11.10) should relate to the form of (11.7). A natural problem to solve for the differential equation is the initial value problem. What initial distribution could lead to a solution of the form of (11.7). This is a normal distribution with variance

$$\sigma^2 = 2\tilde{D}t,$$

so that, if followed back to time $t = 0$, we should recover a distribution with zero variance, that is, a point atom at $x = 0$. This is a delta function. We need a way of approximating the delta function by smooth functions.

Let K be a smooth function that satisfies

$$\int_{-\infty}^{\infty} K(x)\, dx = 1.$$

(K can be a PDF, see Chapter 6) If we define

$$K_\epsilon(x) = \frac{1}{\epsilon} K\left(\frac{x}{\epsilon}\right),$$

it has the following properties:

$$(1) \quad \int_{-\infty}^{\infty} K_\epsilon(x)\, dx = 1, \qquad\qquad (11.19)$$

$$(2) \quad \int_{|x|>\delta} |K_\epsilon| \to 0, \quad \epsilon \to 0, \qquad (11.20)$$

$$(3) \quad f * K_\epsilon \to f, \quad \epsilon \to 0. \qquad (11.21)$$

EXERCISE 9 Show these properties. (*Hint:* A simple change of variables.)

Such a function is called an *approximation to the identity*. This is because it approximates the delta function, and the delta function is the identity with respect to convolution.[3]

EXERCISE 10 Show that the following functions give rise to such approximations.

---

[3]Functions used in integrals in this way are referred to as *kernels*.

(i)
$$K(x) = \frac{1}{\pi(1 + x^2)}$$

(ii)
$$K(x) = \frac{1}{\pi}\left(\frac{\sin x}{x}\right)^2$$

(iii)
$$K(x) = \frac{1}{\sqrt{\pi}}e^{-x^2}$$

The last function in this exercise is called the Gaussian kernel. Notice that if we choose
$$\epsilon = \sqrt{4Dt}$$
in that kernel we arrive at the normal probability distribution (11.7). This illustrates the scaling of the characteristic length. The resulting distribution,
$$H(x,t) = \frac{e^{-x^2/4Dt}}{\sqrt{4\pi Dt}},$$
solves the differential equation
$$\frac{\partial u}{\partial t} = D\frac{\partial^2 u}{\partial x^2}, \quad t > 0.$$

EXERCISE 11 Show this explicitly

The solution to the initial-value problem for the diffusion equation,
$$u_t = Du_{xx} \qquad\qquad -\infty < x < \infty, \quad t > 0, \qquad\qquad (11.22)$$
$$u(x,0) \;=\; f(x), \qquad\qquad\qquad\qquad\qquad\qquad (11.23)$$
is
$$H * f.$$
To see this, compute
$$
\begin{aligned}
u_t = (H * f)_t \;&=\; \int_{-\infty}^{\infty} \frac{\partial H}{\partial t}(x - y, t)f(y)\,dy \\
&=\; \int_{-\infty}^{\infty} D\frac{\partial^2 H}{\partial x^2}(x - y, t)f(y)\,dy \\
&=\; D(H * f)_{xx}.
\end{aligned}
$$
And, because H is an approximation to the identity,
$$\lim_{t \to 0} H * f = f.$$

Signal Distortion

What is behind the distortion of signals in transmission lines? The voltage satisfies the telegraph equation,

$$lcv_{tt} - v_{xx} + (lg + rc)v_t + rgv = 0. \tag{11.24}$$

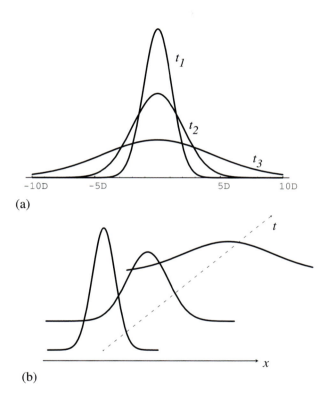

(a)

(b)

Figure 11.4

We will consider a line with negligible inductance. Then, if $g = 0$, [4] the equation (11.24) reduces to the diffusion equation,

$$v_t = \frac{1}{rc}v_{xx}, \tag{11.25}$$

with diffusivity

$$D = \frac{1}{rc}.$$

---

[4]in Problem 9 you will show that we can always reduce to this case

If an impulsive signal is sent at $x = 0$ at time $t = 0$, then, using our solution from the previous section, the signal at any subsequent time is

$$v(x, t) = \sqrt{\frac{rc}{4\pi t}} e^{-rcx^2/4t}. \tag{11.26}$$

The signal spreads, or diffuses (Figure 11.4a). If any inductance is present then there is a definable wave speed but the signal diffuses (Figure 11.4b).

3   Steady-state Diffusion.Boundary Conditions

Imagine particles diffusing along a one-dimensional channel. Suppose that particles are supplied in a steady stream at one end (labeled a) and just as steadily removed at the other end (labeled b). If this has

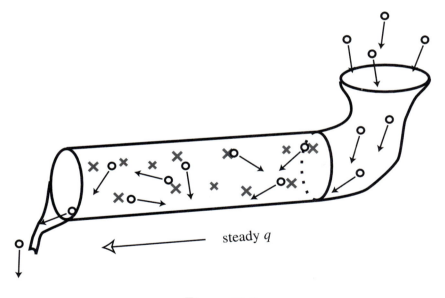

steady q

Figure 11.5

been going on for a long time, then there is no reason to think that the solution is changing in time (Figure 11.5). When we set $c_t = 0$ in the concentration equation, we get the equation for the steady state:

$$c_{xx} = 0.$$

The general solution to this equation is

$$c(x) = k_1 x + k_2.$$

Note that the flux is constant everywhere:

$$q = -D\frac{\partial G}{\partial x} = -Dk_1.$$

EXERCISE 12 What is a probabilistic interpretation of this? What is a micro-mechanical interpretation of this?

If particles are removed from the b end as soon as they reach it, then the condition at that end is

$$c(b) = 0.$$

If the particles are supplied at a constant rate Q at the a end then we have the system of equations for the constants, k_1, k_2,

$$
\begin{aligned}
k_1 &= -Q/D, \\
k_2 &= -k_1 b = bQ/D,
\end{aligned}
$$

and the solution is

$$c(x) = \frac{Q}{D}(b - x). \tag{11.27}$$

Note that we can determine the equilibrium concentration at the a end:

$$c_0 = c(a) = \frac{Q}{D}(b - a).$$

Thus, we can write

$$D = \frac{QL}{c_0},$$

which expresses the diffusivity in terms of a characteristic flux, length, and concentration.

On the other hand, if we keep the concentration at a constant,[5]

$$c(a) = c_0,$$

then the solution is found from

$$
\begin{aligned}
c_0 &= c(a) = k_1 a + k_2, \\
0 &= c(b) = k_1 b + k_2
\end{aligned}
$$

---

[5] This is a typical boundary condition in solidification problems.

and is just the obvious one:

$$c(x) = c_0 \frac{b - x}{b - a}.$$
(11.28)

The solution (11.28) doesn't depend on the diffusivity, but in order to keep the concentration a constant, we have to supply particles at a rate that depends on the diffusivity:

$$Q = \frac{c_0}{b - a}.$$

What if particles are supplied at an interior point? Then the solution depends on how they are supplied: if in such a way as to keep the concentration constant at that point, then there cannot be a smooth solution on the whole interval and the solution is found piecewise on each subinterval (Figure 11.6). It is seen at a glance that the flux at a boundary is inversely proportional to the distance from the source of particles. This can be useful in uncovering certain things about diffusion. We know from the law of large numbers that a particle undergoing a random walk stays near the origin, but only relatively speaking: the uncertainty in the actual position grows as $\sqrt{t}$.

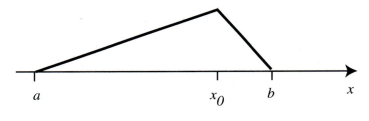

Figure 11.6

What is the probability that the particle will return to the origin at some time? Suppose that the particle reaches some point x_0. What is the probability of returning to the origin from there? We will attempt to find this probability as a function of x_0. This probability is found by considering all the possible outcomes where the particle starts from x_0 and hits the origin and dividing by the number of all possible outcomes. Let us take the limit of a bounded problem. Suppose that particles are supplied at x_0 and removed at 0 and at b. The solution looks like Figure 11.6, with $a = 0$. What is the probability that a particle will be removed at b before being removed at 0, or vice versa. That is, we want the conditional probability of removal at the origin, given that it is removed at one of the end points. We just use the ratio of fluxes:

$$P = \frac{Q_0}{Q_0 + Q_b} = \frac{\frac{1}{x_0}}{\frac{1}{x_0} + \frac{1}{b-x_0}} = 1 - \frac{x_0}{b}.$$

Notice that

$$P \to 1$$

as

$$b \to \infty.$$

Thus, in the one-dimensional random walk a particle returns to the origin with certainty (if one waits long enough). This is perhaps not too surprising: if we know that the particle paths must reflect the maximum entropy, then there must be just as many going toward the negative side as the positive side, and in one dimension they must cross the origin. This intuitive argument doesn't work in two dimensions where there is no such constraint on the topology. Maybe in two dimensions the probability of return is less. And whatever is the case in 2D should also hold in 3D. *Not at all.*

The analogous problem in 3D is to solve the steady-state diffusion equation,

$$\nabla^2 c = 0, \tag{11.29}$$

in the region between two spheres,

$$a < r < b.$$

The equation (11.29) is known as the Laplace equation.

EXERCISE 13 What is the analogous problem for 2D?

If the concentration is kept constant at $r = r_0$, then we must solve the radially symmetric Laplace equation,

$$\frac{1}{r^2} \frac{\partial}{\partial r} (r^2 \frac{\partial c}{\partial r}) = 0. \tag{11.30}$$

EXERCISE 14 What is the radially symmetric Laplace equation in 2D?

The general solution to (11.30) is

$$c(r) = -\frac{k_1}{r} + k_2.$$

EXERCISE 15 What is the general solution for the analogous equation in 2D?

Using the boundary conditions for the interval $a < r < r_0$,

$$c(a) = 0,$$
$$c(r_0) = c_0,$$

we arrive at the solution,

$$c(r) = \frac{c_0}{1 - a/r_0}(1 - a/r).$$

Likewise, on the interval, $r_0 < r < b$, we have the solution (Figure 11.6)

$$c(r) = \frac{c_0}{1 - b/r_0}(1 - b/r).$$

The total flux out at the inner shell is

$$Q_a = \int\int_{r=a} D\frac{\partial c}{\partial r}\, dr = \frac{Dc_0}{1 - a/r_0}a^{-1}(4\pi a^2) = Dc_0\frac{4\pi a}{1 - a/r_0}.$$

Likewise, the total flux out at the outer shell is

$$Q_b = Dc_0\frac{4\pi b}{b/r_0 - 1}.$$

The appropriate probability is

$$P_3 = \frac{Q_a}{Q_a + Q_b} = \frac{a(b/r_0 - 1)}{a(b/r_0 - 1) + b(1 - a/r_0)} = \frac{a}{r_0}\frac{b - r_0}{b - a}.$$

If $b \to \infty$, then

$$P_3 \to a/r_0;$$

in other words, the probability of hitting an inner sphere is proportional to the radius of that sphere. Thus, as $a \to 0$, the probability of returning to the origin is, in fact, 0.

EXERCISE 16 Show that in 2D we get

$$P_2 = \frac{\ln b - \ln r_0}{\ln b - \ln a}.$$

Notice that, as $b \to \infty$,

$$P_2 \to 1$$

independently of a. Thus, in 2D, as in 1D, the particle will always return to the origin.

The practical results of all this is that diffusion is really only efficient as a transport mechanism in dimensions 1 and 2. This will be explored in the problems.

4 Melting and Freezing. Moving Boundary Problems

We will consider the growth of a sheet of ice on a sheet of metal (Figure 11.7). This is really a four-phase problem, but we can reduce it to one phase. Clearly, the most important boundary is the ice–water interface, because that is where the freezing takes place; we can assume that the liquid layer exists mainly through surface tension and, since the density of water is constant, the thickness of this layer is constant. Thus, we can ignore the air phase altogether. The temperature field is varying in extent over the three other phases, but we will show that, for a first model, we can reduce it to a consideration of the the temperature only in the ice phase.

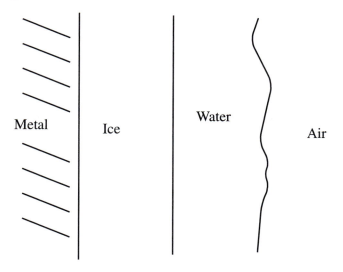

Metal   Ice   Water   Air

Figure 11.7

What are the conditions that hold between the phases? We will make a working assumption that

$$\kappa_{\text{Fe}} \gg \kappa_{\text{ice}} \gg \kappa_{\text{water}}. \tag{11.31}$$

If we compare the conductivities of ice and iron from Appendix C, we see that there is a difference of two orders of magnitude. This means that we can assume that heat is conducted in the iron fast enough to keep it at a constant temperature, θ_F. That takes care of the metal phase.

Freezing takes place at $0°C$. Thus, the temperature is zero on both sides of the ice–water interface.

It would seem that we have enough data for a well-posed boundary-value problem for the temperature field in the ice phase, except for the fact that the position of the ice–water interface is not known and must be calculated as part of the model. What happens at the phase front, the air–water interface, to make it move? If we look at the heat conservation, we see that across this interface there is a jump in the material parameters, the density, the specific heat, and the conductivity. As a result, this interface is the analogue of the shock that we studied in Chapter 10. If we carry out the exact same analysis as is done there for a moving discontinuity,

$$x = \xi(t),$$

we arrive at the same result, that the velocity of the moving front is the ratio of the jump in flux to the jump in energy density,

$$[e]\dot{\xi} = [q]. \tag{11.32}$$

The jump in flux is due to the difference in conductivities between water and ice:

$$[q] = q_r - q_l = -\kappa_{\text{water}}\frac{\partial \theta}{\partial x} + \kappa_{\text{ice}}\frac{\partial \theta}{\partial x} \tag{11.33}$$

The jump in the heat energy across the interface is fundamentally due to the difference in entropies between the liquid and solid phases and so doesn't depend on the temperature or heat flux. This is usually called the *latent heat* and denoted by L. Thus, the condition at the ice–water interface is

$$L\dot{\xi} = -\kappa_{\text{water}}\frac{\partial \theta}{\partial x} + \kappa_{\text{ice}}\frac{\partial \theta}{\partial x} \tag{11.34}$$

This boundary condition couples in the determination of the temperature field in the water phase. At this point, we can call on (11.31) to claim that the first term on the right side of (11.34) is negligible. In fact, on the face of it, it is not quite obviously true: the conductivity of ice is only about six times greater than that of water. But this assumption does lead to an analyzable model. The boundary conditions and the interface condition, along with the heat conduction equation,

$$c\rho\theta_t = \kappa_{\text{ice}}\theta_{xx},$$

represent the model. One would expect to need an initial temperature distribution, but not if we assume a steady-state solution:

$$\theta = \alpha x + \beta.$$

If α and β are constants, then the boundary conditions cannot allow a time-varying ξ, but they can if they are allowed to vary. So we will allow them to vary a little; this is referred to as the *quasi-steady-state* assumption. The left boundary condition fixes β:

$$\beta = \theta_0.$$

The right boundary condition,

$$0 = \theta(\xi, t) = \alpha(t)\xi + \theta_0,$$

expresses α in terms of the position of the boundary:

$$\alpha(t) = -\theta_0/\xi(t).$$

But the interface condition (11.34) gives a differential equation for the boundary:

$$\dot{\xi} = \frac{\kappa_{ice}}{L}\frac{\partial\theta}{\partial x} = \frac{\kappa_{ice}}{L}\alpha(t) = -\frac{\kappa_{ice}}{L}\frac{\theta_0}{\xi}.$$

This can be integrated to give

$$\xi^2 = -\frac{\kappa_{ice}}{L}\theta_0 t + \xi(0)^2.$$

The case $\theta_0 < 0$ corresponds to freezing, and the boundary moves to the right; the case $\theta_0 > 0$ corresponds to melting, and the boundary moves to the left.

5 Problems and Recommended Reading

1. *That open bottle of perfume.* Given that the diffusion coefficient in air is on the order of 0.1 cm s^{-1}, calculate the time for diffusion of perfume molecules in a reasonable-sized room. Does Mr. Science's demonstration illustrate diffusion? What else could be at work? (See [56] or [77].)

2. *Cream in the coffee.* What law of cooling is Mr. Science calling upon? Is it necessarily valid? Could it depend on how the cream is added, and what kind of mixing takes place?

3. A ultracentrifuge is a device (Figure 11.8) that can supply a constant force of 1000 to 3000 g. If a molecule with molecule mass M is placed in the centrifuge show that the ratio of radial (drift) velocity to centrifugal acceleration is

$$\frac{dv}{dr}/\omega^2 r = \frac{M}{N_A \eta}(1 - \frac{\rho_0}{\rho}) = S,$$

where N_A is Avogadro's number, η is the frictional coefficient, ρ is the density of the molecule, and ρ_0 is the density of the diffusional medium. The number S is called the Svedborg number. For hæmoglobin,

$$S = 4.4 \text{ s}, \quad M = 67,000.$$

Estimate D in various media.

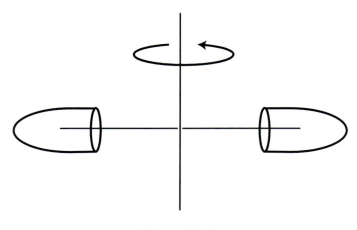

Figure 11.8

4. Evaluate the validity of the sedimentation record for determining the form of the magnetic field during reversals. What information do you need? Do you need to know the strength of the magnetic field?

5. *Nernst Potential* The Nernst law says that the electrical potential across a cell membrane is proportional to the logarithm of the relative concentration (Chapter 1) of the relevant ion species. This law fails to be valid for very small external concentrations of potassium ions (Figure 11.9). Account for the form of this

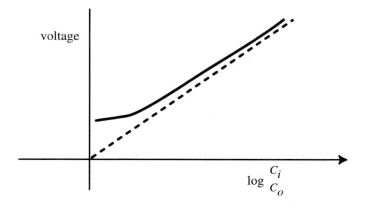

Figure 11.9 Then Nernst potential is not quite log-linear.

graph and the Nernst approximation by a force balance across the membrane. Can you arrive at the exact numerical constant? (*Hint:* You will need to study the derivation of the Einstein-Smoluchowski relation. Usually, for membranes, permeabilities are reported instead of diffusivities. They are related by

$$\text{permeability } = p = \frac{D\beta}{l},$$

where β and l are defined as in Figure 11.10.)

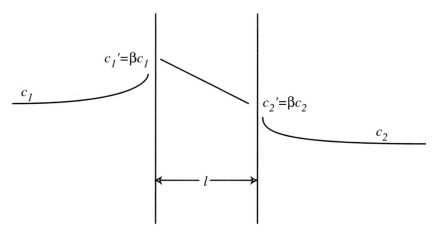

Figure 11.10 Local concentrations across the nerve cell membrane.

6. Show that the Laplace equation reduces to (11.30) when you assume a spherically symmetric solution.

7. *Cooling of a rod* Suppose that a rod of length L has a temperature distribution $f(x)$, $\quad 0 < x < L$, along it and that the ends are placed at $0°$. How will the rod cool? The heat flux in the material is characterized by the thermal conductivity:

$$q = -\kappa\theta_x.$$

Find a change of variables, $\xi = x/\bar{x}$, $\tau = t/\bar{t}$, so that the problem to solve is the boundary-value problem:

$$v_\tau = v_{\xi\xi} \qquad 0 < \xi < 1, \quad \tau > 0$$
$$v(\xi,0) = g(\xi), \qquad v(0,t) = 0 = v(1,t).$$

Show that a characteristic time scale is $\bar{t} = L^2/\kappa$. Solve the boundary-value problem by superposition of infinite modes (Fourier series). That is, represent the solution as a series

$$v(\xi,\tau) = \sum_{m=1}^{\infty} \phi_m(t) \sin m\pi\xi.$$

Show that each mode undergoes simple relaxation. Show that the fundamental mode ($m = 1$) decays much more slowly than the harmonics ($m > 1$). What is the relaxation time for this fundamental mode?

8. In the text we derived the fundamental solution of the diffusion equation by appeal to the probability context. It can be derived solely on dimensional grounds. Show this by taking $x/\sqrt{2Dt}$ as the a dimensionless quantity, finding a dimensionless quantity involving the dependent variable (temperature in the heat equation), and then using the Π-theorem to find the functional form of the independent solution. (For other techniques of solving partial differential equations, see [25].)

9. What is the optimal size and placement of heating elements of a car rear-window defroster?

10. Show that signaling along a transmission line is always by diffusion when $l = 0$. (The existence of a nonzero leakage g only changes the relaxation rate.) (*Hint:* Look for a solution of the form $V = f(t)W$. What equation must f satisfy?)

11. *Electrotonic spread* Note that in the solution to the diffusion equation (11.26) the distribution reaches a single maximum in time. Show how the time to maximum at a particular point can be used to measure the electrical properties of tissue.

12. Suppose that in a electrically conductive medium Ohm's law is satisfied,
$$\mathbf{J} = \sigma\mathbf{E}.$$

    Use Maxwell's equations to show that the magnetic induction, $\mathbf{B}$, satisfies a diffusion equation with diffusivity
$$D = \frac{1}{\sigma\mu}.$$

    Interpret this diffusive decay of magnetic fields.

13. *Flow of a viscous fluid* Suppose that an incompressible viscous fluid fills the region of three-dimensional space, $z > 0$, bounded by $z = 0$. Before $t = 0$ everything is at rest, and at $t = 0$ the plane $z = 0$ moves forward with the constant velocity $U\mathbf{i}$. The resulting motion is a *diffusion of momentum*. To see this, note that because of the symmetry of the problem the only important quantity is the x-momentum, and the only direction of variation is the z-direction. Conservation of x-momentum in a vertical layer, $a < z < b$, is
$$\frac{d}{dt} \int_{z=a}^{z=b} \rho u \, dz = \phi(a, t) - \phi(b, t),$$

    where ϕ is the momentum flux. What is the momentum flux? Since there is no motion of material in the z-direction, there is no convective transport. Due to viscosity, stresses in the fluid are acting as restoring forces to spread out the x-momentum: think of thin sheets of fluid moving parallel (see Figure 3.11) The restoring force will depend only on the difference in velocity between the fluid layers and will tend to average out their velocities; thus, we are led to the constititive law
$$\phi = -\mu\frac{du}{dz},$$

    where μ, the characteristic constant of the restoring force, is the viscosity. Note that ϕ does not depend on the density, since it is a surface force, not a body force. In the equation of motion, the viscosity will appear in the following combination,

$$\nu = \frac{\mu}{\rho},$$

termed the dynamic viscosity. Write down the resulting boundary-value problem. Look for a solution of the form

$$u(z,t) = f(\frac{z}{\sqrt{\nu t}}).$$

Solve this with the help of tables or a computer. Estimate the thickness of the boundary layer.

14. *Models of the atmosphere* The *geostrophic wind* is a horizontal velocity field $\mathbf{v} = (u, v)$ that satisfies

$$fv = \frac{1}{\rho}\frac{\partial p}{\partial x}, \quad fu = -\frac{1}{\rho}\frac{\partial p}{\partial y},$$

where $f = 2\Omega \sin \phi$, ϕ being the terrestrial latitude and Ω the planetary angular velocity. When is the geostrophic wind a good approximation to the true wind? If $\mathbf{u} = (u, v, w)$ is the velocity vector, then the equation of motion for the horizontal component is

$$\frac{\partial \mathbf{v}}{\partial t} + \mathbf{v}\cdot\nabla\mathbf{v} = f\mathbf{k} \times (\mathbf{v}_g - \mathbf{v}) + \nu\nabla^2\mathbf{v} + \nu\nabla_2(\text{div }\mathbf{v}),$$

where $\mathbf{v}_g$ is the geostrophic wind and $\nabla_2 = (\frac{\partial}{\partial x}, \frac{\partial}{\partial y})$. Using characteristic scales for the horizontal and vertical velocities and lengths, and a characteristic time scale, carry and out a nondimensionalization of the equation of motion. Identify which terms in the resulting must be small in order that the geostrophic approximation is valid? Can you assign physical significance to the appropriate dimensionless coefficients? (See [16].)

15. *Solitary traveling wave* Burger's equation has been intended for a simple model of a viscous fluid:

$$u_t + uu_x = \nu u_{xx}, \quad -\infty < x < \infty, \quad t > 0.$$

When $\nu = 0$, this is the scalar nonlinear conservation law discussed in Chapter 10. Any initial condition that is a decreasing function of x will lead to a shock solution. However, we know that the diffusion term on the right side has the effect of smoothing out any extreme gradients such as the shock. It may be that the two tendencies present in the equation will combine to allow a smooth traveling wave. Show that such a *soliton* is a possible solution. [*Hint:* Look for a solution of the form,

$$u(x, t) = f(x - ct), \quad \lim_{x \to \pm\infty} f'(x) = 0.]$$

16. *Populations* We have seen that a useful model for population growth is the logistic equation

$$\frac{dp}{dt} = rp(1 - p/k).$$

Suppose that you take into account the spatial distribution of the population, which reacts to population gradients by diffusion. Show that there can be traveling waves in the population, that is, look for a solution of the form,

$$p(x, t) = f(x - ct).$$

17. Can you find a soliton solution to the Nagumo telegraph equation (Chapter 10)? (See [61]).

18. *Diffusion in crystals* Some atoms can fit in the spaces between the lattice atoms in a crystal. These are called interstitial atoms. In response to gradients in their concentration, they will diffuse. Model this as a random walk on a cubic lattice, and show that the diffusivity, D, can be written as

$$D = -\frac{1}{6}\Gamma\alpha^2,$$

where Γ is the rate of jumping, and α is the lattice spacing. It is noticed that D is an increasing function of concentration for large concentrations; what could be causing that? (For further details, see [36].)

19. *Percolation* Some chemical reactions go faster in the presence of a catalyst. The problem is always to get the reacting species to the site of a catalyst. A *zeolite* is a porous particle that has a high effective surface area. It can act as a catalyst for certain chemical reactions useful in the energy-producing and chemical industries.

Physically, a zeolite is a small particle sporting a labyrinthine surface where reacting particles can wander through, finding the many catalytical sites. A simple model of a zeolite is the purposeful wandering of a particle on a "random grid." On the grid, each site is marked "open" with probability p (and "closed" with probability $1 - p$). Simulate this by first laying down such a grid and then having a particle move in a constant direction until it hits a closed cell, and then takes the smallest possible change in direction. Show that when p is less than some value there is virtually no movement of a particle. How does this compare to the random walk (which is random evolution on a fixed grid)? In particular, does the variance grow proportionally to the time of passage? The times of "covering" are very important in industry.

20. *Mean time to capture*   The differentiation of cells to prepare them for their everyday work is aided by the action of repressor molecules who turn off a particular gene on the strand of DNA. Such a molecule has a strong affinity for the site of that gene. How long, on average, does such a molecule take to reach its binding site? Model this as a random walk along the DNA molecule, with grid parameters, Δx and Δt. Let $W(x)$ be the mean time to be trapped. Justify

$$W(x) = \Delta t + \frac{1}{2}[W(x + \Delta x) + W(x - \Delta x)].$$

Derive a continuum model (differential equation with boundary conditions).

21. *Slime molds*   Amœbae are unicellular organisms. In the presence of food, slime mold amœbae feed and multiply by mitosis. If food is not present, they go through a period where they migrate. They tend to move toward high concentrations of a chemical (attractant). In this way, due to fluctuations in the attractant, there can be localized aggregations of the mold, which can give rise to striking wavy patterns. Model this by starting with the conservation law for the mold. There will be two flux terms: one is self-diffusion of the amœbae, the other is an antidiffusion in response to the gradient of the attractant. The coefficient of this antidiffusion is proportional to the amœba concentration. This must be coupled to the conservation equation for the attractant which is undergoing self-diffusion. There will be a source term

that is proportional to the amoeba concentration and also a sink term that is proportional to attractant concentration, which represents the enzymatic decay of the attractant. Show that there exists a uniform (constant) steady state. Show that this steady state can be unstable. Can you identify a characteristic stability parameter? Can you identify a characteristic wavelength? (For details, consult [45].)

22. Diffusion can sometimes, via instabilities, lead to spatial patterns in chemical reactions. Write down the ordinary differential equations for the species X and Y that undergo the following reactions:

$$A \xrightarrow{k_1} X,$$
$$B + X \xrightarrow{k_3} Y + D,$$
$$2X + Y \xrightarrow{k_2} 3X,$$
$$X \xrightarrow{k_4} E.$$

Now add diffusion terms for the two species. Show that the resulting system of partial differential equations has a uniform solution. Investigate the stability of this solution.

23. Diffusion in three dimensions is not a very efficient mode of transport, so if diffusion must occur for the working of a mechanism some way must be provided to bring the diffusant to a surface or in contact with some tubules. Nature has found many clever ways of doing this. Oxygen must diffuse across membranes separating air and blood, but the bronchial branching is so efficient that these surfaces are so small and thin that diffusion is almost transparent. Muscles are much more difficult to supply with nutrients and thus are highly limited in their rate of working. Certain muscle tissues are, however, phenomenally active. Such is the case with insect wings. These muscles are sometimes bigger than 1 cm. Is there a limit on the activity of large flying insects? Look at some of the anatomical drawings in [81], where such diffusion is discussed in detail. Other examples of diffusion can be found in [27] and [77].

24. *Nonlinear heat equation* While deriving the one-dimensional heat equation

$$c\rho\theta_t = (\kappa\theta_x)_x,$$

it was mentioned that the thermal conductivity, κ, generally depends on the temperature. Suppose that

$$\kappa = \frac{a}{c\rho}\theta^n.$$

Look for a solution of the form

$$\theta = At^{-k}f\left(\frac{x}{Bt^m}\right).$$

What is the relation between k and m?

25. *Delay response* In the constitutive law for heat conduction, there may be a delay in the response of the heat current to the temperature gradient:

$$q = -\int_{-\infty}^{t} K(t-s)\frac{\partial}{\partial x}\theta(x,s)\,ds.$$

Here, the kernel K satisfies the causality condition

$$K(t) = 0, \quad t < 0.$$

Note that if $K(t-s) = \kappa\delta(t-s)$ then we recover the usual law of conduction. A one-side approximation to the delta is provided by

$$e^{-t}.$$

Thus, a typical kernel would be

$$K(t-s) = \frac{\kappa}{\tau}e^{(t-s)/\tau},$$

where τ is a relaxation time. Derive the resulting equation for heat conduction. Notice that it is a wave equation. Interpret the solutions. Show that it reduces to the usual heat equation in the limit of instantaneous relaxation,

$$\tau \to 0.$$

Does this mean that the solutions of the wave equation (you've derived) are diffusion-like when $\tau \approx 0$?

26. *Oxidation* Metals will react chemically with oxygen and thereby form a thin layer of an oxide. This layer can act as a protectant and insulator. Thicker layers than the "native" oxide layers are

sometimes needed for industrial purposes. How can one get a thicker layer? Construct a model similar to that of the freezing problem in the text. Note that the oxide layer forms by the migration of oxygen atoms to the metal–oxide interface, where the oxidation reaction takes place. This is a heterogeneous reaction occurring only at the surface. The reaction rate is proportional to the local concentration of oxygen. This reaction rate will appear as a source term in the interface condition. It can be eliminated as a variable by noticing that the reaction rate must be the rate of change of the total amount of the oxide. The resulting solution is asymptotically similar to the freezing interface (Section 4) but there is a short time period during which the interface moves at a constant rate. (A more complete account, with references, of the oxidation of silicon which is important to the semiconductor industry, can be found in [21].)

27. *Tarnishing* Sometimes the oxidation process can be controlled by the use of an electric field. In this case there is migration of electrons, as well as ions and transport due to the field, as well as diffusion. Write down the constitutive laws for the electrons and ions. Show that they can be combined in a single nonlinear flux law. What are the parameters influencing the growth of the layer of oxide? A similar process is the *carburization of steel*. Steel is iron that has been processed to remove much of the carbon; however, the carbon makes it more wear-resistant. Thus, it is good for steel to have a carbon-rich outer layer. To form this layer, steel is placed in gas containing CO and CH_4, and the concentration of steel is maintained at a certain level on the surface. The layer then grows by diffusion. The diffusivity, D, here depends on the concentration of carbon. Why? How can this detail change the growth rate of the carburized layer? (Many details on oxidation of metals can be found in [39].)

Appendix A

Electromagnetism

The ideas of electric charge and electic field are fundamental concepts in classical electrodynamics. An electric field is a vector field defined in space. The electric charge is a scalar quantity, positive, negative, or zero, and is a property of matter. The basic fact is that when an object with charge q is placed in an electric field, $\mathbf{E}$, it will experience a force,

$$\mathbf{F} = q\mathbf{E}. \tag{A.1}$$

What is the origin of the electric field? It turns out that *charges* can generate an electric field. Gauss's law of flux states that the flux of the electric field through a closed surface is proportional to the amount of charge enclosed:

$$\int\int_{\partial B} \mathbf{E} \cdot d\mathbf{S} = Q/\epsilon.$$

(The value of ϵ depends on the material; the value in a vacuum, ϵ_0, is given in appendix B.) If the charge is distributed over space according to a continuous density, ρ, this can be written in the differential form

$$\operatorname{div} \mathbf{E} = \rho/\epsilon, \tag{A.2}$$

or, as it is sometimes written,

$$\operatorname{div} \mathbf{D} = \rho, \tag{A.3}$$

where $\mathbf{D}$ is the so-called displacement vector (electric flux density). (It may happen in some materials that $\mathbf{D}$ does not depend linearly on $\mathbf{E}$, or that ϵ is a tensor.)

Is $\mathbf{E}$ a conservative vector field? If so, then we can define a potential: ϕ,

$$\phi(b) - \phi(a) = \int_a^b \mathbf{E} \cdot d\mathbf{l}.$$

This potential difference,

$$V = \phi(b) - \phi(a),$$

is called a voltage and is what is measured by a voltmeter. If we use the differential form $\mathbf{E} = \nabla\phi$, then, together with (A.2), we get the Poisson equation:

$$\nabla^2\phi = \rho/\epsilon.$$

If we solve this with a delta function density (charge concentrated at a point) and insert the field into (A.1), we recover Coulomb's inverse-square law (see Chapter 3).

It turns out that $\mathbf{E}$ is not always conservative. Its integral around a closed path is nonzero whenever something called magnetic flux *changes*. Thus,

$$\oint \mathbf{E} \cdot d\mathbf{l} = -\frac{d\Phi}{dt}, \tag{A.4}$$

where Φ is the magnetic flux, which is understood in terms of a vector field, $\mathbf{B}$, called the flux density. The differential form is

$$\operatorname{curl} \mathbf{E} = -\frac{\partial \mathbf{B}}{\partial t}. \tag{A.5}$$

The magnetic flux is *not* produced by something akin to charges. In fact, as far as we know, there are no magnetic charges (monopoles):

$$\operatorname{div} \mathbf{B} = 0. \tag{A.6}$$

Magnetic flux is generated by currents. A current is a flow of charges. The law states that

$$\oint_C \mathbf{H} \cdot d\mathbf{l} = I_{\text{enc}}, \tag{A.7}$$

where $\mathbf{H} = \mathbf{B}/\mu$ is the magnetic field strength,[1] and I_{enc} is the total current enclosed by the loop C. Current can be given in terms of a current density, $\mathbf{J}$, where

$$I = \int\int \mathbf{J} \cdot d\mathbf{S}.$$

To the right side of (A.7) we must also include a "displacement current" due to a changing electric field; thus,

$$I_{\text{enc}} = \int\int_S (\mathbf{J} + \frac{\partial \mathbf{D}}{\partial t}) \cdot d\mathbf{S}.$$

The differential form is

$$\operatorname{curl} \mathbf{H} = \mathbf{J} + \frac{\partial \mathbf{D}}{\partial t}. \tag{A.8}$$

---

[1] We are ignoring magnetic materials for which one defines a density of magnetic moment, $\mathbf{M}$, and the appropriate relation is

$$\mathbf{B} = \mu\mathbf{H} + \mathbf{M}.$$

As the electric field is generated by charges and affects charged objects, the magnetic field is generated by currents and also affects currents. Currents flow through conductors. The charges in current-carrying conductors are affected by the electric field, but the conductor itself can experience a force even in the absence of the electrical field due to a magnetic flux. If $\mathbf{v}$ is the velocity of a moving charge, q, then it experiences a force orthogonal to its direction of motion and to the direction of the magnetic flux:

$$\mathbf{F} = q(\mathbf{v} \times \mathbf{B}). \tag{A.9}$$

Starting from this, we can calculate the forces exerted on each other by two conductors (Biot–Savart law).

The equations (A.2, A.5, A.6, A.8) are called Maxwell's equations, and, along with the two force laws (A.1 and A.9), are the basis of classical electrodynamics.

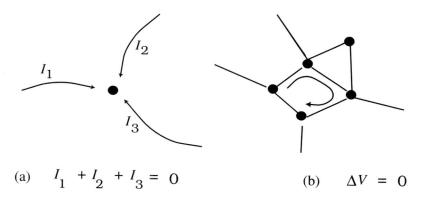

(a) $I_1 + I_2 + I_3 = 0$ (b) $\Delta V = 0$

Figure A.1 The circuit laws.

From Maxwell's equations we can derive the charge conservation law:

$$\frac{\partial \rho}{\partial t} + \operatorname{div} \mathbf{J} = 0. \tag{A.10}$$

If there are no time-varying charges, then this is

$$\operatorname{div} \mathbf{J} = 0$$

and leads to the first of Kirckhoff's circuit laws:

K1) at any node of a circuit, the in-currents must equal the out-currents.

The other circuit law,

K2) the voltage around any closed loop is zero,

follows directly from (A.4) in the case where there is no changing flux (Figure A.1).

If an electric field is turned on in a perfect conductor and there are any free charges present, they will instantaneously reposition themselves. In real materials the charges cannot move freely but suffer collisions, and there is only a mean drift along field lines with a mean drift speed (see Chapter 11). In the simplest model of such behavior the current density is proportional to the applied field:

$$\mathbf{J} = \sigma \mathbf{E},$$

where σ is the conductivity. This is Ohm's law. It is easy to see the relation between σ and the *resistance*, R, in a simple circuit by considering current flowing along a rod of length l and cross-sectional area A:

$$R = \frac{l}{\sigma A}.$$

Capacitance is the amount of charge that must be "placed" on an object in order that it is at a potential (voltage) of unity with respect to some reference. It obviously depends only on the size and shape of the object.

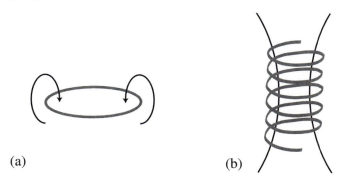

(a)   (b)

Figure A.2

When a current is flowing in a loop, it generates a magnetic field, and thus a flux, by (A.8) (see Figure A.2a). If this current is changing, so is the flux, and this gives rise to an EMF (voltage). And this voltage can be what generates the current! Thus, we have

$$V = L\frac{dI}{dt},$$

where the proportionality constant is called the *inductance*. This induction effect can be enhanced in real circuits by making the current go many times around the "same" loop (Figure A.2b).

Appendix B

The SI System of Units

Fundamental Units

Dimension	Unit	Symbol
Length	meter	m
Mass	dilogram	kg
Time	second	s
Electrical current	ampere	A
Temperature	kelvin	K
Light intensity	candela	cd
Amount	mole	mol

Multiples and fractions of a fundamental unit are usually formed with a prefix. Thus, $1/1000$ m $= 1$ mm, and 1000 m $= 1$ km.

Prefix	Symbol	Value
giga-	G	10^9
mega-	M	10^6
kilo-	k	10^3
centi-	c	10^{-2}
milli-	m	10^{-3}
micro-	μ	10^{-6}
nano-	n	10^{-9}
pico-	p	10^{-12}

Tables of physical properties are usually given in the cgs system, or other systems, of units. Some conversions:

$$1\text{dyn} = 10^{-5}\text{N}, \quad 1 \text{ atm} = 101,325 \text{ Pa},$$

$$1 \text{ cal} = 4.1868 \text{ J}, 1 \text{ erg} = 10^{-7} \text{ J}, 1 \text{ eV} = 1.60219 \times 10^{-19} \text{ J}$$

Some Derived Units

Dimension	Unit	Symbol
Area	square meter	m^2
Volume	cubic meter	m^3
Density		$kg\ m^{-3}$
Velocity		$m\ s^{-1}$
Acceleration		$m\ s^{-2}$
Force	newton	$N = kg\ m\ s^{-2}$
Pressure	pascal	$Pa = kg\ m^{-1}\ s^{-2}$
Energy, work	joule	$J = kg\ m^2\ s^{-2}$
Power	watt	$W = J\ s^{-1} = kg\ m^2\ s^{-3}$
Entropy		$J\ K^{-1}$
Electrical potential	volt	$V = kg\ m^3\ s^{-3}\ A^{-1}$
Electrical charge	coulomb	$C = A\ s$
Electrical capacitance	farad	$F = kg^{-1}\ m^{-2}\ s^4\ A^2$
Electrical resistance	ohm	$\Omega = kg\ m^2\ s^{-3}\ A^{-2}$
Electrical field		$V\ m^{-1}$
Magnetic flux	weber	$Wb = kg\ m^2\ s^{-2}\ A^{-1}$
Magnetic flux intensity	tesla	$T = kg\ s^{-2}\ A^{-1}$
Inductance	henry	$H = kg\ m^2\ s^{-2}\ A^{-2}$

Some Physical Constants

Speed of light in vacuum, c	$2.997925 \times 10^8\ m\ s^{-1}$
Avogadro's number, N_A	6.022252×10^{23}
Universal gas constant, R	$8.3143 \times 10^3\ J\ kmol^{-1}\ K^{-1}$
Boltzmann constant, k_B	$1.38054 \times 10^{-23}\ J\ K^{-1}$
Elementary charge, e	$1.60210 \times 10^{-19}\ C$
ϵ_0	$8.8544 \times 10^{-12}\ F\ m^{-1}$
μ_0	$1.25664 \times 10^{-6}\ H\ m^{-1}$

Appendix C

Some Physical Properties of Materials

Unless otherwise noted, properties are at room temperature and pressure.

Viscosity

Material	Density (g cm^{-3})	Viscosity (g cm^{-1} s^{-1})
Acetone	0.791	3.3×10^{-3}
Benzene	0.879	6.73×10^{-3}
Ethyl alcohol	0.790	1.24×10^{-2}
Ethyl bromide	1.463	4.03×10^{-3}
Glycerin	1.260	$1.393 \times 10^{+1}$
Mercury	13.551	1.566×10^{-2}
Olive Oil	0.915	9.98×10^{-1}
Toluene	0.866	6.02×10^{-3}
Water	0.999	1.06×10^{-2}
Air		1.7×10^{-4}

Thermal Conductivity

Material	Conductivity (cal K^{-1} cm^{-1} s^{-1}) C
Brass	2.7×10^{-1}
Cork	10^{-3}
Glass	1.566×10^{-3}
Ice	6×10^{-3}
Naphthalene	9×10^{-5}
Porcelain	2.5×10^{-3}
Steel	1.11×10^{-1}

Bibliography

[1] Mathematical modeling for instructors. Report IMA-1254, Institute for Mathematics and Its Applications, 1994.

[2] A. Albert. A mathematical theory of pattern recognition. *Ann. Math. Stat.*, 34:284–299, 1963.

[3] I. Asimov. *Space Garbage*. Garth Stevens, Milwaukee, 1989.

[4] E. A. Bender. *An Introduction to Mathematical Modeling*. Wiley, New York, 1978.

[5] B. L. Blackford. The hydraulic jump in radially spreading flow: A new model and new experimental data. *Amer. J. Phys.*, 64:164–169, 1996.

[6] G. H. Bower. Application of a model to paired-associate learning. *Psychometrika*, 26:255–280, 1961.

[7] L. Breiman. The Poisson tendency in traffic distribution. *Ann. Math. Stat.*, 34:308–311, 1963.

[8] S. Bullett and J. Stark. Renormalizing the simple pendulum. *SIAM Review*, 35:631–640, 1993.

[9] E. Burmeister and A. R. Dobell. *Mathematical Theories of Economic Growth*. Macmillan, London, 1970.

[10] S. S. Chen. *Flow-induced Vibration of Circular Cylindrical Structures*. Hemisphere, Washington, DC, 1987.

[11] S. Chikazumi. *Physics of Magnetism*. Wiley, New York, 1964.

[12] A. E. Cook and P. H. Roberts. The Rikitake two-disc dynamo system. *Proc. Camb. Phil. Soc.*, 68:547, 1970.

[13] P. Dallos, C. D. Geisler, J. W. Matthews, M. A. Ruggero, and C. R. Steele, editors. *The Mechanics and Biophysics of Hearing*, Springer-Verlag, Berlin, 1990.

[14] M. A. Dinno, A. M. Al-Karmi, D. A. Stoltz, J. C. Matthews, and L. A. Crum. Effect of free radical scavengers on changes in ion conductance during exposure to therapeutic ultrasound. *Membrane Biochemistry*, 10:237–247, 1993.

[15] C. Doering and P. Constantin. Turbulence energy dissipation. *Phys. Rev. Lett.*, 69:1648, 1992.

[16] J. Dutton. *The Ceaseless Wind.* Dover, New York, 1986.

[17] M. Van Dyke. *An Album of Fluid Motion.* Parabolic Press, Stanford, CA, 1982.

[18] W. Feller. *An Introduction to Probability and its Applications*, volume II. Wiley, New York, 2nd edition, 1971.

[19] R. Feynman. *Lectures on Physics.* Addison-Wesley, Reading, MA, 1965.

[20] A. T. Fomenko. *Empirico-Statistical Analysis of Narrative Material and its Application to Historical Dating.* Kluwer, Dordecht, 1994.

[21] R. Ghez. *A Primer of Diffusion Problems.* Wiley, New York, 1988.

[22] R. Gilmore. *Catastrophe Theory for Scientists and Engineers.* Dover, New York, 1981.

[23] S. Goldberg. *Probability in Social Science.* Birkhauser, Boston, 1983.

[24] J. E. Gordon. *The New Science of Strong Materials.* Princeton University Press, Princeton, NJ, 1984.

[25] R. Haberman. *Elementary Applied Partial Differential Equations.* Prentice–Hall, Englewood Cliffs, NJ, 2nd edition, 1987.

[26] J. M. Hammersley and D. C. Handscomb. *Monte Carlo Methods.* Methuen, London, 1964.

[27] F. C. Hoppensteadt and C. S. Peskin. *Mathematics in Medicine and the Life Sciences.* Springer-Verlag, New York, 1991.

[28] K. Inada. On a two-sector model of economic growth: Comments and a generalization. *Review of Economic Studies*, XXX(83):119–127, 1963.

[29] T. Indow and K. Togano. On retrieving sequence from long-term memory. *Psychological Review*, 77(4):317–331, 1970.

[30] W. D. Iwan and R. D. Blevins. A model for vortex induced oscillations of structures. *Journal of Applied Mechanics*, pages 581–586, 1974.

[31] R. C. Jackson. *The theoretical foundations of cancer chemotherapy introduced by computer models.* Academic Press, San Diego, CA, 1992.

[32] D. Jewett and M. Rayner. *Basic Concepts of Neuronal Function.* Little, Brown, Boston, 1984.

[33] S. Karlin. *A First Course in Stochastic Processes.* Academic Press, New York, 1966.

[34] M. Kimmel. A model for the cell proliferation cycle. *Mathematical Biosciences*, 48:211–239, 1980.

[35] M. Kimmel and A. Swierniak. Control and modeling of cancer cell populations. *Applied Mathematics and Computer Science*, 4, 1994.

[36] C. Kittel. *Introduction to Solid State Physics*. Wiley, New York, 1971.

[37] L. Knopoff and J. Mouton. Earthquake sources. In R. DiPrima, editor, *Modern Modeling of Continuum Phenomena*, pages 221–251. American Mathematical Society, Providence, RI, 1977.

[38] R. E. Kronauer, C. A. Czeisler, S. F. Pilato, M. C. Moore-Ede, and E. D. Weitzman. Mathematical modeling of the human circadian system with two interacting oscillators. *Am. J. Physiol.*, 242:R3–R17, 1982.

[39] O. Kubaschewski. *Oxidation of metals and alloys*. Butterworths, London, 2nd edition, 1962.

[40] L. D. Landau and E. M. Lifshitz. *Fluid Mechanics*. Pergamon, London, 1959.

[41] R. Landl. A mathematical model for vortex-excited vibrations of bluff bodies. *J. Sound and Vibration*, 42(2):219–234, 1975.

[42] P. A. W. Lewis, editor. *Stochastic Point Processes*. Wiley–Interscience, New York, 1972.

[43] M. Lieberstein, editor. *Mathematical Physiology: Blood Flow and Electrically Active Cells*. Elsevier, New York, 1973.

[44] J. Lighthill. Energy flow in the cochlea. *J. Fluid Mech.*, 106:149–213, 1981.

[45] C. C. Lin and L. A. Segel. *Mathematics Applied to Deterministic Problems in the Natural Sciences*. SIAM, Philadelphia, 1988.

[46] T. McMahon. Size and shape in biology. *Science*, 179:1201–4, 1973.

[47] T. McMahon. The mechanical design of trees. *Scientific Amer.*, 233(1):92–102, 1975.

[48] A. Mineyev. Trees worthy of Paul Bunyan. *Quantum*, 4(3):4–10, 1994.

[49] M. C. Moore-Ede, F. M. Sulzman, and C. A. Fuller. *The Clocks That Time Us*. Harvard University Press, Cambridge, MA, 1982.

[50] I. Nagumo, S. Arimoto, and S. Yoshisava. Nerve propagation. *Proc. IRE*, 50:2061–2066, 1962.

[51] R. G. Olsson and E. T. Turkdogan. Radial spread of a liquid stream on horizontal plate. *Nature*, 211:813–816, 1966.

[52] J.O. Pickles. *An introduction to the physiology of hearing*. Academic Press, London, 2nd edition, 1988.

[53] A. B. Pippard. *Response and Stability*. Cambridge University Press, Cambridge, 1984.

[54] A. B. Poore, E. J. Doedel, and J. E. Cermak. Dynamics of the Iwan-Blevins wake oscillator model. *Int. J. Nonlinear Mechanics*, 21(4):291–302, 1986.

[55] G. L. Priest and B. Klein. The selection of disputes for litigation. *J, Legal Studies*, XIII:1–55, 1984.

[56] E. M. Purcell. Life at low reynolds number. *Amer. J. Physics*, 45:3–11, 1977.

[57] W. S. Rhode. Observations of the vibration of the basilar membrane in squirrel monkeys using the Mossbauer technique. *J. Acoust. Soc. Amer.*, 49:1218–1231, 1971.

[58] P. Riesz and T. Kondo. Free radical formation induced by ultasound and its biological implications. *Free Radical Biology & Medicine*, 13:247–270, 1992.

[59] K. J. Rose. *The Body In Time*. Wiley, New York, 1988.

[60] D. S. Ross. Two new moving boundary problems for scalar conservation laws. *Comm. Pure. Appl. Math.*, XLI:725, 1988.

[61] M. V. Sataric and G. I. Gogoljev. A solitonlike solution for action potential in nerve axon. *Neural Network World*, 5(5):709–716, 1995.

[62] K. Schmidt-Neilsen. *Scaling: Why Is Animal Size so Important?* Cambridge University Press, Cambridge, 1984.

[63] L. A. Segel. *Mathematics Applied to Continuum Mechanics*. Macmillan, New York, 1977.

[64] L. A. Segel. *Modeling Dynamic Phenomena in Molecular and Cellular Biology*. Cambridge University Press, Cambridge, 1984.

[65] S. S. Shen. *A Course on Nonlinear Waves*. Kluwer, Dordrecht, 1992.

[66] I. M. Sobol. *A Primer for the Monte-Carlo Method*. CRC, Boca Raton, FL, 1994.

[67] A. M. Starfield and A. L. Bleloch. *Building Models for Conservation and Wildlife Management*. Macmillan, New York, 1986.

[68] A. M. Starfield, K. A. Smith, and A. L. Bleloch. *How to Model It: Problem Solving for the Computer Age*. McGraw-Hill, New York, 1990.

[69] G. I. Taylor. The mechanism of plastic deformation of crystals. *Proc. Roy. Soc. A*, 145:362–87, 1934.

[70] G. I. Taylor. The formation of a blast wave by a very intense explosion. II. The atomic explosion of 1945. *Proc. Roy. Soc. A*, 201:175–186, 1949.

[71] G. I. Taylor. Disintegration of water drops in an electric field. *Proc. Roy. Soc. A*, 281:383–397, 1964.

[72] H. N. V. Temperley and D. H. Trevena. *Liquids and Their Properties.* Ellis Horwood, Chichester, 1978.

[73] T. Thedeen. A note on the Poisson tendency in traffic distribution. *Ann. Math. Stat.*, 35:1823–1824, 1964.

[74] P. N. V. Tu. *Dynamical Systems.* Springer, Berlin, 1992.

[75] H. Uzawa. On a two-sector model of economic growth. *Rev. Econ. Studies*, XXIX(78):40–47, 1961.

[76] H. Uzawa. On a two-sector model of economic growth ii. *Rev. Econ. Studies*, XXX(83):105–118, 1963.

[77] S. Vogel. *Life's Devices.* Princeton University Press, Princeton, NJ, 1988.

[78] L. Voldrich. Mechanical properties of the basilar membrane. *Acta Otolaryngol.*, 86:331–335, 1978.

[79] C. von Euler. Central pattern generation during breathing. *Trends in Neuroscience*, pages 275–277, 1980.

[80] A. J. Walton. *The Three Phases of Matter.* Oxford University Press, Oxford, 1983.

[81] T. Weis-Fogh. Diffusion in insect wing muscle, the most active tissue known. *J. Exp. Biol.*, 41:229–256, 1964.

[82] T. E. Wheldon. *Mathematical Models in Cancer Research.* Adam Hilger, Bristol, 1988.

[83] A. G. Wilson. *Catastrophe Theory and Bifurcation.* Croom Helm, London, 1981.

[84] J. P. Wilson and D. T. Kemp, editors. *Cochlear Mechanisms*, Plenum, New York, 1989.

[85] A. T. Winfree. *The Geometry of Biological Time.* Springer, Berlin, 1990.

[86] J. M. Yeomans. *Statistical Mechanics of Phase Transitions.* Clarendon Press, Oxford, 1992.

Index